Communications
in Computer and Information Science 730

Commenced Publication in 2007
Founding and Former Series Editors:
Alfredo Cuzzocrea, Xiaoyong Du, Orhun Kara, Ting Liu, Dominik Ślęzak,
and Xiaokang Yang

More information about this series at http://www.springer.com/series/7899

Janet Liebenberg · Stefan Gruner (Eds.)

ICT Education

46th Annual Conference of the Southern African
Computer Lecturers' Association, SACLA 2017
Magaliesburg, South Africa, July 3–5, 2017
Revised Selected Papers

 Springer

Editors
Janet Liebenberg
North-West University
Potchefstroom
South Africa

Stefan Gruner
University of Pretoria
Pretoria
South Africa

ISSN 1865-0929   ISSN 1865-0937 (electronic)
Communications in Computer and Information Science
ISBN 978-3-319-69669-0   ISBN 978-3-319-69670-6 (eBook)
https://doi.org/10.1007/978-3-319-69670-6

Library of Congress Control Number: 2017957564

Printed on acid-free paper

This Springer imprint is published by Springer Nature
The registered company is Springer International Publishing AG
The registered company address is: Gewerbestrasse 11, 6330 Cham, Switzerland

Preface

This volume of CCIS contains the revised selected papers of SACLA 2017, the 46th Annual Conference of the Southern African Computer Lecturers' Association, held in the picturesque town of Magaliesburg (Republic of South Africa), July 3–5, 2017. The theme of this conference was: "Keeping Education Relevant: Infinite possibilities."[1]

The goal of the annual meeting of the Southern African computer lecturers is to provide participants with an opportunity to share ideas, while maintaining a high level of academic input from all involved. The accepted papers reflect current trends in teaching and learning in computer science and information systems in tertiary education.

SACLA 2017 called for papers describing educational research projects, classroom experiences, teaching techniques, curricular initiatives, or pedagogical tools. Relevant contemporary topics mentioned in the call for papers included:

- Keeping up with the continuous changes in the fields of computer science and information systems
- Keeping up with changes in technology[2]
- Adapting to new generations of students
- Keeping relevant by collaboration with industry
- Keeping relevant by "regionalizing" curricula
- Relevance of program/curriculum design
- Didactics and methods of teaching and assessment
- Ethical problems
- Transition of newly graduated school pupils
- Transition of newly graduated students into industry
- International comparability of degrees and levels of knowledge and skill
- Separation or combination of teaching and research
- General matters: best/worst practices, success/failure experiences

The Program Committee comprised 53 members, 30 of whom were from outside South Africa. Each submitted paper was reviewed by three members of the Program Committee in a rigorous, double-blind mode, whereby especially the following criteria were taken into consideration: novel research contribution, methodological soundness, theoretical framing and reference to related work, quality of analysis, and quality of writing and presentation. Of the three reviews for each submission, at least one was provided by an international reviewer (from outside South Africa). After the international dissemination of our call for papers, 63 submissions were initially received and carefully reviewed; 40 of them were chosen for presentation at the SACLA 2017 conference,[3] 39 of which were actually presented. Of these submissions, 22 revised

---

[1] http://sacla2017.nwu.ac.za.

[2] For comparison, see LNET, http://www.springer.com/series/11777.

[3] The entire program can be obtained via http://sacla2017.nwu.ac.za/shortprogram/.

selected papers (plus the extended abstract of the invited keynote lecture by Grandon Gill) were finally included in this volume of CCIS as the conference's most noteworthy contributions.

The overall paper acceptance rate for this book is thus 35%, which shows our commitment to high academic quality.

We were fortunate to have a keynote speaker from the USA, Grandon Gill; the extended abstract of his invited lecture is included in this volume. Moreover, the paper by Anwar Parker and Jean-Paul van Belle (see Table of Contents) received the conference's Best Paper Award.

Affiliated with our conference was a local workshop on the topic: The South African Computing Accreditation Board (SACAB): Implementation and Documentation.

We extend our thanks and appreciation to the conference's Organizing Committee, colleagues, and friends who contributed to the success of SACLA 2017. On behalf of the SACLA community, we also wish to express our deepest appreciation to our sponsors: AdaptIT, IITPSA, IBM, NWU Potchefstroom campus, and SAP. Thank you to the authors, as well as the invited speaker, for having chosen SACLA 2017 as the forum for communicating their noteworthy insights and interesting thoughts. A further word of thanks goes to the members of the Program Committee, who all provided extensive and insightful reviews. Last but not least, many thanks also to the helpful staff of Springer, who have made this CCIS publication possible.

Throughout the remainder of this book, ICT stands for information and communication technologies, comprising computer science, informatics, information science, and similar (related) areas of studies (which cannot be sharply distinguished from each other).

We wish our readers a fruitful reading experience with this volume of CCIS, and we look forward to the continuation of the SACLA series in the following years.

August 2017   Janet Liebenberg
                                                                      Stefan Gruner

The supporters and sponsors of SACLA 2017 are herewith gratefully acknowledged.

Organization

General Chair

Estelle Taylor   North-West University, South Africa

Local Arrangements and Technical Support

Susan Campher   North-West University, South Africa
Henry Foulds   North-West University, South Africa
Dimi Keykaan   North-West University, South Africa
Anne Mans   North-West University, South Africa

Program Committee Chair and Proceedings Co-editor

Janet Liebenberg   North-West University, South Africa

Publications Chair and Proceedings Co-editor

Stefan Gruner   University of Pretoria, South Africa

Program Committee

Elisabeth Bacon   University of Greenwich, UK
Trudie Benade   North-West University, South Africa
Torsten Brinda[*]   Universität Duisburg-Essen, Germany
André Calitz[*]   Nelson Mandela Metropolitan University, South Africa
Charmain Cilliers[*]   Nelson Mandela Metropolitan University, South Africa

Janis Voigtländer[+]	Universität Duisburg-Essen, Germany
Gottfried Vossen[*]	Westfälische Wilhelms-Universität Münster, Germany
Jim Woodcock[*]	University of York, UK
Olaf Zukunft[*]	Hochschule für angewandte Wissenschaften Hamburg, Germany

PC members marked with a[*] are continuing PC members from the previous year's conference, SACLA 2016 (Springer: CCIS 642), and PC members marked with a[+] had acted as additional reviewers for SACLA 2016.

Additional Reviewer

Nel, Liezel

"The Master spoke: In preparing the governmental notifications, Pi Shan first made the rough draft; Shi Shu examined and discussed its contents; Tse Yu, the manager of foreign affairs, then polished the style; and, finally, Tse Chan of Tung Li gave it the proper elegance and finish"

— Confucius (Kung Fu Tse 551–479BC), ANALECTS: Chapter 14.

Blending IT Research and Teaching to Create a more Locally Relevant Curriculum (Invited Lecture)

Grandon Gill

Department of Information Systems and Decision Sciences,
University of South Florida, Tampa, USA
grandon@usf.edu

Abstract. This short-paper summarises the invited keynote lecture presented to the SACLA' 2017 conference in Magaliesburg on the 4th of July 2017.

Keywords: Locally relevant curricula · Theory of case studies · Invited keynote lecture

Extended Abstract of the Invited Keynote Lecture

Case studies come in many forms [3, 5, 6]:

Research case studies tell a story of cause and effect and are closely tied to theory. They achieve rigor through triangulation — melding multiple types of data sources, analytical approaches, investigators, and theories. They are published in the same type of outlets as other forms of rigorous research.

Teaching cases are ore diverse in their objectives [7]. Examples cases provide complete stories that resonate with students and practitioners. Walkthrough cases allow a facilitator to guide students through complicated real-world procedures. Exercise cases provide students with access to real-world problems.

Discussion cases are the type of case studies pioneered by institutions such as the Harvard Business School. They are designed for a specific purpose: as a basis for classroom discussion [1]. They nearly always begin by laying out a decision that needs to be made by an individual, referred to as the protagonist. They then provide the reader with a rich description of the context in which the decision is being made, supported by exhibits drawn from the site of the case. The 'classic' discussion case can be described as

- **open:** having no 'right' solution but, instead, offering many possible approaches to the decision, some of which make sense, some of which do not;
- **authentic:** based on observations of a real-world situation, as opposed to a fictional organization;
- **detailed:** supplying enough detail to support a discussion of one to two hours.

Discussion cases are particularly appropriate for situations of high *extrinsic complexity* [4]. These types of situations exist where combinations of factors determine the fitness [2] of the options available for a decision, rather than individual attributes. High extrinsic complexity environments tend to be very situation-specific (i.e., context-dependent). In today's world, such situations are encountered more and more commonly. What is needed to address these situations is judgment — which is the skill that discussion cases are most suited to develop.

An inevitable problem encountered when using discussion cases results from their context-dependence. A case study written about a CEO's decision for a major U.S. American corporation is likely to be of little value to an undergraduate coming from rural South Africa. While the need for judgment exists in both settings, it is very unlikely that such a case will resonate with students who cannot perceive its relevance.

There are huge benefits to developing case studies of local organizations, both businesses and governmental. While the development of local case studies will not lead to the types of publication prized by the traditional academic research community, they will serve to build the researchers' skills of observation and interviewing. The development of these cases will also serve to build relationships between university scholars and the local stakeholder community. Such relationships with key stakeholders will be of inestimable value when university budgets are threatened or when the value of academic research is questioned. In conducting research associated with case development, scholars will also become acquainted with the types of activities their graduates will be performing in the workforce, which can inform the curriculum. Most importantly, the preparation and discussion cases will engage students in a way that complements the existing curriculum.

References

1. Gill, T.G.: A protocol online case discussion. Decis. Sci. J. Innovative Educ. **3**(1), 141–148 (2005)
2. Gill, T.G.: Reflections on researching the rugged fitness landscape. Informing Sci. **11**, 165–196 (2008)
3. Gill, T.G.: The complexity and the case method. Manag. Decis. **59**(9), 1564–1590 (2014)
4. Gill, T.G., Mullarkey, M.: Fitness, extrinsic complexity, and information science. Inf. Sci. **20**, 37–16 (2017)
5. Gill, T.G., Mullarkey, M.: Taking a case method capstone course online: a comparative case study. J. IT Educ. Res. **14**, 189–218 (2015)
6. Gill, T.G., Mullarkey, M., Mohr, J., Limayem, M.: Building an information business school: a case study of USF's muma college of business. Inf. Sci. **19**, 1–73 (2016)
7. Webb, H., Gill, T.G., Poe, G.: Teaching with the case method online: pure versus hybrid approaches. Decis. Sci. J. Innovative Educ. **3**(2), 223–250 (2005)

Contents

Computer Programming Education

ICT Courses and Curricula

ICT Students of a New Generation

The iGeneration as Students: Exploring the Relative Access, Use of, and Perceptions of IT in Higher Education

Anwar Parker and Jean-Paul van Belle(✉) ⓘ

Department of Information Systems, University of Cape Town,
Cape Town, South Africa
PRKANW001@myuct.ac.za, jean-paul.vanbelle@uct.ac.za

Abstract. This paper explores iGeneration age students as they encountered e-learning in higher education. The case was that of the University of Cape Town (UCT). The paper examines the iGeneration as a distinct technologically enhanced generation and describes them within the context of first-year students with specific technological capabilities and needs. It was found that, although the literature strongly suggested that their technological preferences would outweigh those provided at tertiary education, they showed positive attitudes towards the use of technology at tertiary level. As such, the expectations associated with a new generation may not necessarily warrant a reform in order to suit the generation. However, the findings suggest a high degree of consensus amongst the majority of responses indicating that, as per the literature, that they may be identified as a homogenous sample. Furthermore, it is believed that these students value the role that technology plays as a facilitator towards education, especially at university.

Keywords: Information technology · Higher education · iGeneration · Generational characteristics · Students · Technology attitude · e-Learning

1 Introduction

Technology has a strong influence on the ways in which computers and the internet are adopted in the home, at schools, and at work for the 'iGeneration' [6]. The integration of technology in the learning may facilitate innovative and learner-focused learning experiences [10]. However, there exists little historical data regarding digital media consumption of iGeneration students [5]. Further investigation into the appropriate technologies and teaching methods may be able to improve the learning experience for specific generations [6,14]. For instance, the rapid exposure to technology has facilitated the development of multi-tasking as a skill that requires constant stimulation [21]. This research premises that teachers and educational institutions have a responsibility to change to the assumed demands of new generations [8].

J. Liebenberg and S. Gruner (Eds.): SACLA 2017, CCIS 730, pp. 3–18, 2017.
https://doi.org/10.1007/978-3-319-69670-6_1

Each student generation has technological preferences when it comes to, for example, communication; for instance, the 'baby boomers' prefer face-to-face and telephonic conversations [21]. However, the iGeneration prefer instant messaging in contrast to making or receiving phone calls [21]. There may, therefore, exist a technical skill gap between iGeneration and their 'digital immigrant' university teachers which may create a disconnect [5]. In brief, *"prior generations have to restructure their traditional teaching methods to suit the later generation's new and technological enhanced methods of learning"* [6] (p. 262). Our research was driven by three primary questions:

1. How do iGeneration students use technology to connect and communicate with various people on campus?
2. What relative importance do iGeneration students attach to having access to more or better technology for themselves?
3. To what extent does iGeneration students' use of technology enable them to understand, demonstrate their understanding, or study on their own or with others?

The research objectives provide guidance towards the answering of the research questions [16]. Considering the questions proposed, the following objectives were pursued:

- to explore PC ownership among first-year students and the use they make of computing facilities;
- to explore student perceptions of ICT-based learning;
- to explore student attitudes towards ICT as a learning tool in higher education;
- to explore what technologies iGeneration students use to connect and communicate with various people on campus;
- to explore how important access to more or better technology is for iGeneration students;
- to investigate how technology has facilitated their understanding (including: demonstrating their understanding) of course content by themselves or with others.

The value derived from the research is justified by providing an overview of the relative access, use of and perceptions towards IT in higher education. Most previous studies tend to focus on a specific technology (e.g., YouTube or mobile phones); this research does not. Thus, it allows for a usable foundation towards further studies based on the findings thereof. Furthermore, the implications may also be used within regard to the development of further academic streams based on the findings. Lastly, *"through the emergence of technology, students have better access to information anytime and anywhere, allowing learning to become more convenient and flexible, enhancing communication and connection with their instructors and amongst their peers, giving better control over when to engage in course activities and offering improved learning overall"* [10] (p. 22).

2   Literature

2.1   The iGeneration

Students of the iGeneration are seen as a distinct new generation of technologically-enhanced learners [15,21]. 'Generations' can be defined as *"bands of time which include the birth years of individuals that share similar characteristics as shaped by events and circumstances surrounding their lives"* [6] (p. 261). In regard to the iGeneration, they have grown up with digital technology and have an information-age mind-set [6]. In other words, due to their exposure to digital technology, they have a lower tolerance to delays, and require information at their fingertips [6]. The name of the iGeneration is *"so called because their use the iPhone, iPod, iPad (and iEverything), who have redefined communication not only by their acceptance of and hunger for new devices, but because of their sometimes overwhelming reliance on technology for being in touch with others and interpreting their world"* [21] (p. 2).

For the iGeneration there are various opinions as to when this generation had begun. According to [6] these individuals were born from 1992 onwards, whereas [5,15,21] also argued that they were born in the 1990s, even though some stated that these students were born from the 2000 s and onwards [8].

There are various other names associated to the iGeneration: [18] identifies this generation as 'Generation Y'. Other names identified by Sternberg include the 'Net Generation', 'Dot-Coms', 'Echo-Boomers', 'Generation X^2', 'Generation-D', the 'Nexters', 'Generation Next', 'Boomlets', and the 'Nintendo Generation'. Finally, [20] identifies them as 'Generation Me'. Although there exists a variation and overlap of naming conventions amongst authors, 'iGeneration' was the most notable one [6].

The iGeneration have high expectations and expect tailored items based on their wants and needs. Within an academic context, this may pose a problem towards their lecturers [6]. In addition, their likes include that of multi-tasking, communication technologies, virtual social worlds, and brands advertising [15,21]. However, the concern is with defining and labelling groups of people is that it can lead to problems of misrepresentation and generalisation; or that these generations (e.g. Net Generation) may not necessarily be homogenous in their use of technology: *"skill and comfort level with technology differs within the generation"* [14] (p. 475).

The iGeneration consumes technology at a rapid pace. In [15] (p. 8) their consumption is described as a *"thirst for any mobile technology"*, whilst [21] (pp. 2–3) compliments this claim with the statement that the iGeneration *"redefined communication not only by their acceptance of and hunger for new devices, but because of their sometimes overwhelming reliance on technology for being in touch with others and interpreting their world"*. The transformation of the cell phone into a media content delivery platform facilitated an explosion in consumption patterns [12].

2.2   Criticisms Against Technology Consumption

Technology consumption has an impact on behavioural development: *"we expect more from technology and less from one another and seem increasingly drawn to technologies that provide the illusion of companionship without the demands of relationship"* [19] (p. 3). In [21] the views of [19] are supported according to which human beings could not do without the immeasurable and instant connectivity associated with technology. The repercussions of this are noted in [5] whereby iGeneration and generations to come may inaptly lose certain social mechanisms such as empathy. This was attributed towards the over-emphasis on left brain activity from which we use to engage in with digital consumption [5]. Additionally, these new technology-driven platforms has also created a new medium by which social abuse can occur [14]. In [21] the association of social defects w.r.t. technology, such as attention-deficit disorder and inattention, is suggested.

2.3   Education and Technology

Technology provides a way in which learning can be personalised and maximised towards the student. Such technology can also be in the form of mobile learning technology that is personalised towards students with cognitive, physical, and sensorial disabilities [3]. In [2] (p. xvi) specifically the close relationship between technology and pedagogy was noted, and how the *"scope and style of pedgogy change as the technology changes"*. However, [2] critiques technologies within the context of the net generation in that technology does not necessarily have an educational capacity. On the other hand, [13] (p. 134) specifically made reference to higher education and their *"well-established trends toward non-adoption of new technologies"*. Pedagogy needs to set the standard and technology would need to adhere to that standard in order for for realisation to occur [1,11]. A mismatch between the required and set standards may lead to a failure in educational technology being utilised [11].

Students of the iGeneration consume technology as it comes, and mould offerings to their personalised needs [21]. For example, learning management technologies have facilitated consumption behaviour to a great extent as e-learning tools [4]. On the other end, the effectiveness of these tools is based on the purpose they serve for the users and the way in which they are implemented [17]. The Digital Age does, however, allow for further usability and flexibility, along with mobility for users to consume technology at a rate at which they please [3]. In other words, technology as a determining power is almost extinguished as it is determined by the needs of the consumers [11]. In brief, technology has become an integral part of education; however, educational offerings have not necessarily been adjusted to current technological trends.

3   Method

A positivist, quantitative, cross-sectional research approach was chosen to answer the research questions in the introduction. The authors acknowledge that this limits the in-depth understanding of the unique, lived experiences of the individual students, and foregoes the richness and understandings of the complexities underlying these realities, but this approach will allow for a hopefully generalizable 'big picture' view.

The research instrument used here was based on a similar study that aimed to explore *"the experiences of first-year students as they encountered university e-learning"* [7] (p. 724). The development of the instrument was developed with reference to [9] as well as the Educause ECAR studies. In addition, the instrument was intended to collect baseline data about keys aspects of students' use of technology in both their social and academic lives [7]. The four sections include demographic characteristics of the respondents; access to technology; use of technology in university; and course-specific uses of technology. Adjustments were made where applicable in order to tailor the instrument to a South African audience, and the final instrument is available from the authors on simple email request. The sample population was students enrolled in three large Faculties at University of Cape Town (UCT), namely commerce, humanities, and science. Within each faculty, the research instrument was limited to first-year students as they were identified as the most appropriate reflection of the iGeneration and least acculturated to the status quo of teaching practice at the university. Participation was voluntary with informed consent, anonymity and confidentiality attended to.

4   Data Analysis and Discussion

The analysis is constructed under four main headings: the demographic characteristics of the respondents, access to technology, use of technology, and the specific use of technology in courses.

4.1   Demographic Analysis

In total, 86 valid responses were received with a predominance of female respondents: two-thirds, i.e. 65%, of the respondents were female, 34% were male, and one respondent did not answer. 32 responses were from the Humanities, 37 from the Science Faculty, and 17 from Commerce.

The students' attitude towards technology was measured using four items. Each item was scored using a 5-point Likert scale response; (note that the mean response in Fig. 1 and subsequent similar type charts is suffixed to Y-axis labels between brackets). Figure 1 indicates that the student attitude towards technology at university is positive with 57 respondents indicating that it would be useful, 42 were confident about using technology at university, 41 were excited about the technology, and 50 disagreed that they were not interested. In all cases, this is statistically highly significantly ($p < 0.001$) different from neutral ('3').

Figure 2 provides an overview of computer access amongst first-year respondents. Of 86 respondents, 82 indicated ownership of a device. Most students (84%) indicated that they owned a laptop computer, whereas only 18% owned

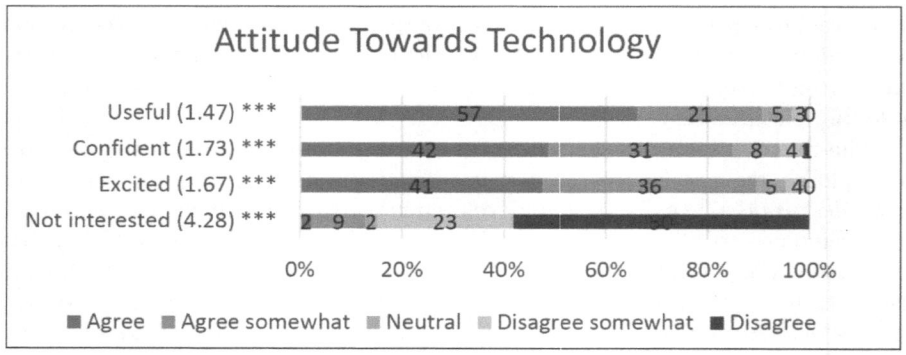

Fig. 1. 1st-year students' attitude towards digital technology

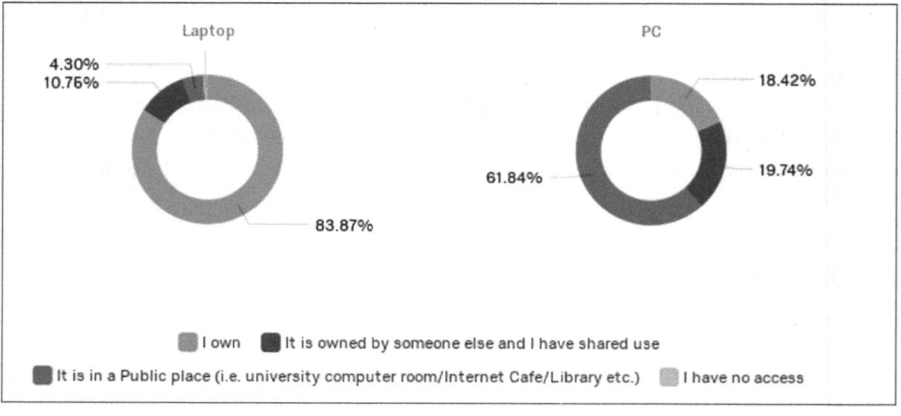

Fig. 2. Distribution of respondents' computer access

a personal computer (desktop PC). However, the contrast between laptop ownership and PC public access becomes evident as 62% of respondents indicated they made use of public resources (i.e. university computer room, internet caf, or library). Thereby, 77% of the respondents said that their PC fully meets their needs. In this case, PC refers to both laptop and personal computer (PC) as highlighted in Fig. 2. The remaining respondents bar one individual (22%) indicated that it mostly met their needs. Of all respondents, 41 (47.7%) indicated their mobile device made use of an 'Android' OS, 32 (37.2%) an Apple 'iOS', 8 (9.3%) as 'Windows', and Other accounted for 5 (5.8%). The majority of the 'Other' specified Blackberry as an operating system. Perhaps, in South Africa, we should speak of the 'A(ndroid)-Generation' instead of the iGeneration?

Figure 3 represents the relative importance of IT for studying, specifically highlighting 3 key use-cases and 3 devices. From the devices the computer scored the highest with a count of 80 for 'Importance', eclipsing even mobile devices (count = 56). The rated importance of the computer appears to be somewhat

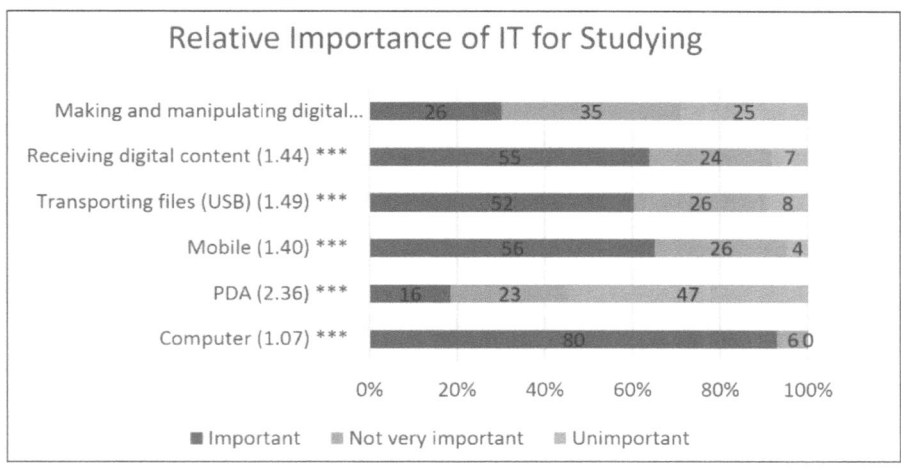

Fig. 3. Relative importance of IT for studying

paradoxical given that making and manipulating digital content was rated the least important IT task. This highlights the fact that receiving digital content is not necessarily always preferred on a mobile device. Interestingly, when asked the same questions, i.e.: about the relative importance of IT uses and devices for leisure, there was a noticeable shift of relative importance in favour to that of the mobile which 79 out of 86 rated 'Important'. Receiving digital content also became more important in the leisure/personal environment than the study context, whereas transporting data was perceived less important in the private sphere.

4.2 Use of Technology

Respondents were asked (by means of a slider) to indicate their average exposure to a computer screen per day. The average of the reported exposures to a computer screen was 6.2 h per day (but with a relatively high standard deviation of 3.7 h). Respondents indicated their average daily connectivity to the internet for studying purposes per day to be just below that with 5.8 h (but an even higher standard deviation of 4.2 h).

Figure 4 indicates how often respondents used specific technologies. Not surprisingly, instant messaging (IM), social media, emailing and texting is a more-than-daily activity for the majority of the students. Streaming audio or video is also done by almost all students—not as frequently though still almost daily—by close to half the students. Surprisingly, most students also use wikis at least weekly, although the question phrasing makes it unclear whether they actually contribute to wikis or just read them. The least frequent activities were shopping online (53 out of 86) and blogging (39 out of 86), which are larger and more time-consuming; however, more than half of the students has never shopped online, which was somewhat of a surprise to us.

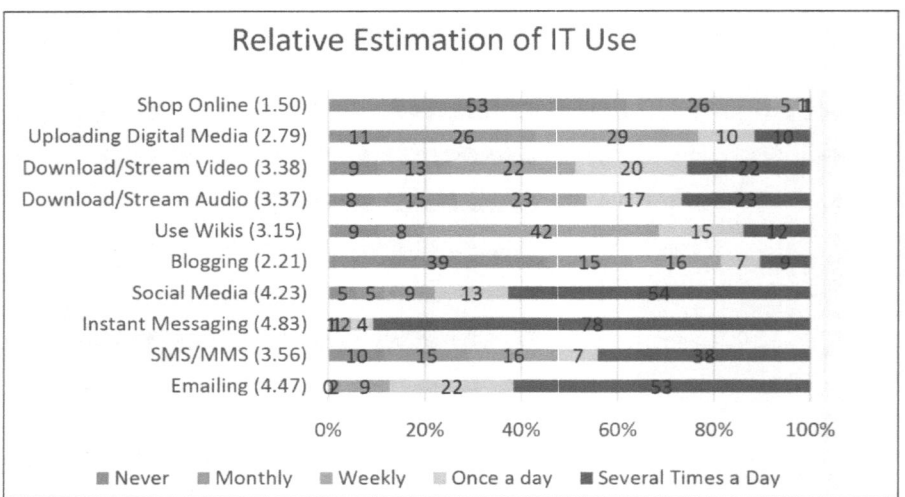

Fig. 4. Relative estimation of IT use

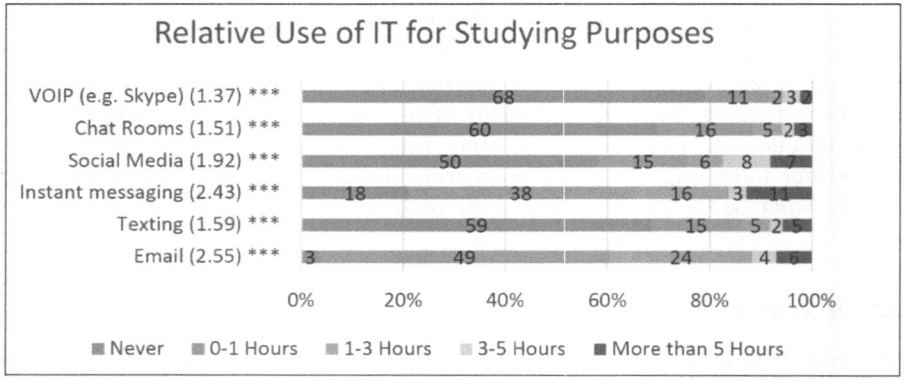

Fig. 5. Relative use of IT for studying purposes

Figure 5 indicates the relative use of IT for studying purposes. Perhaps not surprisingly, tools such VOIP, chat rooms, social media, IM and texting are used only by a small minority for studying purposes. This is in contrast to social life and leisure where half or more of the respondents use VOIP (43 i.e. 50%), social media (95%), IM (96%) and texting (61%). So, interestingly, their use of these tools in the private sphere does not seem to transfer to the study or academic sphere.

When asked about their self-rated efficacy of performing IT-related tasks, it was quite surprising to find that the majority of students do not rate themselves very confident in a number of crucial e-literacy areas (Fig. 6). Although a good majority rated themselves pretty or very confident in the areas of learning

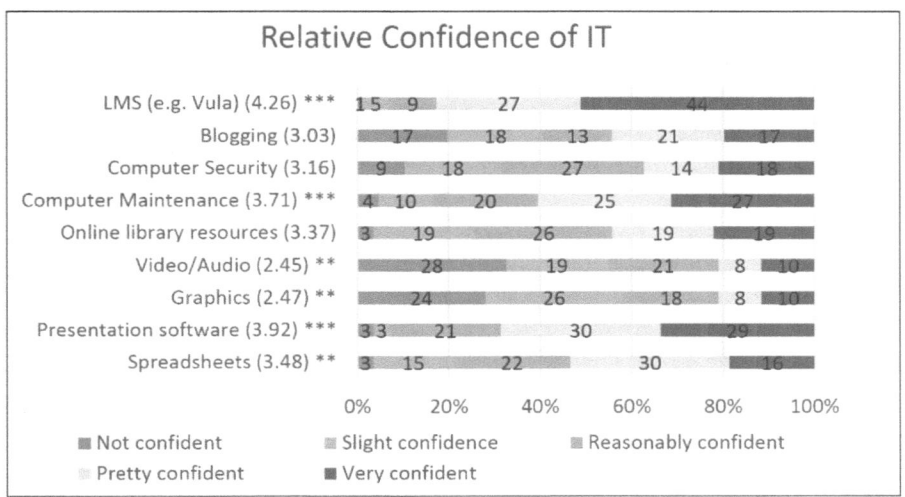

Fig. 6. Relative confidence in IT, with $** = p < .01$ and $*** = p < .001$ different from 'reasonably confident'

management systems (LMS), presentation software, spreadsheets and computer maintenance, the median student only rates him- or herself as reasonably confident in the areas of blogging, online library resources and, quite crucially, computer security. Given the essential nature of the latter two skills in an academic environment, this calls in our opinion for urgent remedial intervention. Also interesting is that, contrary to perceptions, the iGeneration still does not feel comfortable with manipulating multimedia such as graphics, video, or audio with less than half of the students having reasonable confidence in their own ability to do so.

4.3 Course-Specific Use of Technology

Students were asked what type of technology use was required by their courses. Unsurprisingly, the use of UCT's LMS 'Vula' system indicated the highest, almost universal count of 84 with email (72) having the second highest count (Fig. 7). The use of online quizzes and online tests scored as high as the access to course websites, by 55% of respondents. Although the use of electronic books or journals seems low, it must be remembered that these were first-year students; senior students are more likely to be required to use library resources for their course work. Other technologies such as wikis, blogs and social networking sites appear to be required only for a small number of students.

An interesting analysis looks at the iGeneration's preferences of electronic versus paper (i.e. material) format. Course material media preferences were measured across 5 categories. The majority (42%) indicated that they prefer mostly to read course materials on screen, and a further 33% 'always' read on screen. Only a minority of 14% indicated that they mostly 'print out'.

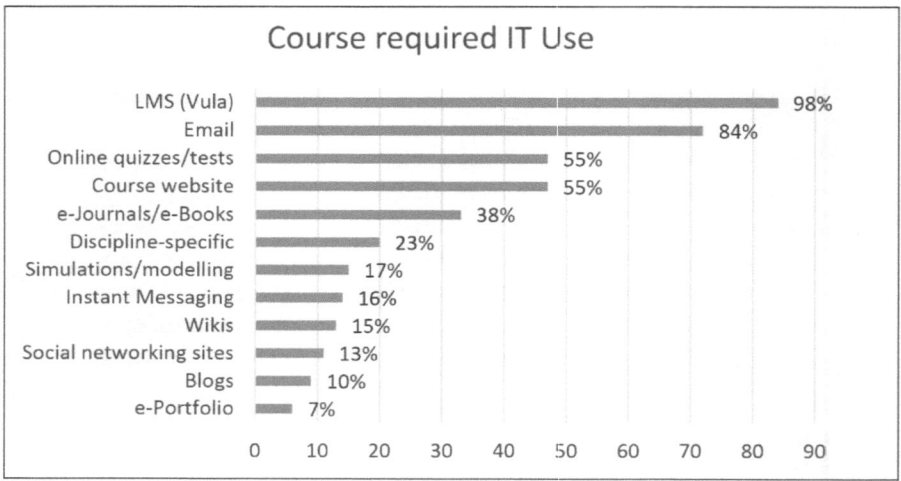

Fig. 7. Distribution of respondents' course-required IT usage

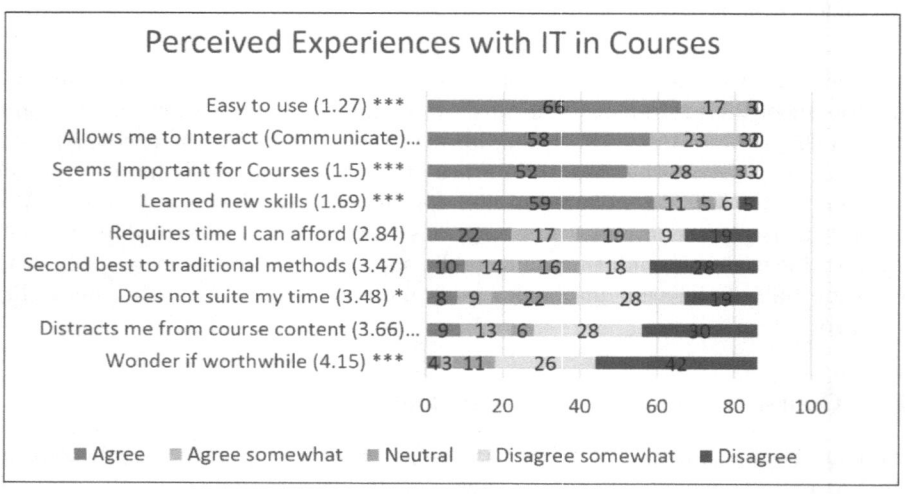

Fig. 8. IT experiences in courses, with $* = p < .05$ and $*** = p < .001$ different from 'neutral'

This attests to the iGeneration being fully accustomed to electronic-only media, although issues of environmental sustainability and financial reasons may contribute to this attitude; further research could validate the motivations.

Finally, the students' perceptions of IT experiences in courses was also probed. Most students agree that IT is easy to use, allows them to interact, seems important for the courses and also taught them new skills (Fig. 8). Generally, they disagreed (statistically significantly) with the statements that it distracted them from the course content or was not worthwhile. However, there was no strong signal that whether IT is second best or not when compared to traditional methods, or even if it requires too much time. On the whole, though, these results indicate a positive attitude towards IT in courses.

5 Findings, Implications, and Limitations

5.1 Summary of Findings

Access to Technology. The majority of respondents (62%) relied on public access (i.e. university computer room, library, etc.) for PC access. On the other hand, the majority of laptop access was through ownership (84%). Within regard to the computer ownership, 77% indicated that their devices fully met their needs. Network access points plays a critical role for access to resources and technology—34% relied on the student or student network for internet access while 28% relied on their wireless mobile connection for internet access. The device ownership was expanded beyond computer to include secondary devices too; of the respondents, all respondents owned a mobile phone and 91% owned a memory stick or card. The relevance of a mobile device became clear when 84% of respondents indicated that this was their most treasured device. Lastly, the relative importance of IT was measured for both studying and leisure using a 3-point scale. Within regard to IT for studying, the respondents indicated that a computer (80) was the most important with the highest count, although their mobile was the most important (79) for their leisure use.

Use of Technology. Mean reported exposure levels to both a computer screen and the internet for studying purposes were 6.22 and 5.79 h respectively, but with great variability in the responses. Respondents were asked to indicate their internet access points across 10 different locations against 'when studying my course' and 'when not studying'. The most notable figures were that of 95% using university computer lab rooms for 'when studying my course' and 92% using the university's library; this indicates a high reliance on university resources for studying purposes. In contrast, the majority of respondents indicated that they utilised their permanent home as an internet access point when not studying with a count of 56 (65%) or the use of their mobile internet with a count of 43 (50%). In addition, they were also asked to indicate their relative estimation of their IT use across 10 categories. Perhaps surprising, most (62%) of the respondents have never done online shopping; predictably, instant messaging (91%), emailing (62%), and social media (63%) were used several times a day. The survey the continued to ask the respondents about their social media activity. Thereby, 98% of students used social media before university; however, 67% indicated that their use of social media has increased while being at university. First-year students are not knowledge creators (yet): 77% have made no contributions towards a blog, and 93% never made a wiki contribution; hopefully life at a higher education institution will instill more of a knowledge creation attitude.

In regard to the relative use of IT for studying purposes, most students do not use the tools they employ frequently in their private (social and leisure) spheres in their academic roles: texting (69%), social media (58%), chat rooms (70%), and VOIP (79%) were never used by students for academic purposes even though the majority of students used them privately. Students were asked to measure their relative confidence of IT across 9 categories. Not surprisingly, they were quite confident of their skills using LMS (e.g. Vula), presentation and spreadsheet software. Disconcertingly, they were not very confident in their own perceived skills in respect of cyber-security and accessing online library resources; this issue would require educational attention. Contrary to expectation, the iGeneration has a very low confidence in manipulating multimedia (audio, images, video) or collaborative knowledge sharing (wikis and blogging).

Course-Specific Use of IT. The use of specific IT for course activities, revealed a surprisingly high use of online quizzes (55%) but low use of e-Books/e-Journals (38%) and even less for wikis (15%) and blogs (10%). The iGeneration no longer relies on printed materials, with a large majority reading and creating their materials in electronic format only—only 14% still prints much of their materials—although the motivation for this was not investigated. On the whole, students displayed a very positive attitude towards the use of technology in their courses, given that the majority of students perceived them to be relevant, useful and important to their courses and imparted new skills.

5.2 Research Questions and Implications

How do iGeneration Students Use Technology to Connect and Communicate with Various People on Campus? The findings suggest that students primarily rely on their access to the university/student network for internet access. Furthermore, although their mobile device was suggested as an important tool for studying, it showed a greater inclination towards that of a personal device rather than a study device. In addition, the relative importance of internet access was justified by the students indicating a high score for communication. This suggests that the majority of communication is digital. Lastly, social media and instant messaging played big roles for communication within a personal capacity, but instant messaging (i.e. WhatsApp) also played a big role as medium of communication for studying purposes.

What is the Relative Importance to iGeneration Students to Have Access to More or Better Technology for Themselves? The majority of respondents indicated positive attitudes towards the use of technology in higher education. This suggested that the technology made available by the university satisfied their needs and had a great relevance for their studies. The majority also indicated that the technology that they encountered as easy to use, opened interaction and communication, and facilitated the development of new IT skills. Lastly, the majority of respondents indicated that their access to certain technologies (i.e. computer) had fully met their requirements thus indicating that there was no need for better technology.

Cross-tabulation analysis suggests that the perceptions of importance do not differ across different cohorts of students. The attitudes and perceptions regarding the relative importance and use of IT within higher education indicates that responses do not vary much across faculty (i.e. Commerce, Sciences, and Humanities) nor across gender. This suggests that their attitudes and perceptions are uniformed as suggested by the literature that iGeneration students are seen as a unit.

To what Extent has iGeneration Students' Technology Use Enabled them to Understand and Demonstrate Their Understanding, or to Tudy on Their Own or with Others? The majority of respondents indicated that technology within courses has not only allowed them to learn and develop more skills, but it also provide a critical role towards course-specific learning tasks as well as communication facilitator as the majority of communication is in that of a digital format. Email and instant messaging are the key mediums of communication that help enable them to study on their own or with others. With respect to their academic studies, students agreed that the role of IT was critical.

5.3   Limitations

Our research was done in a cross-sectional timeframe and at a single higher education institution. The sample, with 86 responses, was also relatively small. Thus care should be taken in trying to extend the findings to the student body for South Africa as a whole. Only gender and faculty affiliation were used as demographic variables, and no significant influence of these demographics was found in the results. Other variables may well have a significant impact, e.g.: learning style, educational or socio-economic background, e-literacy and computer efficacy. Finally, the research was a positivist, quantitative survey. No prior theoretical model was used, so hardly any inferential statistics were possible; analysis was limited mostly to descriptive statistics. No in-depth, nuanced analysis of individual attitudes, differences in perspectives and views, motivations, rationales, expectations or mechanisms was possible; qualitative research approaches would complement the descriptive picture provided.

6   Recommendations

Our research, having followed an quantitative, mostly descriptive strategy, did not however consider the qualitative aspect of exploratory research. This may be expanded in future research to accommodate a better conceptual understanding of student attitudes and perceptions regarding the relative use and importance of IT in higher education, and in doing so having a mix-methods approach.

Our research also considered the relative attitudes and relative importance of IT in higher education from the perspective of one single subgroup, i.e.: first-year students. As per literature, the iGeneration—having been born in 1992 and onwards—could have been studied w.r.t. how their attitudes towards the relative importance and use may vary from first-years to that final-year students, as both could be classified suitably within the iGeneration band of time. In doing so, the variation in perceptions could further validate the results as per similar studies.

A larger sample size, stretching across multiple higher education institutions and across more faculties, would allow for a greater response as well as a greater variation in responses. Based on the initial findings; however, little variation is expected based on the profiling of this particular generation, but this may be explored further.

Lastly, the use of a conceptual model may be established to further guide similar research beyond that of the relatively technology acceptance models in order to better understand the high-level perceptions of technology usage within an academic context.

7 Conclusions

The purpose of our research was to explore the relative use and attitudes towards IT within higher education amongst iGeneration students. The literature review explored the iGeneration as a generation and their technology consumption. This was followed by a section on the role technology plays within the context of pedagogy. A quantitative survey was conducted among first-year students in the three largest faculties of a leading South African university. The analysis provided an insight into the relative use and attitudes. In doing so, communication technologies were deemed important, but they were differentiated based on studying purposes and that of academic application. In addition, it was found that technology played an integral role as an enabler as in the case of a reliance on university infrastructure to facilitate communications. Other aspects included the lack of variation across student cohorts regarding the attitudes and perceptions, but this could be explored further. Lastly, it was found that the technologies utilised proved to be sufficient regarding the needs of the iGeneration and that the majority of the technologies utilised were those provided by the institution rather than personal technologies.

Some findings supported some assumptions about the iGeneration, but debunked some others. For instance, the survey confirmed their dependency on mobile phones using IM, SNS, texting or VOIP for leisure/personal use; however, this is not carried over into their academic context or roles where they actually own and prefer to use their laptop and use email but not IM, SNS or VOIP. On the one side, very few of them rely on printed materials, prefer to consume and create electronically. They felt happy with the relatively high use of and big reliance on technology in the academic context. On the other side, they profess a lack of skills in the use of online library resources or cyber-security. Also surprising was their lack of ability to manipulate multimedia materials as well as their lack of participation in collaborative knowledge creation such as contributing to wikis or blogs. Educators would do well to address these gaps in the university curriculum.

Future research is recommended but it can build on our findings. A qualitative approach may useful to explore some of the findings that were not expected. It will also allow for a more accurate and nuanced representation of student perceptions regarding the relative use and attitude towards IT in higher education. Larger, more representative samples, especially at other tertiary educational institutions, would lend more credibility to the results and allow for more confident generalization of the findings presented here. Finally, other variables regarding the demographics of the respondents should be considered to fully understand how variations may occur based on aspects such as learning styles, or prior technology and education experiences.

References

1. Anderson, T., Dron, J.: Learning technology through three generations of technology-enhanced distance education pedagogy. Eur. J. Open Distance E-learn. **15**(2) (2012)
2. Beetham, H., Sharpe, R.: Rethinking Pedagogy for a Digital Age. Routledge, Abingdon (2013)
3. Fernandez-Lopez, A., Rodriguez-Fortiz, M.J., Rodriguez-Almendros, M.L., Martinez-Segura, M.J.: Mobile learning technology based on iOS devices to support students with special education needs. Comput. Educ. **61**, 77–90 (2013)
4. Hannon, J.: Incommensurate practices: sociomaterial entanglements of learning technology implementation. J. Comput. Assist. Learn. **29**(2), 168–178 (2013)
5. Ives, E.A.: iGeneration: The Social Cognitive Effects of Digital Technology on Teenagers. Doctoral Dissertation, Dominican University of California (2012)
6. Johnston, K.A.: A guide to educating different generations in south Africa. Issues Inf. Sci. Inf. Technol. **10**(36), 261–273 (2013)
7. Jones, C., Ramanau, R., Cross, S., Healing, G.: Net generation or digital natives: is there a distinct new generation entering university? Comput. Educ. **54**(3), 722–732 (2010)
8. Jones, V., Jo, J., Martin, P.: Future schools and how technology can be used to support millennial and generation-Z students. In: Proceedings Ubiquitous Information Technology CUT 2007(B), pp. 12–14 (2007)
9. Kennedy, G., Krause, K.L., Judd, T., Churchward, A., Gray, K.: First year students' experiences with technology: are they really digital natives? Australas. J. Educ. Technol. **24**(1), 108–122 (2006)
10. Mirriahi, N., Alonzo, D.: Shedding light on students' technology preferences: implications for academic development. J. Univ. Teach. Learn. Pract. **12**(1), 6 (2015)
11. Oliver, M.: Learning technology: theorising the tools we study. Br. J. Educ. Technol. **44**(1), 31–43 (2013)
12. Rideout, V.J., Foehr, U.G., Roberts, D.F.: Generation M^2: Media in the Lives of 8- to 18-Year-Olds. Henry J. Kaiser Family Foundation (2010)
13. Roblyer, M.D., McDaniel, M., Webb, M., Herman, J., Witty, J.V.: Findings on Facebook in higher education: a comparison of college faculty and student uses and perceptions of social networking sites. Internet High. Educ. **13**(3), 134–140 (2010)

14. Roodt, S., Peier, D.: Using YouTube in the classroom for the net generation of students. Issues Inform. Sci. Inform. Technol. **10**, 473–488 (2013)
15. Rosen, L.: Welcome to the iGeneration!. Educ. Digest **75**(8), 8–12 (2010)
16. Saunders, M., Lewis, P., Thornhill, A.: Research Methods for Business Students. Pearson Education, London (2009)
17. Schmid, R.F., Bernard, R.M., Borokhovski, E., Tamim, R.M., Abrami, P.C., Surkes, M.A., Woods, J.: The effects of technology use in post-secondary education: a meta-analysis of classroom applications. J. Comput. High. Educ. **26**(1), 87–122 (2014)
18. Sternberg, J.: It's the end of the university as we know it (and i feel fine): the generation Y student in higher education discourse. High. Educ. Res. Dev. **31**(4), 571–583 (2012)
19. Turkle, S.: The flight from conversation. N. Y. Times **22** (2012)
20. Twenge, J.M.: Generational changes and their impact in the classroom: teaching generation me. Med. Educ. **43**(5), 398–405 (2009)
21. Waldron, K.A.: The iGeneration: Technology Guidelines for Parents and Teachers. Trinity University (2012)

A New Generation of Students: Digital Media in Academic Contexts

Daniel B. le Roux and Douglas A. Parry[✉]

Department of Information Science, Stellenbosch University,
Stellenbosch, South Africa
dougaparry@sun.ac.za

Abstract. The growing presence of digital media in the lives of university students signals a change in how use of such media in educational contexts should be viewed. Institutional focus on technologically mediated education and the promotion of blended learning initiatives further serve to encourage media use in academic settings. Scant attention has been afforded to the potential negative consequences arising from heightened media engagement. This is especially the case in areas of study where technological artifacts are often the medium and the subject of interest, for instance the computer and information sciences. In this study a survey was done to investigate students' use of media, as well as the behavioural beliefs, norms and motivators surrounding such use.

Keywords: Media multitasking · Beliefs · Norms · New media · Students · Behaviour

1 Introduction and Research Questions

With the growing 'hype' surrounding *Blended Learning* and the potential benefits of digital technology in tertiary institutions, the prospects of negative implications associated with the use of personal digital media in educational settings are seldom considered. We regard this position to be particularly tenuous in light of the growing body of knowledge suggesting the *potential for negative consequences associated with digital media use in academic settings* [5,11,17,20,28,31]. Mobile digital media devices such as laptops, tablets and smartphones have become ubiquitous on today's university campuses [9]. The ubiquity of media, as well as the characteristics inherent in modern digital media have contributed to the growing prevalence of continuous media use among the current generation of university students. In describing members of this generation, which includes current university students, as the 'net generation' [29] or the 'digital natives' [26], the significant role media plays in shaping their lives is further highlighted. Increasingly, students are engaging in media multitasking behaviour, rapidly switching between numerous ongoing activities. Such behaviour has been shown to *reduce the ability to pay attention to individual*

© Springer International Publishing AG 2017
J. Liebenberg and S. Gruner (Eds.): SACLA 2017, CCIS 730, pp. 19–36, 2017.
https://doi.org/10.1007/978-3-319-69670-6_2

tasks [13,16]. These outcomes suggest that media multitasking implies cognitive costs, impeding the processing and encoding of information into long term memory—key functions necessary for learning to take place [16]. Studies in this area indicate that there exists a negative correlation between media multitasking frequency in academic contexts and academic performance [31]. Owing to the ubiquity of extensive media use and media multitasking behaviour amongst contemporary student populations, the development of a deeper understanding of the beliefs, norms, motivations and nature of students' media use is of substantial importance.

In addition to the ubiquity and presence of media in students' lives, the increased institutional focus on technologically mediated education and the continued promotion of blended learning and e-learning initiatives imply that the opportunities for students to use media whilst in academically focused situations are greater now than ever before. Blended learning typically involves the integration of physical, face-to-face, lecture-based learning opportunities with online learning experiences [14]. Through blended learning, digital media use in academic settings is not specifically discouraged, rather, it is explicitly encouraged. This is especially noticeable with the structural interventions in many academic institutions such as Wifi access points in lecture halls and the increasing prominence of e-learning environments. Furthermore, the use of digital media technologies is particularly commonplace in areas of study where technological artifacts are often the medium *and* the subject of interest, for instance the computer and information sciences. Within these academic subjects students typically conduct assignments and projects on devices such as desktop computers, laptops, tablets and even smartphones.

University students are characterised as being members of the 'net generation' [29], a cohort displaying an unprecedented propensity for engaging and interacting with digital media, both throughout their everyday lives, as well as in the course of their academic experiences [8,17]. Exemplifying this line of reasoning, a recent EDUCAUSE Center for Applied Research (ECAR) study (with $\approx 50,000$ respondents in 11 countries at 161 universities) indicates that 98% of the students posses a mobile computing device (laptop, smartphone or tablet), capable of Internet connection [9]. While one may reason that these statistics are reflective of the students in developed economies, a study of mobile phone use by South African university students [24] found that only 1% did not own a mobile phone, or had not owned one recently. Similarly, an earlier study of low-income South African students shows extensive use of mobile devices [19].

In addition to extensive engagement with media in their general lives, many studies show that media use within academic settings has become increasingly common [6,13,15,20,28]. This is both the case in structured as well as self-regulated academic contexts. For the purpose of this study a 'structured academic context' is defined as a *classroom based environment within which students observe and record material provided by a facilitator.* Correspondingly, a 'self-regulated academic context' is defined as *a student or group of students under-*

taking academic work without direct supervision by a facilitator, either within a personal or public study environment.

From the aforementioned definitions it is clear that structured and self-regulated academic environments are distinguished by different physical properties as well as distinct social, and behavioural norms. In a structured academic context the presence of a facilitator is the key element distinguishing such an environment. The facilitator is responsible for presenting material to the students. In addition to this, the facilitator regulates the behaviour of the students within this context through the establishment and maintenance of order [2]. However, students' behaviour within this environment is not only a function of the facilitator, but is also modulated by the behavioural and social norms established by their peers [3]. While structured lecture contexts might constitute the primary academic environment experienced by students, students spend a far larger amount of their time engaged in informal, self-regulated study environments outside of scheduled class times [21]. These environments are defined as being self-regulated, because the behaviour exhibited within them is not dependent on external rules imposed by a facilitator. For the most part, the nature of these environments is determined by the individual's personal choices [32].

We believe that academic staff in technical domains such as Informatics, Computer Science and Information Science have an important role to play in shaping the outcomes of blended learning initiatives at universities. It is our perception that staff with limited knowledge of the inner-workings of computer-based systems often operate under naïve assumptions of the affordances of digital technologies for learning, leading to misguided application efforts. While we acknowledge the potential of digital media in learning, we argue for a *balanced perspective which acknowledges both the positive and negative consequences of digital media use* in academic contexts.

In this study we investigate data collected through a survey (with $n = 1,678$) conducted at a large, public South African university. The survey concerns, primarily, respondents' online media use in general and in structured academic settings and, in addition, considers the beliefs, norms and motivators surrounding media use. Specifically, our study aims to answer these questions:

1. *Which media do students generally use and how frequently do they use them?*
2. *Which media do students use in structured academic contexts and how frequently do they use them?*
3. *What beliefs do students hold in relation to their media use in structured academic contexts?*
4. *What triggers media use in structured academic contexts?*
5. *Does media use in structured academic contexts influence academic performance?*

We discuss the implications of our findings for teaching and learning at tertiary educational institutions within South Africa, whereby specific attention is given to the implications of our findings for the teaching of computer and information sciences.

2 Literature

2.1 Students' Digital Media Use

Over the preceding decade a number of studies have focused on the growing prevalence of students' media use [4,5,13,16–18,20]. These studies reveal that media use is commonplace for students whether they are engaged in academic study (structured or self-regulated), or in non-academic activities.

A study focusing on the variety and frequency of media used by students [30] classifies students' media use into nine distinct categories. Of these nine categories, the two categories referred to by [30] as 'rapid communication technology' and 'web resources' were used most frequently by a majority of the students surveyed. Combined, these categories include activities such as: calling or texting on a mobile phone, using social networking sites, watching online video, and web-searching.

Using an experience sampling method, [23] shows a real-time examination of students' Internet behaviour and reveals that on average students spent 56 min online per day [23]. This result represents a significantly smaller amount of time than suggested by studies relying on self-reported data: for instance, a large sample of American students w.r.t. their digital media usage habits is discussed in [17]. Results from that survey indicate that on average students spend over two hours per day engaging with online media.

Over and above media use in the course of their everyday lives, numerous studies support the argument that students' media use within academic settings has become increasingly common. In [15], for example, data were gathered about media use during academic study, finding that two-thirds of the sampled students reported media use either while in class or in self-regulated study. Similarly, [28] shows students' media multitasking habits in their own personal study environments, whereby students averaged less than six minutes on task before switching to another task. The most frequent causes of task-switching were observed to be technological distractions such as social media and texting [28].

In [6] we can find an experimental approach to determine the number of media interruptions students experience, the duration of these interruptions as well as the proportion of study time devoted to media multitasking behaviour. It was found that students engaged with an average of 35 distractions of six seconds or longer, with an aggregated mean duration of 25 min [6], whereby cellphone and laptop use constituted the largest frequency and duration of distraction from academic work. Those results are commensurate with other studies in this area such as [10,13,28].

Following a survey of 1,839 students, in-lecture media use was classified in [16] into three categories: 'high', 'moderate' and 'low' frequency media use. Of the categories examined, texting on a mobile phone was found to be the only media activity which could be classified as high-frequency, with 69% of the sample indicating in-lecture texting activity [16].

In a study examining the nature of students' in-class laptop use, it was found that students spend a substantial amount of time multitasking on laptops within a lecture [13], whereby over the 20 week period of the study students reported using their laptops for non-lecture related activities for an average of 17 min out of each 75 min lecture [13].

In another survey study, conducted in the USA [22], it was found that 92% of the respondents used a digital device during a lecture for off-task activities at least once during a typical day. Similar to the results of [16], texting was found to be the most frequently engaged in activity, with email, social networking and web-browsing following. In [22] over 80% of the students indicated that multitasking with a digital device in class caused them to pay less attention.

The use of digital media in university lectures and the use of mobile devices whilst attending lectures was also discussed in [27] where it was found that 66% of the respondents used a mobile device whilst in lectures, and that laptops were primarily used for non-academic purposes [27].

Research in the South African context is generally congruent with the aforementioned international studies. For instance, [24] shows mobile phone usage by South African university students, finding that in a sample of 362 students only 1% did not own a mobile phone. Another study surveys 500 low-income South African students [19] and shows extensive use of mobile devices amongst this demographic sample. That paper shows that for low-income students in South Africa, a mobile device constitutes their primary connection to the Internet, with 83% of participants accessing mobile Internet applications on a daily basis.

Similarly, in another South African university survey [20], it was found that the dominant class of media in use by South African students in a lecture context is non-academic in nature. Specifically, instant messaging and social networking are the most frequently used media during a lecture [20].

In summary, students spend a significant amount of their time engaging with digital media in many varied forms and contexts, in general as well as throughout formal, structured educational settings. Hence, nowadays *students' lives are to a large extent mediated* by the digital technology through which they engage many aspects of the world.

2.2 Students' Beliefs and Behavioural Norms

Research focusing on students' perceptions of media use, social norms and beliefs about media use in academic settings is limited. 'Normative beliefs' are defined as an individuals' perception of social normative pressures on them to perform a certain behaviour [12]. Similarly, behavioural beliefs are described as an individual's beliefs about the consequences arising from a particular behaviour [12]. In the context of students' media use, normative beliefs relate to social pressures to engage with digital media in academic settings. Correspondingly, behavioural beliefs relate to perceptions about the implications of media use for their academic performance.

While there may be limited research within this area, two focus-group based studies provide some level of insight. At a medical university in the United States, all students were required to use laptops for study purposes. A number of focus-groups was employed in [1] to investigate the benefits and drawbacks of the required laptop program at that university. Students in the focus groups reported frequently using laptops in lectures for off-task activities. They explained that they commonly engage in social media and web browsing when they become bored with the lecture [1]. Despite reporting the use of laptops for off-task activities, the students explained that from their perspective, laptops improved their communication abilities, access to learning material as well as increased the flexibility of education [1]. However, the most frequently reported drawback of laptop use was distraction. This outcome is in agreement with [13] where students perceive their own use of a laptop as well as that of those around them to be the single greatest distraction to learning in the classroom setting.

A later series of focus groups was conducted at a South African university [25], focusing on students' behavioural beliefs, the triggers underlying their behaviour and the nature of their behaviour with media in academic contexts. Through a thematic analysis of these focus groups it was found that students are cognizant of the impact that media multitasking has on their cognitive functioning and, as a consequence, their academic performance. This is especially the case when in structured lecture contexts. In terms of social norms, [25] indicates that students believe that the behaviour that they exhibit with digital media is shared by their peer group and that off-task media use in academic settings has become a normal mode of functioning.

3   Method

While many aspects of students' media use have been extensively studied (as discussed above), there remain many unanswered questions in this domain. Few studies have been conducted in the South African higher education context, and little is known of the role demographic factors play in moderating media use. Also the influence of subjective norms and beliefs on media use frequency in structured academic contexts has yet to be investigated. This paper sets out to investigate online media use among university students, both within and outside structured academic contexts. To this end a *survey-based approach* was followed, enabling the collection of data from a large population through a series of *Likert-type questions*. The sections which follow address the survey design and the data collection.

3.1   Survey Design

A web-based survey consisting of three sections was utilised to collect data relating to the research problem. The first section of the survey concerned subjects' media use behaviour in and outside of structured academic contexts (i.e., lectures, practical classes or tutorial classes). Six forms of online media typically used by the target population were selected. These included:

1. Social Networks (SN);
2. Micro-blogs (MB);

3. Encyclopedic (or structured data) browsing (ENC);
4. Instant Messaging (IM);
5. Search (engine) activities (SE);
6. The university's e-learning platform (EL).

For each medium, general use frequency was elicited through a five-point Likert-type question with indicators for 'not at all', 'sometimes (at least once per month)', 'often (at least once per week)', 'at most once a day', and 'many times per day'. For use frequency in structured academic contexts (which we refer to as *in-lecture use*) the indicators were 'not at all', 'once or twice', 'every 10 min', 'every 5 min', and 'constantly'.

The second section of the survey concerned, firstly, subjects' beliefs about social norms regarding the use of smartphones in structured academic contexts and, secondly, their beliefs about the motivations or triggers underlying smartphone use in these contexts.

In the third and final section of the survey demographic variables were elicited. These included age, gender, language, highest qualification of parents and area of study (i.e. the subject's home faculty). In addition to these subjects were asked to report their general level of academic performance through a scale ranging, in 5% intervals, from 'below 40%' to '96%–100%'.

3.2 Data Collection

A single round of invitations to complete the survey was sent to 14,122 undergraduate students across all faculties at a large South African university. Because the survey contained questions concerning subjects' academic performance in the previous academic year, first-year students were excluded from the list of recipients. Completion of the survey was incentivised by offering subjects that have completed the survey a chance to win a gift voucher through a separately managed lucky draw. A total of 1,678 completed surveys were submitted within a three-week period following the invitation.

4 Data Analysis

The data analysis is presented in five parts. The first provides a descriptive overview of the sample population based on the following demographic variables: age, gender, language, highest qualification of parents and area of study (i.e. the subject's home faculty). This is followed by the analysis of the media use patterns (in general and during lectures) in relation to the six demographic variables. We then consider data relating to students' beliefs and norms about in-lecture media use, followed by data concerning the reasons (triggers) underlying instances of use in-lecture use. Finally, we briefly consider the relationship between media use and academic performance.

4.1 Sample Population

Just over 82% of the 1,678 students that completed the survey are between 20 and 23 years of age. Of the remaining subjects 7% did not disclose their age while the rest are older (mostly 24 or 25 years of age). The sample included slightly more female (51%) than male (49%) respondents. In terms of first language, 47.6% of the respondents indicated Afrikaans, while 42.5% indicated English. The remaining 10% indicated isiXhosa (2.3%), Zulu (1.9%), Sepedi (0.8%), other African languages (3%), or other European languages (1.9%). Almost 60% of the respondents indicated that at least one of their parents has a university qualification. This includes Bachelor's degrees (31.1%), Honours and/or Master degrees (24.6%),[1] and Doctorates (4.2%). However, a large section of the respondents' parents' highest qualification is a secondary-school (pre-university) certificate (38.4%). The university's 10 faculties were generally well-represented in the sample with the exception of the faculties of Education (40 respondents), Theology (19 respondents) and Military Science (10 respondents). Table 1 shows a summary of the number of the respondents for each faculty.

4.2 Media Use (in General)

The medium used most frequently among the sample population is instant messaging (IM) with a mean frequency of 4.9 (on a 5–1 Likert scale), whereby 97% of the respondents indicated that they use IM multiple times per day. Only 12 respondents indicated that they do not use IM at all. This is followed by search engines with a mean use frequency of 4.73 and the university's learning platform with a mean frequency of 4.5. Social networks are used only slightly less frequently (4.36) while encyclopedias (2.97) and micro-blogs (2.23) are used with substantially lower frequencies. Comparison of general media use frequencies based on gender reveals that female students tend to use social media more frequently than male students. This is the case for instant messaging (4.96 vs. 4.91), social networks (4.5 vs. 4.21) and micro-blogs (2.49 vs. 1.96). Male students are more frequent users of encyclopedias (3.1 vs. 2.85), search engines (4.77 vs. 4.68) and the university's learning platform (4.54 vs. 4.47). The data offer little evidence that the age difference between subjects in the sample has a substantial influence on media use behaviour. When considering use across all media there is a slight drop in frequency between the ages of 20 and 23. Respondents aged 20 have a mean frequency of 24.0 across the six media while for those aged 21 and 22 it is 23.9 and 23.5 respectively. However, this number increases to 23.6 for respondents aged 23. Much like the respondents' age, the data indicate that neither first language, home faculty or parents' highest qualification influence general media use frequency in any substantial way.

---

[1] For readers from outside South Africa: the South African 'honours' degree is an extension of the classical 'B.Sc.' degree which enables a student to commence with Master-studies thereafter. While already considered 'postgraduate' in South Africa, the 'honours' degree in South Africa is reasonably well comparable to the final study-year in the (longer) U.S.American 'B.Sc.' curriculum.

Table 1. Respondents per home faculty

Faculty	# Respondents	Percent (%)	Cumulative percent (%)
Engineering	402	24.0	24.0
Economics and management	353	21.0	45.0
Medicine and health science	263	15.7	60.7
Arts and social sciences	223	13.3	74.0
Natural sciences	196	11.7	85.7
Agricultural sciences	107	6.4	92.1
Law (Jurisprudence)	63	3.8	95.9
Education (Pedagogics)	40	2.4	98.3
Theology	19	1.1	99.4
Military science	10	0.6	99.9
(Missing)	2	0.1	100
(Total)	1678	100	100

4.3   Media Use During Lectures

The sessions (i.e.: lectures, practical classes and tutorials) at the university where the data were collected are all 50 min in duration. We customised the Likert items and their indicators to elicit a respondent's use frequency during a single session. The resulting five-point scale had indicators for 'not at all', 'once or twice', 'every 10 min', 'every 5 min', and 'constantly'. As is the case for media use in general, IM is the medium used most frequently during lectures with a mean use frequency of 2.82. Thereby, 16% of the respondents reported that they IM constantly during lectures with another 30% indicating that they do so at least every 10 min. Search engines (mean = 2.06) closely followed by social networks (mean = 2.03) were the second and third most used media, while the e-learning platform (mean = 1.96), encyclopaedia (mean = 1.51) and micro-blogs (mean = 1.32) were used less frequently. In terms of gender differences the same pattern observable for general use repeats itself for in-lecture use. Male students reported slightly higher use of search engines, encyclopaedia and the e-learning platform, while female students reported slightly higher use of social media. The mean scores are presented in Table 2. Finally, the data offered no evidence that students' in-lecture media use frequencies differ based on their parents' highest academic qualification. This suggests that socio-economic status is not a determinant of in-lecture use.

Table 2. Mean in-lecture use frequency of different media, by gender

Gender	SN	MB	ENC	IM	SE	EL
Male	1.99	1.27	1.53	2.81	2.14	1.98
Female	2.07	1.36	1.48	2.83	1.98	1.94

With the exception of instant messaging, there is a slight but consistent decrease in in-lecture use of *social* media for *older* respondents. Table 3 presents the mean use frequencies for our four primary age cohorts (20–23 years). The data suggest that older students tend to use social networks and micro-blogs less during lectures, but encyclopaedia more. The mean use across all media, calculated as a sum of the six media, drops from 12.08 for 20-year olds to 11.7 for 23-year olds.

Table 3. Mean in-lecture use frequency of different media, by age groups

Age	SN	MB	ENC	IM	SE	EL
20	2.14	1.37	1.46	2.97	2.08	2.06
21	2.05	1.32	1.47	2.85	2.03	2.01
22	1.95	1.31	1.53	2.77	2.01	1.84
23	1.98	1.26	1.61	2.88	2.14	1.82

As indicated in Table 4, Military Science emerged as the faculty with the highest in-lecture use across all media. However, the number of respondents from that faculty (only 10) is too low to justify conclusions to be drawn. The other nine faculties are bunched between 12.33 (Education) and 10.94 (Law).

A more detailed breakdown of in-lecture use per faculty is given in Table 5 which shows the mean frequencies for each medium.

4.4   Beliefs and Norms

We now turn our attention to students' beliefs and norms regarding in-lecture media use. Respondents were asked to indicate their level of agreement with two statements using a five-point Likert scale. The provided statements (to agree or to disagree with) were:

Table 4. Mean in-lecture use frequency for all media and faculties

Faculty	Mean value	N
Agricultural sciences	11.02	107
Arts and social sciences	12.16	223
Economics and management	11.87	353
Education (Pedagogics)	12.33	40
Engineering	11.40	402
Law (Jurisprudence)	10.94	63
Medicine and health science	12.25	263
Military science	14.60	10
Natural sciences	10.97	196
Theology	11.89	19

Table 5. Mean in-lecture use frequency of different media, by faculty

Faculty	SN	MB	ENC	IM	SE	EL
Agricultural sciences	1.95	1.18	1.46	2.77	1.90	1.77
Arts and social sciences	2.13	1.49	1.59	2.79	2.10	2.05
Economics and management	2.19	1.32	1.37	2.93	2.05	2.01
Education (Pedagogics)	2.03	1.35	1.53	3.10	2.20	2.13
Engineering	1.97	1.23	1.44	2.79	2.04	1.94
Law (Jurisprudence)	2.14	1.25	1.29	2.84	1.70	1.71
Medicine and health science	2.03	1.41	1.75	2.90	2.23	1.93
Military science	2.30	1.50	2.10	2.90	3.10	2.70
Natural sciences	1.77	1.26	1.49	2.63	1.92	1.90
Theology	1.95	1.21	1.63	2.47	2.32	2.32

1. *It is acceptable to use my phone during lectures.*
2. *In-lecture mobile phone use negatively affects my studies.*

Close to 30% of the respondents either agreed (20%) or strongly agreed (9%) that it is acceptable to use their phones during lectures. 39% of the respondents felt neutral about the statement while the remaining students either disagreed (24%) or strongly disagreed (8%). Stronger agreement was found in relation to the second statement with 44% of the respondents either agreeing (27%) or strongly agreeing (17%) that in-lecture phone use negatively impacts their studies. Only 15% of the respondents disagreed, and 9% strongly disagreed with this statement. In the case of both statements students' beliefs were found to correlate significantly with their in-lecture use frequencies. Correlation tests revealed that belief about the acceptability of in-lecture phone use is a stronger predictor of use (with $\rho = 0.33, p < 0.01$) than belief about the impact of in-lecture phone use on academic performance (with $\rho = -0.10, p < 0.01$). These correlations are illustrated in Fig. 1.

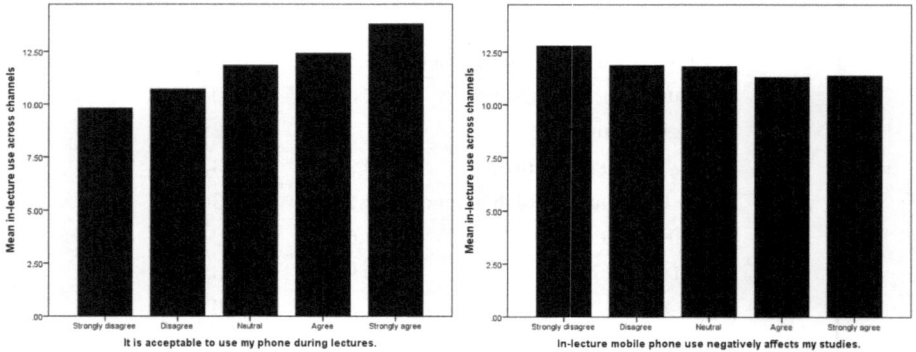

Fig. 1. Frequency of in-lecture media use in relation to beliefs about its acceptability (left-hand-side), and frequency of in-lecture media use in relation to beliefs about its effect on academic performance (right-hand-side)

Despite the existence of these correlations it is worth noting that students often act in contradiction to their beliefs. For example, 339 respondents indicated that they disagreed that it is acceptable to use phones during lectures. However, 257 of these respondents indicated that they used their phones at least once or twice during lectures to send instant messages. Similarly, 60 of the respondents who indicated that they used their phones once or twice per lecture to access a social network strongly disagreed with the acceptability of in-lecture phone use. The same pattern is also observable for beliefs about in-lecture phone use and academic performance. Of the 285 respondents who strongly agreed that in-lecture phone use negatively affects their studies, 49 constantly used IM during lectures and another 74 did so at least every 10 min. Only 24 of the 285 indicated that they did not use IM at all during lectures. The respondents were also asked to indicate how frequently in-lecture media use is discouraged by lecturers and peers. A five-point Likert scale ranging from 'never' to 'always' was provided in relation to the following two statements:

1. *Lecturers discourage mobile phone use during lectures.*
2. *My classmates discourage mobile phone use during lectures.*

33% of the respondents reported that lecturers always discourage in-lecture phone use (12.4%) or do so most of the time (20.5%). The largest portion of the respondents (39%) indicated that this happens sometimes. 57% of the respondents reported that their classmates never discourage in-lecture phone use, and only 19% indicated that this happens sometimes (16%), most of the time (2%), or always (1%). Our analysis revealed that neither of these items are predictors of the level of in-lecture use frequency reported.

4.5   Reasons for Use

The survey tested for five reasons for use (RU_i) using five-point Likert questions with frequency indicators ranging from 'never' to 'always'. The five items tested for were:

RU_1: *Use as a response to boredom.*
RU_2: *Use to chat with friends.*
RU_3: *Use to stay up to date with current events.*
RU_4: *Use to find information relevant to the lecture.*
RU_5: *Use after receiving a notification.*

All in all, the respondents indicated that the receipt of notifications is the most frequent trigger of in-lecture phone use (3.39), followed by experiences of boredom (3.03) and the desire to chat with friends (2.9). Use triggered by the desire to find information related to the lecture (2.73) or about current events (2.69) were reported to occur less frequently. Our analysis further indicated that the demographic variables tested for did not influence these numbers in any noteworthy manner.

4.6   Use and Academic Performance

Now we briefly consider the variables tested for in relation to the respondents' self-reported academic performance.[2] Our data indicate, firstly, that media use in general does not correlate with academic performance across the sample. However, media use during lectures (calculated as the sum of frequency of use of all media tested for) was found to correlate negatively with self-reported academic performance (with $\rho = -0.07, p < 0.01$).

A *counter-intuitive finding* was made upon analysis of the relationship between the general use of individual media and academic performance. The only medium for which correlation was found is the university's e-learning platform and, interestingly, this correlation is *negative* ($\rho = -0.09, p < 0.01$). Explanation of this finding requires further investigation. However, we believe it may suggest that students who consider the e-learning platform as a replacement of lectures and, as such, over-rely on it, tend to perform worse in assessments.

Analysis of the relationship between in-lecture use of individual media and academic performance revealed that social networks (with $\rho = -0.07, p < 0.01$), and instant messaging (with $\rho = -0.06, p < 0.01$) were the media responsible for the negative correlation observed between in-lecture use and academic performance reported above.

---

[2] The scope of this paper does not allow for a detailed analysis of the relationship between media use and academic performance. Our findings in this regard shall be published separately.

5   Discussion

In the section which follows we discuss our findings in relation to five themes. Firstly, we consider our findings in the context of the recent 'hype' around 'blended learning'. Then we consider demographic variables as predictors of media use. This is followed by considerations of our findings regarding the students' beliefs about media use and, thereafter, the role of institutions in managing this behaviour. Finally, we briefly reflect upon the findings made regarding media use and academic performance.

It is our opinion that the recent popularity of 'blended learning' initiatives in higher education environments have, perhaps inadvertently, *cultivated social norms around media use in structured academic contexts that may be obstructive*, as opposed to conducive, to learning. This, we believe, is primarily a result of naïvety about the distractions contemporary technology introduce to learning environments. Our data, accordingly, clearly indicate that high levels of off-task media use has become the norm in structured academic contexts. While there is evidence that students still perceive this as unacceptable behaviour, the greatest majority engage in multiple instances of off-task media use in a single 50-min lecture.

These findings should be contextualised within the rapid development of both mobile devices themselves and the applications they afford. Students have, at their fingertips, an increasingly wide range of entertainment and social attractions. Institutional infrastructure like WiFi-enabled lecture halls not only enables in-lecture media use but serves to establish such behaviour as the norm among students. In turn, *lecturers are placed in the difficult position of having to compete with media for students' attention*. Their challenge is exacerbated by, what seems to be, a decrease in the value students attribute to lectures as learning opportunities [25].

While we acknowledge their value and strongly encourage the continuation and expansion of blended learning initiatives, our findings raise a 'red flag' regarding the range of implications resulting from their implementation. The successful harnessing of this double-edged sword requires thoughtful management. As the attention economy becomes more competitive facilitators of learning should emphasise and enable attentional control among students, something which might imply limiting rather than maximising media use in learning environments. This is perhaps best illustrated by the finding that higher levels of use of the e-learning platform correlates with lower academic performance. Though this finding requires further investigation, it should serve to encourage careful, balanced deliberation about the role of technology learning.

In terms of demographics our findings suggest that high levels of media use in and outside academic contexts are not limited to particular subpopulations. However, slight variations in use behaviour are detectable. Female students use social media slightly more frequently than their male counterparts, while male students use encyclopedia and search engines more frequently. Likewise, older students tend to be more frequent users of encyclopedia and search engines and less frequent users of social media. We emphasise that these differences are not substantial and provide little ground for conclusions to be drawn.

In accordance with prominent technology-use theories [12], our data indicate that students' subjective norms concerning in-lecture media use influence their in-lecture media use frequency. Similarly, beliefs about the negative effects of use on academic performance were also found to be a statistically significant predictor of in-lecture use frequency. It should be noted that our data did not afford an in-depth understanding of the role beliefs and norms play in this regard. While [25] has made notable progress in this area, we encourage future research to provide, particularly through qualitative data, insight into the subjective factors that determine media use.

Our data suggest that lecturers at the institution where the study was performed do, at least sometimes, discourage in-lecture media use. This, however, does not seem to be a broadly accepted (enacted) policy. Furthermore, students' perception of how regularly this occurs was found not to be a predictor of their in-lecture use frequency. In accordance with these findings it is our position that institutions will achieve little success if they attempt to manage in-lecture media use through top-down control mechanisms.

Finally, our study confirmed previously reported findings regarding the negative correlation between in-lecture media use and academic performance. Of course, this correlation may be interpreted in a variety of ways and does not necessarily suggest that in-lecture media use impedes learning. A more detailed analysis of this phenomenon is forthcoming. However, it is our proposition that the high levels of media use observed among students encourage a 'culture of distractibility' [7] which has, under the guise of 'blended learning', become ingrained in structured academic contexts at higher learning institutions.

Several of the more important *limitations* require attention. First, this study used a self-report survey as its data gathering technique. While yielding a large number of responses, there exist potential limitations in the nature of the findings that can be generated through this technique, potentially limiting the insights possible. The second limitation of this study is due to the use of discrete media activities in the survey design. This design potentially limits the reporting of other media (either on or off-task) engaged in by students. This may imply that students' use of media is in fact under-reported. Third, by limiting the scope of students' responses, through the use of Likert scales, the potential for further qualitative insight into their beliefs and behaviour is limited. However, despite limitations such as these, the study findings are of interest because of the large sample size, the unique context, and the novel insights into students' behaviours and beliefs uncovered.

6   Conclusion

The current generation of students, as members of the 'net generation', display a propensity for continuous media use throughout their lives, as well as in structured academic contexts. This behaviour is further encouraged by increased institutional moves towards blended learning and e-learning strategies. This is especially the case for the fields of computer science and, information science, where these digital artifacts are often the medium and the subject of interest.

The findings presented in this study indicate that demographic factors are irrelevant as predictors of media use. For all students, using media for off-task purposes during lectures has become the normal mode of functioning. This finding, in combination with the finding that no link has been found between institutional attempts to curb off-task media use and changes in students' behaviour, suggests that such behaviour is so widespread and entrenched that it has come to define the current generation of students in higher education.

Extending from this study are a number of potential *future research* directions. This study primarily focused on the behaviour and beliefs displayed by students in regards to their use of media. Future research should focus on the role media play in fostering a distracted state of mind in both students, as well as in the general population. In addition to this, we conjecture that the issue of students' beliefs and perceptions of behavioural norms is nuanced to such a degree that further research in this area would benefit from adopting a qualitative method. In particular, we encourage those educators undertaking blended learning initiatives to carefully and continuously monitoring student behaviour through the collection of qualitative data. This feedback should reveal differences between the desired behaviour and the actual behaviour of learners in terms of media use. We propose, accordingly, that researchers should be sensitive to the emergence of unforeseen behavioural patterns in which digital technology obstructs rather than aids learning. In addition to this, these findings should be shared with colleagues at the appropriate institutional fora.

The pervasive ubiquity, and extensive media use within educational institutions continues to raise profoundly important questions about the impact of such media within these contexts. We acknowledge the potential of digital media in learning. However, we argue for a balanced perspective which acknowledges both the positive and negative consequences arising from digital media use in academic contexts. In this regard, we hold the view that those with technical knowledge have an important role to play in acknowledging the changing nature of the student population and, how academic institutions propose to meet this challenge.

References

1. Annan-Coultas, D.L.: Laptops as instructional tools: student perceptions. Tech Trends **56**(5), 34–42 (2012)
2. Bain, K.: What the Best College Teachers Do. Harvard University Press, Cambridge (2004)
3. Berkowitz, A.D.: The social norms approach: theory, research, and annotated bibliography. Technical report (2004)
4. Blackburn, K., Lefebvre, L., Richardson, E.: Technology interruptions: technological task interruptions in the classroom. Fla. Commun. J. **41**(2), 107–116 (2013)

5. Burak, L.: Multitasking in the University classroom. Int. J. Scholarsh. Teach. Learn. **6**(2), 8 (2012)
6. Calderwood, C., Ackerman, P.L., Conklin, E.M.: What else do college students do while studying? an investigation of multitasking. Comput. Educ. **75**, 19–29 (2014)
7. Carr, N.: The Shallows: How the Internet is Changing the Way we Think, Read and Remember. Atlantic Books, London (2010)
8. Cotten, S.R., McCullough, B., Adams, R.: Technological influences on social ties across the lifespan. In: Fingerman, K.L., Berg, C., Smith, J., Antonucci, T.C. (eds.) Handbook of Life-Span Development, pp. 647–671. Springer, New York (2011)
9. Dahlstrom, E., Bichsel, J.: ECAR study of undergraduate students and information technology. Technical report (2014)
10. David, P., Kim, J.H., Brickman, J.S., Ran, W., Curtis, C.M.: Mobile phone distraction while studying. New Media Soc. **17**(10), 1661–1679 (2014)
11. Dietz, S., Christopher Henrich, C.: Texting as a distraction to learning in college students. Comput. Hum. Behav. **36**, 163–167 (2014)
12. Fishbein, M., Ajzen, I.: Belief, attitude, intention, and behavior: an introduction to theory and research. Philos. Rhetoric **10**(2), 130–132 (1977)
13. Fried, C.B.: In-class laptop use and its effects on student learning. Comput. Educ. **50**(3), 906–914 (2008)
14. Garrison, D.R., Kanuka, H.: Blended learning: uncovering its transformative potential in higher education. Internet High. Educ. **7**(2), 95–105 (2004)
15. Jacobsen, W.C., Forste, R.: The wired generation: academic and social outcomes of electronic media use among university students. Cyberpsychol. Behav. Soc. Netw. **14**(5), 275–280 (2011)
16. Junco, R.: In-class multitasking and academic performance. Comput. Hum. Behav. **28**(6), 2236–2243 (2012)
17. Junco, R., Cotten, S.R.: A decade of distraction? how multitasking affects student outcomes. In: A Decade in Internet Time Symposium on the Dynamics of the Internet and Society, Oxford (2011)
18. Kay, R.H., Lauricella, S.: Exploring the benefits and challenges of using laptop computers in higher education classrooms: a formative analysis. Can. J. Learn. Technol. **37**(1), 1–18 (2011)
19. Kreutzer, T.: Internet and online media usage on mobile phones among low-income urban youth in cape town. In: ICA Proceedings International Communication Association Conference, pp. 1–21 (2009)
20. Leysens, J.L., le Roux, D.B., Parry, D.A.: Can i have your attention, please? an empirical investigation of media multitasking during University lectures. In: Proceedings South African Institute of Computer Scientists and Information Technologists, paper #21, ACM (2016). doi:10.1145/2987491.2987498
21. Lomas, C., Oblinger, D.G.: Student practices and their impact on learning spaces. In: Learning Spaces, Educause (2006)
22. McCoy, B.R.: Digital distractions in the classroom: student classroom use of digital devicees for non-class-related purposes. J. Media Educ. **4**(4), 5–12 (2013)
23. Moreno, M.A., Jelenchick, L., Koff, R., Eikoff, J., Diermyer, C., Christakis, D.A.: Internet use and multitasking among older adolescents: an experience sampling approach. Comput. Hum. Behav. **28**(4), 1097–1102 (2012)
24. North, D., Johnston, K., Ophoff, J.: The use of mobile phones by South African university students. Issues Inf. Sci. Inf. Technol. **11**, 115–138 (2014)
25. Parry, D.A.: The digitally-mediated study experiences of undergraduate students in South Africa. Stellenbosch University, Doctoral Dissertation (2017)

26. Prensky, M.: Digital natives, digital immigrants. On the Horiz. J. Serv. Manag. **9**(3), 245–267 (2001)
27. Roberts, N., Rees, M.: Student use of mobile devices in University lectures. Australas. J. Educ. Technol. **30**(4), 415–426 (2014)
28. Rosen, L.D., Carrier, L.M., Cheever, N.A.: Facebook and texting made me do it: media-induced task-switching while studying. Comput. Hum. Behav. **29**(3), 948–958 (2013)
29. Tapscott, D.: Grown up Digital. McGraw-Hill, New York City (1998)
30. Thompson, P.: The digital natives as learners: technology use patterns and approaches to learning. Comput. Educ. **65**, 12–33 (2013)
31. van der Schuur, W., Baumgartner, S.E., Sumter, S.R., Valkenburg, P.M.: The consequences of media multitasking for youth: a review. Comput. Hum. Behav. **53**, 204–215 (2015)
32. Zimmerman, B.J.: Investigating self-regulation and motivation: historical background, methodological developments, and future prospects. Am. Educ. Res. J. **45**(1), 166–183 (2008)

Students' Learning Approaches: Are They Changing?

Estelle Taylor[(✉)] and Kobus van Aswegen

North-West University, Potchefstroom, South Africa
{estelle.taylor,kobus.vanaswegen}@nwu.ac.za

Abstract. Some researchers believe that not much has changed in terms of how students prefer to study, whilst others believe that tertiary institutions have to adapt to a new generation of students. The aim of this paper is to determine if the learning approaches of students are changing. This will be done by comparing the learning approaches of students in 2016, with that of students of 10 years ago (2006). Questionnaires were distributed electronically to students in 2006, and again in 2016. It appears that learning approaches of these students have not changed much. Computer anxiety has increased, as well as the number of students unsure of their own chances of success. There is a decrease in students preferring to solve problems themselves rather than using available solutions, and more students prefer to work under supervision. These results show that it is essential that teaching and learning are designed according to the needs of the specific group of students it is meant for, not a mythical 'new generation' student.

Keywords: Learning approaches · Changes

1 Background and Problem Statement

According to [3] most students enrolled in colleges and universities at that point in time (2007) were members of both the Generation X and Millennial groups, and that these students, according to research, had both positive and negative qualities. On the positive side studies show that they were hard working, they were accustomed to working to schedules and following rules, they were used to being assessed, receiving focused feedback, and being goal-directed, they were team-oriented, and they have mastered the ability to multi-task. On the negative side studies show that they were often exceedingly close to their parents, they may have been over-reliant on communications technology — which may have stunted their interpersonal (face-to-face) skills, and being routinely engaged in multi-tasking behaviours, enabled in part through the use of technology, may have shortened their collective attention span. Ten years later there are (still) different opinions on the concept of a 'new generation' of students, and whether there is actually any difference in the way today's students approach learning, compared to the way previous generations of students approached learning.

© Springer International Publishing AG 2017
J. Liebenberg and S. Gruner (Eds.): SACLA 2017, CCIS 730, pp. 37–47, 2017.
https://doi.org/10.1007/978-3-319-69670-6_3

There are those that argue that, since the members of this generation have been immersed in a networked world of digital technology, they behave differently, have different social characteristics, different ways of using and making sense of information, different ways of learning, and different expectations about life and learning [2,4]. The suggestion is made in literature that today's learners are becoming impatient with traditional modes of teaching, due to the fact that they have grown up digitally.

Comparing modern and traditional students, [13] states that there are enormous differences between these two groups, and the main reason for these differences is the digitalization of the modern age. Accordingly, today's students can find information themselves wherever or whenever they choose and therefore, the characteristics of today's students are different from the characteristics of traditional students. According to [9], teachers have to be aware that today's students have different needs because they have been exposed to technology since birth. Moreover, digital natives prefer to communicate through symbols and pictures rather than through words [11], and their thoughts and behaviours are very different from the thoughts and behaviours of the previous generations. They are more visual learners than auditory learners; they like to learn by doing and not by being told how to do; further, they are self-starters, self-motivators and self-learners.

On the other hand: research conducted in different countries and at various different institutions shows that while the use of digital technology is growing, and that young people tend to use it more than older people, it also shows that the issues are not defined by generation and that the implications for education are far from clear [1,4].

According to [2] the idea that the generation born after 1982 is fundamentally different than previous generations has become so firmly entrenched that it is treated as a self-evident truth. However, a comprehensive review of the research and popular literature on the topic suggests that there are no meaningful generational differences in how learners say that they use ICT, or in their perceived behavioural characteristics [2]. In another study, interviews with 25 students revealed that the majority of them believed that their generation is not as good at learning as the pre-ICT generation [6]. Yet another study found no evidence to support previous claims suggesting that the current generation of students adopt radical learning styles, exhibit new forms of literacies, use digital technologies in sophisticated ways, or have novel expectations from higher education [8]. According to [8], the attitudes to learning appear to be influenced by lecturers' teaching approaches and students appear to conform to traditional pedagogies. In fact, students' emphasised that they expected to be 'taught' in traditional ways. On this basis, previous claims of a growing and uniform generation of young students entering higher education, with radically different expectations about how they will learn, seem unwarranted.

"Claims about the Net generation fall into three categories: the widespread and intensive use of digital technologies; the impact of this use on how this generation accesses and uses information, how they interact socially, how they

learn; as well as the unique behavioural characteristics and learning styles of this generation. With the exception of the first category (widespread and intensive use) our review of the popular and academic literature shows that there is no empirical support for the most prevalent claims in the other two categories. Furthermore, other literature reviews confirm our findings and the methodologically sound research tends to contradict the claims" [2].

In [5] we can find arguments against the opinion that universities and academic staff would have to 'adjust' to 'accommodate' a new Net Generation of Digital Native students. It is important to realize that the so-called 'new generation' of students is not homogenous, nor is it articulating a single clear set of demands. Universities and academics are, as always, faced with choices about how to change, as well as the need to be better informed about the kinds of students that are entering their institutions [5].

Moreover, there is not sufficient knowledge on generational differences to be used as a solid conceptual framework; thus more substantive studies in this area are still necessary [10]. Generational differences have been discussed, literally for generations, and various authors with interest in a particular aspect of different generations have used diverse perspectives to describe their area of concentration. Despite a growing interest in this topic there has been no clear consensus in this area.

According to [7], 'generation' is not a determining factor in students' use of digital technologies for learning, nor has generation had a radical impact on learning characteristics of higher education students. Results of this study show little difference in learning characteristics or technology-use between generations. Furthermore, findings from this study do not support the notion of a unique learning preference for the current generation of young people [7].

This brings us to the research question of this paper: Are the learning approaches of students changing? This we will try to answer by comparing the way students approach learning today, to that of students 10 years ago.

2   Method

The first questionnaire (in 2006) was answered voluntarily by 2031 of a possible 2767 students in their first year at a South African university. This gives us a participation rate of 73.4%. The initial questions were formulated after an extensive literature study as well as interviews [12]. A questionnaire containing many of the same questions was distributed electronically to first year students at the same university in 2016. This questionnaire was completed by 1391 students out of a possible 3589 which gives as a participation rate of 38.8%. Both groups of students were enrolled in a compulsory first year computer literacy module. This group of students were chosen for two reasons, namely: due to the nature of this module (which is a computer-based e-learning module), we had access to a large population of students with different cultural and educational backgrounds; the data obtained had the potential to provide a good indication of the computer background, online skills, learning approaches (etc.) of South African students at the start of their university studies. Quantitative research was done as we wanted to measure certain variables. To do this statistical analysis was done on questions pertaining to the students' learning preferences. Qualitative research was also done, as we wanted to understand the students' own perceptions on possible changes in approaches to learning. The answers to two open questions were analysed interpretively.

3   Results

3.1   Statistical Analysis of Data

In this sub-section we will compare some of the results of the two questionnaires (2006 and 2016). In 2016, as in 2006, about 95% of the students that completed the questionnaire indicated that they have used a computer before starting their studies at university. There has been an increase in regular email usage from 81% (2006) to 88.5% (2016). The ages at which the respondents started using computers on a regular basis, can be seen in Table 1, and it appears that there is a slight increase in the percentage of students exposed to computer before the age of 10 (from 21.3% to 27.5%).

Respondents were asked to indicate whether they think they will use computers in their other modules, with almost a 6% decrease from 2006 to 2016. Another interesting comparison is that 88.3% of students were not afraid of appearing incompetent when having problems using a computer in 2006, compared to a much lower 64.9% in 2016. Students are also more apprehensive about making mistakes when using computers with 36.3% indicating apprehension in 2016, compared to 29.0% in 2006.

We wanted to determine if there is a case for learning content being delivered on tablet devices. The results indicated that more than 50% of the students who answered the questionnaire in 2016 do not own tablet devices. Of the 43.8% of the students who own a tablet, 41% (<19% in terms of the total number of respondents) use it regularly. Respondents were asked to what extent they agree to statements regarding their online skills. The results can be seen in Table 2. It is clear that most students agree that they have the necessary online skills, but there are, however, still some students who do not feel confident about their online skills.

Table 1. Age group and regular computer usage

Age group (years)	2006	2016
under 5	5.1%	6.0%
6 to 9	16.3%	21.5%
10 to 14	37.3%	32.4%
15 to 17	34.9%	30.2%
18 or older	6.6%	9.9%

Table 2. Online skills of students, with SD = strong disagreement, WD = weak disagreement, WA = weak agreement, SA = strong agreement

Statement	SD	WD	WA	SA
I have the basic skills for finding my way around the Internet (e.g., using search engines, entering passwords)	7.1%	10.8%	22.4%	59.8%
I can send an email with a file attached	2.7%	4.2%	6.5%	86.6%
I think that I would be comfortable using a computer several times a week to participate in a course	6.0%	11.5%	26.0%	56.6%
I think that I would be able to communicate effectively with others using online technologies (e.g., email, chat)	4.2%	5.0%	12.9%	77.9%
I think that I would be able to schedule time to provide timely responses to other students and/or the instructor	5.3%	26.6%	27.3%	40.8%

We also wanted to compare the learning approaches of students; the results are listed in Tables 3 and 4. For the questions reported on in Table 3, the students were asked to choose between two contrasting statements and these results are paired together.

From the results in Table 3 it can be seen that there were no drastic changes between the learning approaches of 2006 and 2016. Students still prefer (by a slight margin) for things to stay the same, rather than changing frequently. They prefer learning new information in picture format (graphical learning style rather than text), and they prefer having information explained to them

Table 3. Learning approaches (Part A)

Statement	2006	2016
I prefer it when things around me change regularly	43.7%	43.5%
I prefer it when things around me stay the same	56.3%	56.5%
When someone is showing me information, I prefer charts, schemas and graphs (pictures)	64.2%	63.1%
When someone is showing me information, I prefer text (words) summarizing the results	35.8%	36.9%
I prefer it if someone explains new information to me	77.5%	71.3%
I prefer new information in written format	23.0%	28.7%
I prefer to solve new problems by myself	61.6%	59.5%
I prefer to use available solutions	38.4%	40.5%
The best way to master new information is to reflect about it and to scrutinise it	20.8%	15.9%
The best way to master new information is through practice and action	79.2%	84.1%
I believe that I can successfully complete anything I embark on	84.1%	80.8%
Many times I am unsure about my chances of success	15.9%	19.2%

Table 4. Learning approaches (Part B). (N = no, Y = yes, U = unsure)

Question	N 2006	N 2016	Y 2006	Y 2016	U 2006	U 2016
Do you work better under supervision?	67.2%	55.1%	18.0%	29.2%	14.8%	15.8%
Do you plan in advance how to manage your time to ensure that you fulfil all your duties?	26.8%	16.1%	62.0%	69.6%	11.2%	14.3%
When you plan, do you find it difficult to follow your own plan?	45.4%	61.8%	26.3%	30.2%	28.3%	8.1%

(verbal learning style, rather than written), although there is a slight increase in students preferring information in written format. There is a slight decrease in students preferring to solve problems themselves rather than using available solutions, but also a slight increased realization that the best way to master new information is through practice rather than reflection. There is also a slight increase in the number of students unsure of their own chances of success.

For the questions reported on in Table 4, the students were given three options to choose from, namely 'no' (N), 'yes' (Y), and 'unsure' (U).

As it can be seen from the results in Table 4, the minority of students indicated that they work better under supervision, but this percentage increased from 18.0% in 2006 to 29.2% in 2016. This is one of the biggest changes found in these results. The majority of the students indicated that they plan in advance how to manage their time, to ensure that they fulfil all their duties. This percentage has increased from 62.0% (2006) to 69.6% (2016). There is also a change in the results for the question on the students' ability to follow the plans they make, with an increase in the number of students who indicate that they do not find it difficult to follow their plans, from 45.4% in 2006 to 61.8% in 2016. These results will be discussed in the conclusion section of this paper.

3.2 Interpretive Analysis of Open Questions

In this sub-section the results of the open questions will be presented.

1st Question: *Do you think students today learn differently to students of the past, and how?*

Of the 1232 students that answered this question, 1136 indicated that, according to them, students do learn differently than students of the past. Only 96 indicated that students today do not learn differently than students of the past. The main reason for this difference, as stated by 800 of the students, was technology (computers and internet). Some interesting answers are quoted as follows:

- *"Yes, because the technology changes us"*;
- *"Yes, students of today do not necessarily need to be in classes (at university) to get information from lectures and they also do not need to rely on books from the library as some of the information is available on the internet"*;

- *"Yes, students depend on technology; they also form their own methods of learning and understanding rather than just 'parroting' the work"*;
- *"Yes, most students watch online lectures instead of studying the textbook"*.

Some students (56) mentioned that students today are exposed to a greater variety of methods for learning. Examples of answers were:

- *"Yes, they do. Students today are more exposed to different methods of learning compared to just having a lecturer stand in front of the class in the past"*;
- *"Yes, in the past education was focused on teacher-centred and now it is verging more towards learner-centred where the learner is actively participating"*.

Some students (39) who indicated that students do learn differently saw this difference as negative, mentioning reasons such as students being lazy. Examples of answers were:

- *"Yes, firstly I do not think students of today learn. In the olden days the students had their textbooks which they have studied and did extra work, today students just read through the work and are lazy when it comes to extra work"*;
- *"Yes, students back then did not have many distractions like students of today and they were dedicated towards their studies despite the few resources they had"*;
- *"Yes, we have definitely become lazy with certain things, for example we would rather watch a video on something than read about it"*;
- *"Yes, we do not learn to understand like they did"*;
- *"Yes, in the past more students actually wanted to learn"*;
- *"Yes, they were more motivated and have excellent literature skills because they had to use hardcopies"*.

Of the 96 students indicating that, according to them, students today do not learn differently than students of the past, few mentioned why they felt this way. Reasons mentioned were that students still need books, notes and repetition. For example:

- *"No, I still prefer a book and paper to work through and learn from. I prefer personal interaction and explanation than staring at a screen"*.

2nd Question: *How do you feel lecturers can improve and adapt their style of lecturing to better suit this generation of students?*

Of the 1391 students who completed the questionnaires, 886 answered this question. Some of the answers referred to more than one aspect. Of these students, 427 mentioned better use of technology, for instance multimedia, Internet, online tests and videos. Some interesting answers are quoted as follows:

- *"They can help students in the comfort of their own homes by posting videos online"*;
- *"They can use more multi-media etc. to keep the people interested and keep the classes short and sweet, because students these days do not concentrate for more than a few minutes"*;

- *"The use of media such as internet videos. I feel that some lecturers should be instructed well in the use and implementation of technology in teaching students"*;
- *"By using the latest technology during lectures"*;
- *"Provide smartphone-friendly versions of their presentations"*.

Some students (107) felt that no improvement is necessary. Examples of such answers were:

- *"None, lecturers of the present day are already caught up with modern strategies of breaking through to students"*;
- *"I think our lecturers are even more adapted than some of the students. Our university is quite well adapted"*.

Other students (62) wanted more or clearer explanations. Examples of such answers were:

- *"To explain the work more and in words that we can understand. Most of the lecturers speak to high and expect that you have to know the work already"*;
- *"Not to just read off the slides but to actually explain the work"*;
- *"They can explain it so that you can understand and not just for you to know it"*.

Some students (55) mentioned more interesting or creative classes. For example:

- *"They could make the lectures a little more fun and interesting"*.

Students (50) also mentioned the attitude of the lecturers, wanting more understanding lecturers with an attitude of wanting to help. Examples of such answers were:

- *"Engage with the student and not let technology replace them"*;
- *"Some need to be easier to approach"*;
- *"They can be friendlier. With this generation, if someone is rude to them, they completely ignore that person"*.

Students (44) wanted more visual teaching, for instance using more pictures and diagrams. For example:

- *"Using more graphical features during lectures will accommodate each student as they only use words currently. Not all students understand the information in that manner"*.

Students (34) also recommended using more 'real-world' examples:

- *"Relate their lessons with things that happen in everyday lives so that students can gave a better understanding of what the lecture is trying to say"*;
- *"More examples of how something is in life"*.

Other factors mentioned were: more exercises and practices (25), language problems (16), lecturers putting more effort into preparing for classes (9), wanting more tests (7) or fewer tests (1), more lectures (5), more group work (2) or less group work (1). Also 'cultural' issues were mentioned more than once, as can be seen from the following answers:

- *"Be more patient because students come from different backgrounds"*;
- *"I feel that would be an unfair disadvantage to those who are from poor backgrounds"*;
- *"Yes, simply because we come from different backgrounds"*;
- *"They should treat us equal then black students would participate more"*;
- *"Patience is critical for African child, and those around you, they will feel comfortable"*;
- *"Lecturers are also reminded not to expect too much from their students"*;
- *"Not to expect us to know everything in the textbook all the time. We do have lives"*.

4   Discussion and Conclusion

As the old adage goes, the more things change, the more they stay the same, which can be seen in a large proportion of the statistics in this paper.

It is surprising that the number of students who think they will use computers in their other modules has decreased with almost 6% from 2006 to 2016. More research should be done on the reasons for this. One can speculate that some of their assignments, class tests, etc., can be done with tablets or other mobile devices and they do not always need to use a computer for certain tasks.

It appears that computer anxiety has increased, with 23.4% more students (than in 2006) indicating that they are afraid of appearing incompetent, when having problems with using a computer. There is also an increase in the number of students feeling apprehensive about making mistakes when using computers. This may be due to the fact that students feel that there is an expectation that everybody should, in this day and age, know how to use a computer. This must be taken into account when considering an increase in the use of technology in teaching and learning.

As can be seen by the questions on tablet ownership and use, less than half of these students own a tablet, and it will be very difficult to use tablets as a medium for delivering educational content. For such a strategy to be considered, the cost of issuing each and every student with a tablet will have to be included in the cost of education. Overall the response from the students, to questions regarding their online skills, have been very affirmative, with close to 90% of them agreeing to positive statements regarding their online skills. It appears that the students will have sufficient skills to participate in an online and/or mobile learning course.

Many of the learning preferences have stayed the same over the last 10 years. This is compatible with what was found in studies mentioned above in the literature review section this paper. Students still prefer (by a slight margin) for things to stay the same, rather than changing frequently. They show a preference towards a graphical learning style, as well as a verbal learning style. This matches the findings of [11], namely, that nowadays students prefer to communicate through symbols and pictures rather than through words and that they are more visual learners than auditory learners. However, Srinivasan also writes that they like to learn by doing and not by being told how to do, as well as that they are self-starters, self-motivators and self-learners. The results of this study show that there is a slight decrease in students preferring to solve problems themselves, rather than using available solutions and this should be of some concern. Further research should be done on the reasons for this, as well as ways to amend this. Can this, for instance, be the results of the current secondary school system most of these students attended *before* university?

There is also a slight increase in the number of students unsure of their own chances of success.

More students (with an increase of 12%), prefer to work under supervision than 10 years ago, which is a reason for concern as education is moving toward self-paced, self-driven, asynchronous learning. Again, research should be done on the reasons for this tendency and ways to change this.

Although the statistical analysis did not show significant change in the learning approaches of students the past 10 years, an overwhelming 92% of the students, who completed the questionnaire in 2016, indicated (in answers to the open questions) that according to them, approaches to learning have changed over the years. Most of them stated technology as the main reason for this change. Another reason mentioned was that students today are exposed to a greater variety of methods for learning. Other students were of the opinion that approaches to learning have not changed, because students still need books, notes and repetition.

Suggestions from students on how lecturers can improve and adapt their style of lecturing, to better suit this generation of students, included: better use of technology (for instance multimedia, Internet, online tests and videos); providing more or better explanations; more interesting or creative classes; more understanding lecturers with an attitude of wanting to help; more visual teaching, for instance using more pictures and diagrams; using more 'real-world' examples; more exercises and practices; and lecturers putting more effort into preparing for classes. As always, some conflicting suggestions were also offered, for instance writing more tests versus writing fewer tests, and doing more group work versus doing less group work.

If institutions, and/or individuals responsible for teaching and learning, are leaning towards a new age of technology-driven, digital delivery for teaching and learning, then these results may be seen as cautionary. Students are not necessarily, as occasionally assumed, completely proficient in the use of computers and technology, and they may be less prepared for solving problems, working on their own (without supervision), and taking on more responsibility than they were 10 years ago. Secondary and tertiary institutions should do more to prepare students to be more self-reliant and self-confident.

These results show that it is essential that teaching and learning is designed according to the needs of the specific group of students it is meant for, not a mythical 'Net generation' student. In [1] we can find important suggestions accordingly: use technologies that are programme-relevant; do not assume that all your students have access to the latest technologies or are proficient in their use, because research is showing that there is a continuum of access, use and comfort with digital technologies.

Future research can focus on the post-millennial era to see how these learners and students compare to the previous generation, in terms of technology proficiency and learning preferences.

References

1. Bullen, M., Morgan, T.: Digital learners not digital natives. La Cuestión Universitaria **7**, 60–68 (2016)
2. Bullen, M., Morgan, T., Qayyum, A.: Digital learners in higher education: generation is not the issue. Can. J. Learn. Technol. / La Revue Canadienne de l'Apprentissage et de la Technologie **37**(1), 1–24 (2011)
3. Elam, C., Stratton, T., Gibson, D.D.: Welcoming a new generation to college: the millennial students. J. Coll. Admission **195**, 20–25 (2007)
4. Huang, R., Yang, J.: Digital learners and digital teachers: challenges, changes, and competencies. In: Spector, J.M., Ifenthaler, D., Sampson, D.G., Isaías, P. (eds.) Competencies in Teaching, Learning and Educational Leadership in the Digital Age, pp. 47–56. Springer, Cham (2016). https://doi.org/10.1007/978-3-319-30295-9_4
5. Jones, C., Ramanau, R., Cross, S., Healing, G.: Net generation or digital natives: is there a distinct new generation entering university? Comput. Educ. **54**(3), 722–732 (2010)
6. Kolikant, Y.B.-D.: Digital natives, better learners? Students' beliefs about how the internet influenced their ability to learn. Comput. Hum. Behav. **26**(6), 1384–1391 (2010)
7. Lai, K., Hong, K.: Technology use and learning characteristics of students in higher education: do generational differences exist? Br. J. Educ. Technol. **46**(4), 725–738 (2015)
8. Margaryan, A., Littlejohn, A., Vojt, G.: Are digital natives a myth or reality? University students' use of digital technologies. Comput. Edu. **56**(2), 429–440 (2011)
9. Morgan, H.: Maximizing student success with differentiated learning. Clearing House J. Educ. Strat. Issues Ideas **87**(1), 34–38 (2014)
10. Oh, E., Reeves, T.C.: Generational differences and the integration of technology in learning, instruction, and performance. In: Spector, J., Merrill, M., Elen, J., Bishop, M. (eds.) Handbook of Research on Educational Communications and Technology, pp. 819–828. Springer, Newyork (2014). https://doi.org/10.1007/978-1-4614-3185-5_66
11. Srinivasan, R.: Emerging shifts in learning paradigms – from millenials to the digital natives. Int. J. Appl. Eng. Res. **11**(5), 3616–3618 (2016)
12. Taylor, E.: n Model van die Faktore wat die Sukses van Onderrigleer van Tegnologie-gebaseerde Onderwerpe beïnvloed. Doctoral dissertation, North-West University, South Africa (2007)
13. Uygarer, R., Uzunboylu, H., Ozdamli, F.: A piece of qualitative study about digital natives. Anthropologist **24**(2), 623–629 (2016)

The Effect of Using YouTube in the Classroom for Student Engagement of the Net Generation on an Information Systems Course

Sumarie Roodt[⊠], Nitesh Harry, and Samwel Mwapwele

Department of Information Systems, University of Cape Town,
Cape Town, South Africa
{sumarie.roodt,hrrnit002,mwpsam001}@uct.ac.za

Abstract. This paper aims to determine the effect on the engagement of 'Net Generation' students from using 'YouTube' in the classroom. The education and engagement of the Net Generation students are a growing challenge among institutions of higher learning. Net Generation arrival overlaps with the advent of digital technology. Thus, this explains why the students have dissimilar styles of learning due to their comfort with and use of digital technology. Literature on educating and engaging the Net Generation asserts incorporating the Web 2.0 elements; YouTube inside and/or outside classroom. The target sample includes Net Generation students in their 2nd-3rd year at the University of Cape Town in the Commerce faculty enrolled in an Information Systems course. The research instrument was a questionnaire. Two samples were included, the first consisted of students currently enrolled in the course and the other, students previously enrolled in the course. The result shows that the use of YouTube had a positive effect on the engagement of Net Generation students.

Keywords: YouTube · Net generation · Millennials · Digital natives · Student engagement · Information systems education · Higher education · Technology-enhanced learning · South Africa

1 Introduction

The emergence of digital technologies such as the internet and PCs has resulted in a new generation of technically literate individuals called the 'Net Generation' [25]. Due to the technical literacy of these individuals, their learning styles differ from previous generations. The use of 'YouTube' and other Web 2.0 technologies in education has been suggested as a tool to engage Net Generation students [11,26]. The use of YouTube in education is a relatively new field of study and not much literature has been published regarding the subject [30].

'IT in Business' (INF2004F) is a second-third year undergraduate course offered by the Department of Information Systems (IS) at the University of Cape Town (UCT). The course is compulsory for all students majoring in accounting

J. Liebenberg and S. Gruner (Eds.): SACLA 2017, CCIS 730, pp. 48–63, 2017.
https://doi.org/10.1007/978-3-319-69670-6_4

and finance. The course builds on from a first year level IS course, which lays the basic foundations of Information Systems. The course consists of a theoretical component as well as a practical component (Excel and Pastel). A new aspect to teaching this course had been adopted through the use of YouTube videos in the class. These videos brought a new approach to teaching and it is believed that the use of YouTube as a teaching tool could have an effect on the level of student engagement.

The purpose of our research was to discover if the use of YouTube in the class had an effect on student engagement. Towards the end of the semester, students enrolled in the course were given a questionnaire asking what effect the use of YouTube videos in class had on their level of engagement. Furthermore, students previously enrolled in the course were given a similar questionnaire asking them about their level of engagement with the course. By conducting a comparative analysis, we aimed to discover what effect, if any, the use of YouTube in the class had on student engagement.

2   Literature

2.1   The Net Generation

The Net Generation, also known as the Millennials, Generation Y and the Digital Natives, is the term used to describe the generation born between 1982 and 2003 [5]. This group, which will be referred to as the Net Generation throughout this paper, grew up with the digital technology that arrived in the last decades of the 20th century [25]. The term Net Generation was first mentioned in [32]. The term Net Generation comes from the fact that members of this generation's birth coincided with the emergence of the Internet and digital technology [5].

Members of the Net Generation have grown up with computers and the Internet and are said to have a natural aptitude and high skill levels when using new technologies [15–18]. Prensky used the term 'Digital Natives' to describe this group as he stated that members of this generation were so accustomed to using digital technology that they 'speak' the 'digital language' [25].

Berk noted that defining and labelling groups of people and ascribing characteristics to them can lead to problems of misrepresentation and generalization [5]. Various studies have shown that the Net Generation is not homogenous in their use of technology and thus some of the assumptions made about the Net Generation are not entirely true [7,18,19,21,22].

2.2   YouTube

YouTube is a well-known video sharing website where users can upload, view and share video clips [11]. YouTube was launched in 2005 and is a repository for user-generated content. Content on YouTube includes music videos, TV clips and personal videos uploaded by users, who are mainly members of the public [29]. Videos can be viewed by anyone with an internet connection; however, in order to upload videos, a free user account is required [6]. Various organisations such as businesses, television broadcasters, universities, political parties and non-governmental organisations have set up YouTube channels in order to deliver their message to a wider audience [3,9,14]. Thereby, YouTube has shifted from mainly user-generated content to professionally generated content as well [20].

The use of video clips embedded in multimedia presentations to improve learning in higher education classes was suggested in [4]. Accordingly, videos can have a strong effect on the mind and senses [4]. A list of 20 potential learning outcomes to consider when using videos in the was also provided [4]. These include using videos to: grab students' attention, focus students' concentration, generate interest in the class, draw on students' imagination, improve attitude towards content and learning, and to make learning enjoyable [9,27,28]. One of the methods of using videos in the classroom is using YouTube.

A study into the use of YouTube as a tool to support collaborative learning can be found in [26]. From a sample of 185 students, it was found that the use of YouTube had a positive impact on the students [26]. Moreover, YouTube was perceived as an innovative learning technology by the majority of students [26]. The use of YouTube videos to help explain key ideas in a sociology course was discussed in [31]; accordingly the use of YouTube videos helped the students and was seen as an effective way of supporting their learning [1,8,31].

2.3   Student Engagement

Student engagement, according to [2], refers to *"how involved or interested students appear to be in their learning and how connected they are to their classes, their institutions, and each other"*. The roots of student engagement theory can be traced back as far as 70 years ago to Tyler's research conducted in the 1930s into the amount of time students spent on work and its effect on learning [2]. Student engagement theory also has its roots in research conducted in the 1960s and 1980s, which focused on quality of effort [12], as well as in research from the 1980s which focused on students' involvement [2,24].

In this context, three different dimensions of student engagement were discussed in [13], namely behavioural, emotional, and cognitive engagement:

Behavioural engagement entails positive conduct, involvement, effort and participation [10,13]. Additionally, behavioural engagement is action on the part of the student that could be observed [2].

Emotional engagement refers to students' affective reactions such as interest, enjoyment or a sense of belonging or comfort [13].

Cognitive engagement stresses investment in learning and involves self-regulation and being strategic [13]. Moreover, cognitive engagement is when students go beyond the minimal requirements; it can be used to facilitate learning of complex material [12]. Going beyond the minimal requirements includes getting clarification for concepts by asking questions, persisting with difficult tasks, and reading extra material over and above the prescribed material [12].

3   Materials and Methods

3.1   Research Method

The underlying 'paradigm' of this paper was interpretive. We attempted to measure the change in the level of student engagement through the use of YouTube in the classroom. This was done through measuring the level of engagement with YouTube and comparing it with the level of engagement without YouTube. The underlying purpose of this research was exploratory, as our research was attempting to discover what the effect of using YouTube in the classroom has on student engagement. The research was done in the South African context and sought new insights into the use of YouTube in education.

3.2   Research Objectives and Questions

The main objective of this research was to determine if using YouTube videos in the classroom had an effect on the engagement of Net Generation students. In order to meet this objective a number of research questions were asked. These were:

RQ₁: *Were the students Net Generation students?*
RQ₂: *Did the use of YouTube in class have an effect on overall student engagement and the different types of engagement?*
RQ₃: *How did students feel about the use of YouTube in the classroom?*

3.3   Sampling Techniques

As our research was conducted as comparative study, data was collected from two samples. The first sample was of students currently enrolled in the course INF2004F. Students in this class were potentially part of the Net Generation. Furthermore, these students had been exposed to the use of YouTube in the classroom, thus making them the ideal sample. This sample was referred to as the 'current sample' throughout this paper.

The second sample was of students who had previously been enrolled in the course INF2004F. Students from this course were the highest population of previously enrolled INF2004F students available to us. As with the students currently enrolled in the course, students from this sample were potentially part of the Net Generation. However, as far as we know, those students were not exposed to the use of YouTube in INF2004F lectures. This sample is referred to as the 'previous sample' throughout this paper.

3.4   Data Collection

Data were collected from the two sample groups in the form of an online questionnaire. The questionnaire was conducted on Vula, an education system used at the University of Cape Town. The questionnaire was preceded by a declaration assuring the respondents that data would be kept confidential and anonymous. The questionnaire was split up into four sections, covering the four main themes of this paper. Three of the sections (Net Generation, YouTube and Other Web 2.0 tools) in the questionnaire were the same across both samples, while the fourth (Student Engagement) differed slightly.

The data that the questionnaire obtained were quantitative and qualitative in nature. 5-point Likert scales were used throughout the questionnaire. Open ended questions were also included to allow respondents to express ideas that were difficult to place on a Likert scale. The questionnaire for the first sample of students was put up on the INF2004F Vula site towards the end of the first semester. The questionnaire received 156 responses of 616 students, which showed a response rate of 25%. The questionnaire for the second sample of students was put up on a specially created Vula site in the middle of the second semester. The participants of this site were students enrolled in the course 'ACC3009W' (Financial Reporting III). A class list was obtained from the ACC3009W course convener and the students added to the Vula site. The questionnaire received 85 responses of 488 students, which showed a response rate of 17%.

3.5   Data Analysis

Once the data had been collected from the sample groups, the results were exported into 'Microsoft Excel' for the quantitative data, and text files for the qualitative data. Before the data could be analysed the data needed to be cleaned. This included removing incomplete responses and responses that were clearly filled in incorrectly. Once the data had been cleaned, the current sample consisted of 104 responses ($nc = 104$) and the previous sample consisted of 70 ($np = 70$). Thus, the total sample consisted of 174 responses ($N = 174$). Furthermore, the quantitative data needed to be changed from words into numbers before any statistical tests could be run. Each scale used in the questionnaire was given a numerical legend and each question turned into a variable name. The data was analysed using tables to compare the figures. A few t-tests were used to test the difference in means. All t-tests were done with the 'Statistica' tool. However, before the t-tests could be carried out, the data needed to be checked for normality. All variables used in the statistical analysis were skewed within three standard deviations of the mean.

3.6   Limitations

The sample size, which was a total of 174 across two samples, was not a large sample size given the total number of students in both target groups. Furthermore, the use of YouTube in INF2004F was a new intervention and, thus, the full effect of using YouTube in the classroom might not yet be discovered by this investigation.

4   Data Analysis

For the analysis, the responses from each of the following questions were tabulated and discussed. Where questions were the same amongst both samples, the responses were not tabulated separately. Rather, the responses were tabulated together and marked as 'Current' for the current sample of students, and as 'Previous' for the previous sample of students. Furthermore, when responses showed little or no variation in responses, no table was used. *All data tables are shown in* the **Appendix.**

4.1   The Net Generation

The following questions had been asked.

Were you born between 1982 and 1994? All respondents in both samples answered that they were born between 1982 and 1994.

What is your gender? The gender of the respondents in both samples was almost equally split between male and female. Only one respondent, from the previous sample, did not disclose the gender.

Are you digitally literate, i.e.: do you interact with technological devices such as cell phones, computers, laptops or tablet PCs on a regular basis? In both samples, all respondents answered that they were digitally literate.

How many hours per week do you spend interacting with technological devices? In both samples, more than half of the respondents said that they spent more than 15 h per week interacting with technological devices. Over 33% in both samples said that they spent more than 21 h a week with such interactions.

Are you connected to a technological network (cell phone network or the internet)? In the sample of current students, all 104 responded that they were connected to a technological network. Of the 70 previous students, all but one responded that they were connected to a technological network.

Are you a member of a Social Network (Facebook, Twitter, etc.)? Of the 104 current students, 102 answered that there were a member of a social network. Of the 70 previous students, 67 answered that there were a member of a social network.

Characteristics and learning preferences: Current students. In the current sample, most of the respondents agreed to some extent that they multitasked and expected frequent responses. Most agreed to some extent that they preferred a practical learning environment and preferred to perform tasks with a known set of guidelines and rules. Concerning learning and working in teams, most of the current sample was neutral or disagreed to some extent that they preferred to learn and work in teams. Within the current sample, most respondents agreed to some extent that they preferred interactive and visual material. Furthermore, slightly less than 80% of respondents said they preferred to learn about 'things that matter'.

Characteristics and earning preferences: Previous students. In the previous sample, most students agreed to some extent that they multi-tasked and expected immediate responses from technology. Furthermore, most respondents agreed to some extent that they preferred a practical learning environment and performing tasks with a known set of guidelines. Many students were neutral to learning and working in teams while most disagreed to some extent that they preferred to learn and work in teams. Within the current sample, most respondents agreed to some extent that they preferred interactive and visual material. Furthermore, slightly fewer than 80% of respondents said they preferred to learn about 'things that matter'.

4.2   YouTube and Student Engagement

The following questions had been asked.

Were YouTube videos used in the classroom? In the current sample more than 90% of the respondents said that YouTube was used in the classroom to some extent. For the current sample, most respondents claimed that YouTube was used occasionally. Within the previous sample, nearly 43% of respondents claimed that YouTube was never used in class. Nearly 30% of respondents claimed YouTube was used occasionally while 27% were not sure.

In lectures where YouTube videos were not used, were you engaged? When YouTube videos were not used in class, most respondents in the current sample were engaged occasionally. Approximately 23% were engaged frequently while just fewer than 7% were always engaged when no videos were used in class. Within the previous sample, 30% of respondents were occasionally engaged and nearly 33% were engaged frequently when no videos were used. Slightly more than 7% were always engaged while slightly over 17% were never engaged when no videos were used in class.

Did the YouTube videos engage you in the classroom? (Current students). In the current sample 64% of respondents said that they were at least frequently engaged by the YouTube videos in class. 25% of respondents were occasionally engaged while 7% were never engaged.

The effect of the use of YouTube videos on engagement (Current students). In the current sample, approximately 30% of the respondents said that the use of YouTube videos increased their attendance of lectures to some extent. Slightly more than 31% of students disagreed to some extent that the use of YouTube increased their attendance of lectures. Furthermore, slightly more than 71% of students agreed to some extent that the use of YouTube increased their attention in class with approximately 13% disagreeing to some extent. The use of YouTube increased effort of students in class by 38% to some extent. Almost half the respondents were neutral on whether the use of YouTube increased their effort in class. In the current sample, 55% of respondents agreed to some extent that the use of YouTube had increased their interest in the coursework. Approximately 27% of respondents were neutral while 17% disagreed to some extent. With regards to increasing enjoyment towards the coursework, almost 61% of respondents agreed to some extent that the use of YouTube had increased their enjoyment. Again, approximately 27% were neutral while slightly fewer than 13% disagreed to some extent. In terms of increasing the sense of comfort towards the coursework, 45% of respondents agreed to some extent that the use of YouTube increased their sense of comfort. Slightly over 44% were neutral while 11% disagreed to some extent. With regards to increasing the willingness to learn, 48% of the respondents agreed that YouTube had increased their willingness to learn by some extent. Slightly fewer than 9% disagreed to some extent while 43% were neutral. In terms of aiding in establishing learning goals, 42% of respondents agreed to some extent that YouTube had aided them in establishing their learning goals. Approximately 44% were neutral while 13% disagreed to some extent.

How students feel about the use of YouTube in the classroom **(Current students).** In the current sample, slightly more than half the respondents agreed to some extent that the use of YouTube was successful in class. Furthermore, 65% of the respondents agreed that they would recommend the use of YouTube in other courses.

5   Discussion of the Findings

5.1   RQ$_1$: Were the Students Net Generation Students?

The literature reviewed defined the Net Generation as the generation born between 1982 and 2003 [5]. Thus, looking at the findings of Q$_1$ of the questionnaire, all respondents across both samples are members of the Net Generation according to the definition in [5]. Furthermore, a list of characteristics of the Net Generation can be found in [23].

With regards to being digitally literate, connected and social, the findings confirmed that all respondents across both samples described themselves as being digitally literate [25]. Furthermore, all the respondents in both samples were connected to some form of technological network and a vast majority were members of social networks.

In terms of being immediate and experiential, a majority of the respondents across both samples answered that they did multi-task and expected immediate responses from technology [33]. In addition, a majority of students across both samples agreed to some extent that they preferred a practical learning environment as opposed to a theoretical learning environment [35].

With respect to teamwork most of the students across both samples answered that they did not prefer to learn and work in teams [25]. This is an interesting finding as it is contrary to the literature reviewed and is consistent amongst both samples of students surveyed [23]. A possible reason for this could be the student's field of study. However, before any conclusions can be drawn, further research will need to be conducted to investigate this claim.

In terms of the preference for structure, the findings indicated that across both samples, a majority of the respondents relied on guidance from both lecturers and technological tools to aid in their studies. Furthermore, most of the respondents across both samples preferred to perform tasks with a known set of guidelines.

Lastly, in terms of engagement and experience, visual and kinesthetic and things that matter, the findings indicated that across both samples, respondents adhered to the characteristics noted in [23].

Thus, in answering RQ_1, we found that the students surveyed are Net Generation students. All of the respondents were born after 1982. Furthermore, the majority of the students possessed almost all of the characteristics noted in [23].

5.2   RQ₂: Did the Use of YouTube in Class Have an Effect on Overall Student Engagement and the Different Types of Engagement?

In trying to achieve the first objective of the research, the discussion has established that the students surveyed were Net Generation students and that YouTube had been used in class. What remains to be established is whether the use of YouTube in class had an effect on overall engagement and the different types of engagement. Furthermore, if the use of YouTube did have an effect on student engagement, what types of engagement were affected?

Looking at the finding, the largest proportion, 58%, in the current sample said that they were occasionally engaged in lectures were no YouTube videos were used. Nearly 23% said that they were frequently engaged while only 6% said that they were never engaged when no YouTube videos were used. When comparing the findings for the current sample, one can see that 64% of respondents said that the YouTube videos had engaged them at least frequently. Thus, on the surface, it appears that the use of YouTube videos had a positive effect on engagement. However, in order to more rigorously test whether the use of YouTube had an effect on engagement, a t-test for difference in means was used to compare between the current and previous samples. The output of the t-test can be seen in the screen-snapshot of Fig. 1.

The test compared the means from both samples. From the output of the test, one can see that the p value is smaller than 0.05 and thus there exists a statistically significant difference in the engagement of students. Furthermore, from the output one can see that the mean for the current sample is higher than the mean of the previous sample. Thus, one can conclude that the use of YouTube had an effect on student engagement and that this effect was a positive effect.

The findings indicate that the use of YouTube had a statistically significant positive effect on attention in class as well as enjoyment towards the course work. For both of these factors, the effect was positive. The effect of YouTube

Variable	T-tests; Grouping: CURRENT (Data All Final) Group 1: Current Students Group 2: Previous Students										
	Mean 1	Mean 2	t-value	df	p	Valid N 1	Valid N 2	Std.Dev. 1	Std.Dev. 2	F-ratio Variances	p Variances
ENG_YT_ENGAGED	4.673077	4.314286	2.09055	172	0.038039	104	70	1.118618	1.097333	1.039170	0.873371

Fig. 1. T-test output for difference in level of engagement across the samples

| Variable | T-tests; Grouping: CURRENT (Data All Final) Group 1: Current Students Group 2: Previous Students | | | | | | | | | | |
	Mean 1	Mean 2	t-value	df	p	Valid N 1	Valid N 2	Std.Dev. 1	Std.Dev. 2	F-ratio Variances	p Variances
ENG_ATTENDANCE	3.009615	3.071429	-0.35222	172	0.725101	104	70	1.101589	1.183478	1.154201	0.504618
ENG_ATTENTION	3.750000	3.214286	3.43254	172	0.000749	104	70	0.962904	1.075321	1.247125	0.306935
ENG_EFFORT	3.269231	3.128571	1.02204	172	0.308196	104	70	0.803279	1.006089	1.568699	0.037798
ENG_INTEREST	3.471154	3.185714	1.83304	172	0.068525	104	70	0.974995	1.053536	1.167598	0.471532
ENG_ENJOYMENT	3.567308	2.985714	3.96908	172	0.000106	104	70	0.889687	1.028477	1.336333	0.180626
ENG_COMFORT	3.394231	3.457143	-0.43838	172	0.661659	104	70	0.817506	1.072525	1.721208	0.012246
ENG_WILLINGNESS	3.451923	3.242857	1.44504	172	0.150265	104	70	0.811086	1.095918	1.825669	0.005491
ENG_EST_GOALS	3.317308	3.514286	-1.38285	172	0.168504	104	70	0.884213	0.974201	1.213901	0.369215

Fig. 2. T-test output for difference in factors of engagement across the samples

on the rest of the factors was not statistically significant. Furthermore, using YouTube had an insignificant negative effect on attendance of lectures, sense of comfort towards the coursework and establishing learning goals. Using YouTube had an insignificant positive effect on effort in class, interest in the coursework and willingness to learn. As this t-test was not conclusive in determining the effect of YouTube on the three types of engagement, another t-test was done. This time, the separate factors in each type of engagement were combined. The output of this t-test is shown in the screen-snapshot of Fig. 2.

From the output one can see that the p-values for attention and enjoyment were less than 0.05. Thus, the use of YouTube had a statistically significant positive effect on attention in class as well as enjoyment towards the course work. For both of these factors, the effect was positive. The effect of YouTube on the rest of the factors was not statistically significant. Furthermore, using YouTube had an insignificant negative effect on attendance of lectures, sense of comfort towards the coursework and establishing learning goals. Using YouTube had an insignificant positive effect on effort in class, interest in the coursework

| Variable | T-tests; Grouping: CURRENT (Data All Final) Group 1: Current Students Group 2: Previous Students | | | | | | | | | | |
	Mean 1	Mean 2	t-value	df	p	Valid N 1	Valid N 2	Std.Dev. 1	Std.Dev. 2	F-ratio Variances	p Variances
ENG_BEHAVIOURAL	3.342949	3.138095	1.480562	172	0.140553	104	70	0.822364	0.993526	1.459590	0.081022
ENG_EMOTIONAL	3.477564	3.209524	2.016917	172	0.045259	104	70	0.794757	0.948210	1.423445	0.103246
ENG_COGNITIVE	3.384615	3.378571	0.045259	172	0.963953	104	70	0.788974	0.964725	1.495139	0.063504

Fig. 3. T-test output for difference in types of engagement across the samples

and willingness to learn [30]. As this t-test was not conclusive in determining the effect of YouTube on the three types of engagement, another t-test was carried out. This time, the separate factors in each type of engagement were combined. The output of this t-test is shown in Fig. 3.

In terms of behavioral and cognitive engagement, the use of YouTube had a statistically insignificant positive effect. In terms of emotional engagement, the use of YouTube in the classroom had a statistically significant positive effect.

5.3   RQ$_3$: How Did Students Feel About the Use of YouTube in the Classroom?

In terms of the students' feelings towards the success of using videos in the classroom, a majority of students across both samples agreed to some extent that the use of videos in the classroom was successful. Furthermore, in terms of whether students would recommend the use of videos in other courses, a majority of students across both samples agreed to some extent that they would recommend the use of videos in other courses.

The feelings of the respondents in this study were consistent with the characteristics of the Net Generation noted by Oblinger as well as other authors' claims that videos could be used in higher education [4, 11, 23, 26, 31, 34].

6   Conclusion

We found that the students surveyed were a fair representation of the Net Generation according to the literature reviewed. Furthermore, the extent of the use of YouTube in the classroom between the two samples was significantly different. Lastly, we found that the use of YouTube in the classroom had a positive effect on overall engagement as well as on behavioural, emotional and cognitive engagement. The students surveyed in this research felt that the use of YouTube in class was successful, and many students recommended the use of videos in other courses.

Appendix: Data Tables

In all of the following tables, SA = 'strongly agree', A = 'agree', N = 'neutral', D = 'disagree', SD = 'strongly disagree' on the 5-point Likert scale (Tables 1, 2, 3, 4, 5, 6, 7, 8 and 9).

Table 1. Gender of respondents

Gender	Current sample: n (%)	Previous sample: n (%)
Male	51 (*49%*)	21 (*30%*)
Female	53 (*51%*)	48 (*69%*)
No answer	0 (*0%*)	1 (*1%*)
(Total)	104 (*100%*)	70 (*100%*)

Table 2. Interaction time with technological devices

Hours/week	Current sample: n (%)	Previous sample: n (%)
Not sure	13 (*12%*)	9 (*13%*)
0–5	8 (*8%*)	1 (*1%*)
6–10	9 (*9%*)	5 (*7%*)
11–15	18 (*17%*)	10 (*14%*)
16–20	21 (*20%*)	20 (*29%*)
21++	35 (*34%*)	25 (*36%*)
(Total)	104 (*100%*)	70 (*100%*)

Table 3. Social network membership

Membership	Current sample: n (%)	Previous sample: n (%)
Yes	102 (*98%*)	67 (*69%*)
No	2 (*2%*)	3 (*4%*)
(Total)	104 (*100%*)	70 (*100%*)

Table 4. Characteristics and learning preferences of current students, with total number of respondents $n = 104$ (*100%*) in every row

Question	SD	D	N	A	SA
Do you multitask frequently and do you expect immediate responses from technology?	1 (*1%*)	5 (*5%*)	15 (*14%*)	54 (*52%*)	29 (*28%*)
Do you prefer a practical learning environment as opposed to a theoretical one?	0 (*0%*)	2 (*2%*)	20 (*19%*)	49 (*47%*)	33 (*32%*)
Do you prefer to learn and work in teams?	10 (*10%*)	26 (*25%*)	44 (*42%*)	19 (*18%*)	5 (*5%*)
Do you prefer to perform tasks with a known set of guidelines and rules as opposed to following your own approach?	2 (*2%*)	6 (*6%*)	25 (*24%*)	41 (*39%*)	30 (*29%*)
Are you more comfortable with interactive material (e.g. Videos) instead of static material (slides, notes)?	4 (*4%*)	16 (*15%*)	30 (*29%*)	35 (*34%*)	19 (*18%*)
Are you more comfortable with visual teaching material (e.g. Images) instead of static material (slides, notes)?	5 (*5%*)	10 (*10%*)	34 (*33%*)	34 (*33%*)	21 (*20%*)
Do you like to learn about things that matter, such as environmental and economic concerns?	0 (*0%*)	6 (*6%*)	17 (*16%*)	50 (*48%*)	31 (*30%*)

Table 5. Characteristics and learning preferences of previous students, with total number of respondents $n = 70$ (*100%*) in every row

Question	SD	D	N	A	SA
Do you multitask frequently and do you expect immediate responses from technology?	0 (*0%*)	1 (*1%*)	4 (*6%*)	36 (*51%*)	29 (*41%*)
Do you prefer a practical learning environment as opposed to a theoretical one?	0 (*0%*)	4 (*6%*)	14 (*20%*)	26 (*37%*)	26 (*37%*)
Do you prefer to learn and work in teams?	11 (*16%*)	18 (*26%*)	28 (*40%*)	10 (*14%*)	3 (*4%*)
Do you prefer to perform tasks with a known set of guidelines and rules as opposed to following your own approach?	0 (*0%*)	8 (*11%*)	18 (*26%*)	29 (*41%*)	15 (*21%*)
Are you more comfortable with interactive material (e.g. Videos) instead of static material (slides, notes)?	1 (*1%*)	13 (*19%*)	19 (*27%*)	23 (*33%*)	14 (*20%*)
Are you more comfortable with visual teaching material (e.g. Images) instead of static material (slides, notes)?	1 (*1%*)	16 (*23%*)	22 (*31%*)	23 (*33%*)	8 (*11%*)
Do you like to learn about things that matter, such as environmental and economic concerns?	0 (*0%*)	6 (*9%*)	18 (*26%*)	27 (*39%*)	19 (*27%*)

Table 6. Use of YouTube in the classroom

YouTube usage	Current sample: n (%)	Previous sample: n (%)
Not sure	8 (*8%*)	19 (*27%*)
Never	1 (*1%*)	30 (*43%*)
Occasionally	60 (*58%*)	20 (*29%*)
Often	31 (*30%*)	1 (*1%*)
Always	4 (*4%*)	0 (*0%*)
(Total)	104 (*100%*)	70 (*100%*)

Table 7. Extent of engagement when *no* YouTube videos were used in class

How engaged without YouTube	Current sample: n (%)	Previous sample: n (%)
Not sure	7 (*7%*)	9 (*13%*)
Never	6 (*6%*)	12 (*17%*)
Occasionally	60 (*58%*)	21 (*30%*)
Often	24 (*23%*)	23 (*33%*)
Always	7 (*7%*)	5 (*7%*)
(Total)	104 (*100%*)	70 (*100%*)

Table 8. Extent of engagement *with* YouTube videos in class for the current students

How engaged with YouTube	Current sample: n (%)
Not sure	4 (*4%*)
Never	7 (*7%*)
Occasionally	26 (*25%*)
Often	45 (*43%*)
Always	22 (*21%*)
(Total)	104 (*100%*)

Table 9. Effects of YouTube (YT) in the classroom on the current students, with total number of respondents $n = 104$ (*100%*) in every row

Question	SD	D	N	A	SA
YT increased my attendance of lectures	9 (*9%*)	24 (*23%*)	39 (*38%*)	21 (*20%*)	11 (*10%*)
YT increased my attention in class	2 (*2%*)	12 (*12%*)	16 (*15%*)	54 (*52%*)	20 (*19%*)
YT increased my effort in class	1 (*1%*)	15 (*14%*)	48 (*46%*)	35 (*34%*)	5 (*5%*)
YT increased my interest in the course work	3 (*3%*)	15 (*14%*)	28 (*27%*)	46 (*44%*)	12 (*12%*)
YT increased my enjoyment towards the course work	2 (*2%*)	11 (*11%*)	28 (*27%*)	52 (*50%*)	11 (*11%*)
YT increased my sense of comfort towards the course work	2 (*2%*)	9 (*9%*)	46 (*44%*)	40 (*38%*)	7 (*7%*)
YT increased my willingness to learn	2 (*2%*)	7 (*7%*)	45 (*43%*)	42 (*40%*)	8 (*8%*)
YT aided in establishing my learning goals	4 (*4%*)	10 (*10%*)	46 (*44%*)	37 (*36%*)	7 (*6%*)
I feel that the use of YT was successful in class	3 (*3%*)	6 (*6%*)	29 (*28%*)	50 (*48%*)	16 (*15%*)
I would recommend the use of YT also in other courses	4 (*4%*)	6 (*6%*)	24 (*23%*)	44 (*42%*)	26 (*25%*)

References

1. Agazio, J., Buckley, K.: An untapped resource: using YouTube in nursing education. Nurse Educ. **34**(1), 23–28 (2009)
2. Axelson, R., Flick, A.: Defining student engagement. Change: Mag. High. Learn. **43**(1), 38–43 (2011)

3. Baxter, G.J., Connolly, T.M., Stansfield, M.H., Tsvetkova, N., Stoimenova, B.: Introducing web 2.0 in education: a structured approach adopting a web 2.0 implementation framework. In: Proceedings 7th International Conference on Next Generation Web Services Practices, pp. 499–504. IEEE (2011)
4. Berk, R.A.: Multimedia teaching with video clips: TV, movies, YouTube, and MTVU in the college classroom. Int. J. Technol. Teach. Learn. **5**(1), 1–21 (2009)
5. Berk, R.A.: Teaching strategies for the net generation. Transform. Dialogues: Teach. Learn. J. **3**(2), 1–24 (2009)
6. Burke, S.C., Snyder, S., Rager, R.C.: An assessment of faculty usage of YouTube as a teaching resource. Internet J. Allied Health Sci. Pract. **7**(1), 1–8 (2009)
7. Chetty, M., Banks, R., Bernheim-Brush, A., Donner, J., Grinter, R.E.: While the meter is running: computing in a capped world. Interactions **18**(2), 72–75 (2011)
8. Churchill, D.: Educational applications of web 2.0: using blogs to support teaching and learning. Br. J. Educ. Technol. **40**(1), 179–183 (2009)
9. Clifton, A., Mann, C.: Can YouTube enhance student nurse learning? Nurse Educ. Today **31**(4), 311–313 (2011)
10. Cole, M.: Using Wiki technology to support student engagement: lessons from the trenches. Comput. Educ. **52**(1), 141–146 (2009)
11. Duffy, P.: Engaging the YouTube Google-Eyed generation: strategies for using web 2.0 in teaching and learning. Electron. J. e-Learn. **6**(2), 119–130 (2008)
12. Finn, J.D., Zimmer, K.S.: Student engagement: what is it? why does it matter? In: Christenson, S., Reschly, A., Wylie, C. (eds.) Handbook of Research on Student Engagement, pp. 97–131. Springer, Boston (2012). doi:10.1007/978-1-4614-2018-7_5
13. Fredricks, J.A., Blumenfeld, P.C., Paris, A.H.: School engagement: potential of the concept, state of the evidence. Rev. Educ. Res. **74**(1), 59–109 (2004)
14. Grosseck, G.: To use or not to use web 2.0 in higher education? Procedia Soc. Behav. Sci. **1**(1), 478–482 (2009)
15. Harris, A.L., Rea, I.A.: Web 2.0 and virtual world technologies: a growing impact on IS education. J. Inf. Syst. Educ. **20**(2), 137–145 (2010)
16. Hazari, S., North, A., Moreland, D.: Investigating pedagogical value of Wiki technology. J. Inf. Syst. Educ. **20**(2), 187–199 (2010)
17. Hrastinski, S., Aghaee, N.M.: How are campus students using social media to support their studies? An explorative interview study. Educ. Inf. Technol. **17**(4), 451–464 (2011)
18. Jones, C., Ramanau, R., Cross, S., Healing, G.: Net generation or digital natives: is there a distinct new generation entering university? Comput. Educ. **54**(3), 722–732 (2010)
19. Kennedy, G., Judd, T., Dalgarno, B., Waycott, J.: Beyond natives and immigrants: exploring types of net generation students. J. Comput. Assist. Learn. **26**(5), 332–343 (2010)
20. Kim, J.: The institutionalization of YouTube: from user-generated content to professionally generated content. Media Cult. Soc. **34**(1), 53–67 (2012)
21. Margaryan, A., Littlejohn, A., Vojt, G.: Are digital natives a myth or reality? University students' use of digital technologies. Comput. Educ. **56**(2), 429–440 (2011)
22. Merlino, N., Rhodes, R.: Technology in the 21st century classroom: key pedagogical strategies for millennial students in university business courses. J. Supply Chain Oper. Manag. **10**(1), 113–130 (2012)
23. Oblinger, D., Oblinger, J.: Educating the Net Generation. EDUCASE (2005)

24. Pike, G.R.: The convergent and discriminant validity of NSSE scalelet scores. J. Coll. Student Dev. **47**(5), 550–563 (2006)
25. Prensky, M.: Digital natives, digital immigrants – part 1. On the Horiz. **9**(5), 1–6 (2001)
26. Roodt, S., de Villiers, C.: Using YouTube as an innovative tool for collaborative learning at undergraduate level in tertiary education. In: Proceedings AIS SIG-ED IAIM Conference (2011)
27. Sherer, P., Shea, T.: Using online video to support student learning and engagement. Coll. Teach. **59**(2), 56–59 (2011)
28. Skiba, D.: Nursing education 2.0: YouTube. Nursing Educ. Perspect. **28**(2), 100–102 (2007)
29. Smith, J.: The YouTube revolution: engagement, perception and identity. In: Proceedings ICPhS XVII, Hong Kong, pp. 100–103 (2011)
30. Snelson, C.: YouTube across the disciplines: a review of the literature. MERLOT J. Online Learn. Teach. **7**(1), 159–169 (2011)
31. Tan, E., Pearce, N.: Open education videos in the classroom: exploring the opportunities and barriers to the use of YouTube in teaching introductory sociology. Res. Learn. Technol. **19**(1), 125–133 (2012)
32. Tapscott, D.: Growing up digital: the rise of the net generation. McGraw-Hill, New York (1997)
33. Virkus, S.: Use of web 2.0 technologies in LIS education: experiences at Tallinn University, Estonia. Program Electron. Libr. Inf. Syst. **42**(3), 262–274 (2008)
34. Williams, J., Chinn, S.J.: Using web 2.0 to support the active learning experience. J. Inf. Syst. Educ. **20**(2), 165–175 (2010)
35. Wilson, M., Gerber, L.E.: How generational theory can improve teaching: strategies for working with the 'Millennials'. Curr. Teach. Learn. **1**(1), 29–44 (2008)

Technology and Gaming in Nowadays Education

Mobile and Game Usage, Gender and Attitude Towards Computing Degrees

Eleftherios A. Nicolau and Lisa F. Seymour$^{(\boxtimes)}$ ⓘ

Department of Information Systems, University of Cape Town,
Cape Town, South Africa
Terry.Nicolau@gmail.com, lisa.seymour@uct.ac.za

Abstract. There is a global demand for graduates with computing skills as well as a global shortage of computing professionals, especially women. In Africa these shortages are even more critical. To resolve this problem, studies in the past have argued for increased computing usage in schools that would lead to increased interest in computing degrees. Recently computing usage in the form of mobile devices and mobile gaming has increased substantially in schools, yet in South Africa interest in studying computing at a tertiary level has not increased substantially. Hence, this paper aims to determine whether mobile and gaming usage impacts school learners' attitude towards computer related degrees. A survey of 292 South African secondary school learners confirms low interest in computing degrees, especially by girls, as well as misperceptions of computing degrees. An IT Usage Attitude Model is proposed and validated in which mobile and gaming usage vary substantially based on gender but usage does not influence attitude towards computing degrees.

Keywords: Gender · Gaming · Mobile usage · Technology in schools · Technology attitudes · Computing career choice

1 Introduction

Information Technology (IT) infrastructure, diffusion and usage has increased dramatically and IT is seen as fundamental to socio-economic development [2,21]. Yet the supply of graduates within computer fields is not fulfilling the demand, with predictions that by 2020 in the USA there will be more than 3 times as many jobs for computing graduates compared to the number of available graduates [9] In Australia, Information Systems (IS) departments are reporting reduced student numbers and increasing demand for IS skills [47]. Computing graduates study Computer Related Degrees (CRD) defined by the Computing Curricula as Computer Science (CS), IS, Computer Engineering (CE), Software Engineering (SE) and IT [1]. There is confusion within the industry, as well as among school learners and parents, as to what these various CRD offer, which is increased by many universities offering degrees in blends of IS, CS and IT [3]. The shortage of computing skills is an international problem and also a problem

© Springer International Publishing AG 2017
J. Liebenberg and S. Gruner (Eds.): SACLA 2017, CCIS 730, pp. 67–82, 2017.
https://doi.org/10.1007/978-3-319-69670-6_5

for the South African (SA) economy [10,43].[1] The SA Grade 12 examination results suggest that it will get even worse: in 2013, the Grade 12 results revealed a drastic decrease in the number of learners choosing mathematics and science (approximately 20% in each subject) which are needed to study CRD [16].

This shortage of computing skills is exacerbated by the under-representation of *women* in computing careers and CRD. While this is a global phenomenon, there are some countries (e.g., Malaysia) where this is not a concern [25]. In SA this situation is of high concern and worsening in that the number of women applying for jobs in the IT sector is decreasing annually, hence the field is becoming more male-dominated [20]. SA's census in 2011 found that 8.1% of males chose to study a CRD, contrasted with only 5.4% of females. This is a decrease from 2001 where 7.8% of females chose CRD [49]. Likewise, the percentage of female members of the Institute of IT Professionals SA (IITPSA) shrank from 18.9% in 2015 to 16.2% in 2017 [37].

The shortage of computing graduates was traced to lack of interest in CRD at school [3] and negative attitudes to IT which could be altered to increase interest in CRD [19]. In response, our study aimed to confirm whether IT usage impacts school learners' attitude towards CRD. Two technologies, mobile phones and mobile gaming, were chosen as they are pervasive in secondary schools.

2   Literature

Learners have negative misperceptions of CRD, the career paths available through studying a CRD, and their own capabilities within computing [14,45, 46]. A negative attitude towards IT is seen to impact attitude towards CRD and is an explanation for the gender imbalance in CRD [54]. Yet this relationship has not been validated. Low interest in CRD particularly amongst girls has also been attributed to shortage of IT usage [32] and lower female self-esteem about computing skills [54]. In addition, when adolescents are faced with making adult decisions, the perception that IT is a domain for males is pronounced [45]. The 'geek' perception of CS, the dumbing down of computing to 'IT literacy', and a lack of initiatives to inspire girls to take up IT dissuade girls [28,36]. With mobile devices and gaming becoming increasingly popular amongst the youth, they should have a noticeable influence on attitude towards those technologies. This could influence their attitude towards CRD.

The uptake of mobile devices in Africa progresses rapidly. In SA mobile internet subscription is predicted to achieve 69% penetration by 2019 [41]. The Mobile Africa 2015 survey of SA, Nigeria, Kenya, Ghana and Uganda found that 40% of internet browsing was via mobile devices [23]. Following these trends, children's use of mobile phones has increased rapidly across sub-Saharan Africa [39]. The advantages and disadvantages of mobile device usage can influence perceptions and hence attitudes. In terms of advantages, mobile devices have increased and improved our ability to obtain, search and process information

---

[1] At the same time, unemployment rates are high in SA among low-skilled citizens.

with speed and ease [4,22]. Mobile devices allow us to reach people we would not have been able to reach before, enabling better decisions, real-world experiences with such people, and/or advice [22]. Conversely, the disadvantaged of usage in schools in Africa is becoming apparent, too [39]. Over-extended use of mobile devices reduces the attention span of learners [47]. Studies have shown a noticeable rise in attention deficit hyperactivity disorder in learners and a decreasing ability to retain information, which have been partly attributed to having information easily available through mobile devices [21]. Mobile devices have also disrupted learners' sleep patterns [44] and are a major distraction in the classroom, causing academic performance to drop and an increase in cheating [4]. Finally, accessing inappropriate content or pornography is damaging to learners and to schools' reputations [4,39]. Cyberbullying, under-age sexting, and meeting strangers online are increasing risks [44].

Computer gaming has become a part of everyday life and organization are even embracing gamified mechanisms [53]. In 2015 a report found that 26% of U.S. gamers are under the age of 18 [52]. Mobile gaming growth has increased exponentially [42], and in Africa the social or casual gaming revenue (which is primarily application-based on mobile devices) is predicted to grow at a strong compound annual rate of 9% in SA and 16% in Nigeria [41]. Gaming has advantages and disadvantages: from the negative perspective, an explanation given for the decreasing interest in CRD by women is the rise of a strongly male-dominated computer gaming culture [20] together with the negative stereotype of gamers as 'geeks' [34]. Yet this stereotype seems to be changing as in 2015 females made up 44% of the U.S. gaming industry [52]. But gaming also has advantages: gaming can be a powerful learning tool [53], can increase social interaction and social skills, and has created a gaming culture that has helped many teenagers to connect socially [26]. Gaming can improve the ability to follow instructions, logic and problem solving skills, hand-eye coordination, as well as fine motor and spatial skills [26].

3   Theoretical Framework

The purpose of our study is to determine whether mobile and game usage has an impact on school learners' attitudes towards CRD. This topic is not yet well researched. In the literature a single theoretical framework was not found; thus a combination of already existing approaches was used to formulate our hypotheses.

Social cognitive theory (SCT) builds on *self-efficacy theory*—see below—however SCT emphasizes how personal, and environmental factors interact to determine motivation and behaviour [7,15]. Human behaviour and functioning is attributed to the interaction of personal factors, behavioural factors, and environmental factors [5]. The personal variable 'gender' dominates in the literature and is our concern, too. Hence we propose

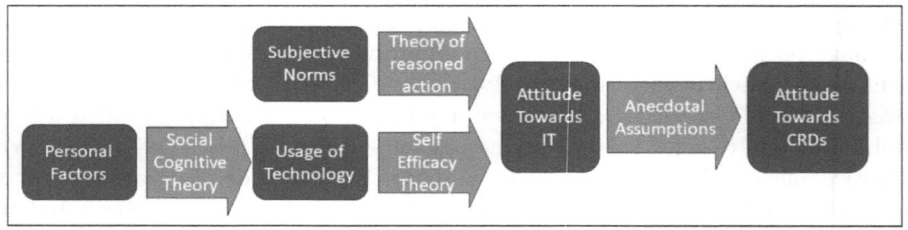

Fig. 1. Initial theoretical framework

H1: *gender influences learners' mobile and gaming usage.*

Self-efficacy theory explains how persons' judgements of how well or able they would be in performing particular tasks has a direct correlation with higher motivation levels; this subsequently has a direct correlation with positive attitude forming [6]. On the basis of this theory we propose

H2: *mobile device usage influences attitude towards mobile devices positively;*
H3: *gaming usage influences attitude towards gaming positively.*

Theory of Reasoned Action (TRA) has also been applied to the question of how major-topics are chosen by students [18]. According to TRA, students' attitudes toward their majors are influenced by individual factors which are separated into two categories: *experiential beliefs* such as perceived difficulty, and *normative beliefs* (referred to as subjective norm) which include the influence of salient others [18]. We include the subjective norm variable 'attitude of friends towards gaming', as it has strong literature support, in

H4: *friends' attitudes towards gaming influences learners' attitudes towards gaming positively.*

Finally, attitude towards IT is assumed to influence attitude towards CRD [33]:

H5: *attitude towards mobile devices influences attitude towards CRD positively;*
H6: *attitude towards gaming influences attitude towards CRD positively.*

Figure 1 shows the initial framework merged from the literature. It includes 5 variable groups: usage of technologies, perceptions and attitudes of the school learners towards these technologies, their personal attributes such as gender, subjective norm factors, and their attitudes towards CRD which would influence their behaviour w.r.t. choosing to study a CRD or not. The relationships, with their theoretical source, are shown by arrows. Social cognitive theory attributes IT usage to personal factors. Self-efficacy theory attributes IT attitudes to IT usage, and TRA attributes IT attitudes to subjective norm factors. On the basis of the literature we assume that the attitude towards CRD are informed by attitude towards IT, and that if the attitude towards CRD improves then the learners will more likely choose to enrol for CRD at university.

Two prior studies have found that learners without access to computers at school are more inclined to study CS than those with access to computers [24, 45]. Learners being exposed to more IT has not translated into more students studying CRD. Hence we must remain self-critical about our own proposed theoretical framework. Some studies on the impact of gaming are inconsistent with other studies finding that computer gaming influences intention to study CS; other studies have found no such relation [40].

4 Method

Related to **H1–H6** of above, we ask the following somewhat more general

Research Question: *does the usage of mobile devices and gaming impact the attitudes that school learners have w.r.t. CRD?*

A deductive strategy in the positivist paradigm with a quantitative survey [8] was chosen. Table 1 lists the relevant hypotheses and their variables. The main dependent variable is attitude toward CRD. All variables are ordinal, except for 'gender' which is nominal.

Participation in the survey-based questionnaire was voluntary.[2] To increase the sample at least one class was requested to complete the survey from Grades 10–12 in each school. In an effort to entice schools into cooperating with us, each school was offered a digitized report of the survey with their learners' responses. In total the survey was taken 430 times. However, 79 incomplete and 63 pilot surveys were discarded, leaving 296 for analysis. The survey included multiple choice, ranking scale, and 5-point Likert scale questions. The target population was limited to Cape Town that has positioned itself as a technological hub of SA. The schools chosen all taught mathematics and science up to Grade 12, which is a requirement for learners before applying to universities for CRD studies. The initial intent was to test for perceptions of IT impact within the school curriculum so only schools offering computing subjects at school were sampled.

Table 1. Hypothesis and variables

Category	Hypotheses	Variables (coding)
Personal factors	**H1**	Gender (G)
Usage of IT	**H2–H3**	Mobile usage (MU), Gaming usage (GU)
Subjective norm factors	**H4**	Friends' attitude towards gaming (FR)
Attitude towards IT and CRD	**H5–H6**	Attitude towards Mobile (MA)
		Attitude towards Gaming (GA)
		Attitude towards CRD (CRD)

---

[2] Approval was acquired from the Education Board, Ethics Committee of our university and the respective schools.

Table 2. Cronbach alpha values and #questions for multiple item constructs (MIC)

MIC	Final α	#Questions	MIC	Final α	#Questions	MIC	Final α	#Questions
MU	0.82	9	GA	0.84	4	MA	0.68	6
GU	0.68	2	FR	0.89	4	CRD	0.86	4

Table 3. Survey response demographics

Demographic	Percentage	Demographic	Percentage
Male	62%	Female	38%
Private all girls' school	7%	Younger than 15	0.3%
Public all girls' school	18%	15 years	2.0%
Private all boys' school	33%	16 years	19.9%
Public all boys' school	18%	17 years	51.7%
Private coeducational school	8%	18 years	24.0%
Public coeducational school	16%	Above 18 years	2.0%

Access to expensive mobile and gaming technology needed to be possible. These criteria excluded poorer schools. For the sake of diversity 6 schools, each of a different school type, were chosen (Table 3).

All variables other than gender were multiple item constructs (MIC) which were then analysed for reliability [48]. Subsequently, factor analysis and Eigenvalue testing were done. After removing some less relevant questions, all constructs passed the tests: see Table 2. Most variables were then averaged and transformed (some reversed) onto a scale between 0 and 1, where 0 represents *no usage* or *negative attitude* whereas 1 represents *very high usage* or *positive attitude*. T-tests were done when a variable was nominal, to test if the difference of means was statistically significant [31]. For correlations between ordinal variables the non-parametric Spearman rank order correlation test was applied (all with the StatSoft Statistica software).

5   Results and Discussion

The demographics are listed in Table 3. The age mode and mean are in the 17-year-old range. The data has several *limitations*. The largest contribution was from the *Private All Boys' School*, which ran the survey amongst the entire Grade 11. The lowest result was from the *Private All Girls' School* as the school asked the girls to do it at home. Hence, the representation from each school type and gender was not equal. The ethnic composition of learners at schools in the Western Cape represents regional (not national) demographics: for example, in 2011 the ethnic demographics of Western Cape school learners was 57% 'coloured' and 11% 'white' compared to the national average of 8% 'coloured' and 5% 'white' [17]. Sufficiently affluent schools were also chosen (as they have better access to smart phones that are enabled for mobile gaming) whereas many other schools in SA do not have suc possibilities. These data limitations impacts the generalisability of our descriptive results. Table 4 lists the correlation test results with the statistical significance of the indicated relationships.

Table 4. Spearman rank order correlation

Variable pair (Hyp.)	Sig.	N	Spearman (R)	$t(N-2)$	p-value
MU and GU (none)	95%	287	0.12	2.02	0.043456
MU and MA (**H2**)	no	286	−0.03	−0.46	0.642979
GU and GA (**H3**)	99%	278	0.31	5.36	0.000000
FR and GA (**H4**)	99%	277	0.60	12.29	0.000000
MA and CRD (**H5**)	no	266	0.06	1.02	0.306870
GA and CRD (**H6**)	99% (rev)	266	−0.16	−2.58	0.010301

5.1 Mobile Access, Gaming Access and Data Availability

Demographic questions related to IT access were included in the survey. All learners had access to at least one mobile device, and all learners had on average access to two. The learners' primary mobile device were smartphones (61% of all learners), feature phones (29%) and tablets (10%). In terms of gaming access (Table 5), most gaming access was through smart phones (83%), followed by laptops and PCs (73%). Even Tablet access (56%) had overtaken specifically designed gaming devices.

Table 5. Gaming access

Device	Gaming access %
Smart phone	83%
Normal PC or laptop	73%
Tablet: iPad, Nexus, Galaxy	56%
3rd generation console: PS3, XBOX 360	40%
Gaming laptop or PC	26%

5.2 Mobile and Gaming Usage

Mobile and gaming usage, in hours per day, is shown in Fig. 2. The learners spent most of their time (an average of 3 h per day) using their mobile devices for instant messaging or listening to music. By contrast, using mobile phones for schoolwork was less than 1.5 h, and making telephone calls less than 1 h. The learners spent an average of 2 h a day gaming on their smart phones, followed by 1.2 h on standard PCs or laptops, and 0.6 h on tablets. Out of 288 learners, 4 did not partake in gaming: as explanation all 4 indicated that they did not see the benefits in gaming, and only 1 learner indicated lack of access. 36% of learners never game in a group or social environment, while 16% game socially at least weekly.

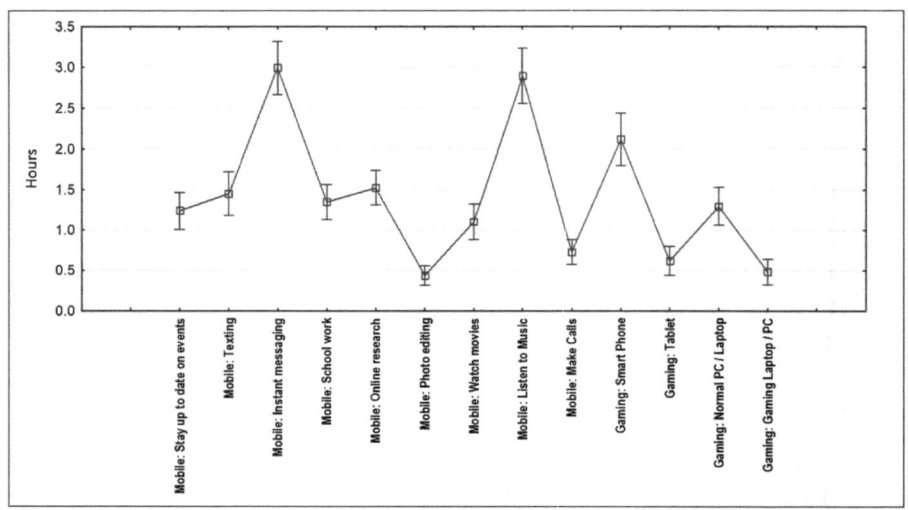

Fig. 2. Gaming and mobile usage in hours

5.3 Attitude Towards Mobile and Gaming

Table 6 summarises some of the attitudinal questions. In the table the 5-point scale was condensed into 'agree', 'neutral' and 'disagree'. The final two questions were not used in hypothesis testing; they were asked to confirm or describe particular attitudes.

Table 6. Learners' attitudes towards mobile and gaming

Agree	Neutral	Question-statements suggested to the respondents
93%	6%	Mobile devices are very useful to my life
93%	6%	Mobile devices have helped me to stay in contact with my friends and family easily
52%	22%	Mobile devices cause a lack of attention within students
55%	22%	Mobile devices contribute to a lack of social skill development: talking via instant messaging and constantly checking phone
48%	32%	Gaming helps to improve a person's ability to follow instructions
58%	28%	Gaming helps to improve a person's ability to solve problems (logic)
55%	26%	Gaming helps to improve a person's hand-eye coordination, fine motor and spatial skills
24%	32%	Gaming helps a person's social skills as it provides a platform for social interaction and community gaming
53%	20%	Gaming is male-dominant
12%	25%	Gaming is for geeks

Generally, the learners and their friends perceived gaming favourably while mobile applications met mixed attitudes (Table 6). The learners stated overwhelmingly that mobile devices are useful and help with staying connected with friends and family. However they also believed that they cause lack of attention and contribute to lack of social skills. Many learners agreed with most advantages of gaming and they generally did not believe that gaming is only for 'geeks'. However, they do not believe that it supports social skills whilst many believe that it is male-dominant.

5.4   Attitude Towards CRD

Our survey sheet explained to the learners the notion (and importance) of 'CRD', and asked how many of the 5 CRD they could explain to family or friends. Learners could not explain 2 CRD to family or friends (mean of 1.5). 22% of the learners believed that one need not study computer programming at school to study a CRD at university. These finding confirm that learners do not know enough about CRD. Furthermore the learners were asked how many of the above five CRD degrees would interest them, with the mean value (0.89) not even reaching 1. The CRD attitudinal questions are shown in Table 7. The learners appear to generally believe that the degrees are useful and interesting but have neutral perceptions of pay and job availability.

Table 7. Learners' attitudes towards CRD

Agree	Neutral	Question-statements suggested to the respondents
45%	31%	CRD are interesting
63%	24%	CRD are useful
38%	40%	In SA there are many open jobs for people with a CRD
36%	45%	In SA, the jobs for people with CRD are highly paid

Subsequently the learners were asked to rate $7°$ in order of personal preference with their most preferred career given a value of '7' and their least preferred career given the value of '1'. Table 8 shows that out of the $7°$ suggested, CRD were rated as the second least preferred (in accordance with the literature). Learners were also asked to rate $7°$ in order of their perception of which career pays the highest, with the lowest paying career being given a value of '7' and the highest paying career being given the value of '1'. Table 8 shows that out of the seven degrees suggested, CRD were rated as the third lowest, with the learners only seeing careers in humanities and marketing as paying lower. CRD also had the highest standard deviation, which shows the highest amount of uncertainty about salary knowledge w.r.t. CRD. Salaries in SA were compared by means of [38] in the same year in which we surveyed the lerners. The listed jobs were searched for and ranked according, and the retrieved ranking was then compared

Table 8. Ranking subjects by interest and pay

Subject, career (our study)	Pref. mean (rank)	Pay mean (rank)	Job title according to 'Payscale' [38]	Median salary p.a. (rank) [38]	Rank diff.
Medical	3.96 (4)	2.77 (1)	General practitioner	R373938 (1)	±0
Accounting	4.81 (2)	3.29 (2)	Accountant	R228 582 (5)	+3
Law	4.04 (3)	3.42 (3)	Attorney/lawyer	R196 527 (7)	+4
Engineering	3.63 (7)	3.53 (4)	Civil engineer	R307 751 (3)	−1
CRD	3.94 (6)	3.77 (5)	Software engineer, developer, programmer IT business analyst senior software engineer, developer, programmer average of all roles above	R233 834 R335 385 R433 351 R334 190 (2)	−3
Marketing	3.98 (5)	5.02 (6)	Marketing Specialist	R256 818 (4)	−2
Humanities	4.82 (1)	6.46 (7)	Bachelor of arts degree	R210 000 (6)	−1

to learners' perceptions. A problem with CRD is the confusingly large variety of job 'titles' in circulation. This might be a competitive advantage for specialized graduates but makes salary comparisons difficult. The median salaries for 3 CRD titles are included; they show that the learners' underestimation of the salaries was strongest for CRD.

5.5   Impact of Gender

Table 9 shows significant gender differences in mobile usage and gaming. **H1** was therefore *accepted*. In accordance with the literature [51] we found (for MU) that girls use their mobile devices 5.8% more than boys do. Although access to smartphones is relatively equal amongst genders, girls are significantly more likely than boys to access the internet using their phones [29]. By contrast, boys partake in gaming (GU) 7% more than girls do. This result was expected and is in agreement with [20] according to which gaming is a male-dominated culture. Our t-tests confirm that girls have a more negative attitude towards CRD than boys have.

Table 9. T-test: gender versus IT usage and attitude towards CRD

Variable	Mean (boys)	Mean (girls)	$t - value$	df	p	N (Boys)	N (Girls)
GU	0.255647	0.222275	2.57383	286	0.0106	182	106
MU	0.380078	0.426504	−3.47354	285	0.0006	182	105
CRD	0.592387	0.529808	2.456954	264	0.0147	162	104

5.6   Usage Influences Learner's Attitude Towards IT

H3 was strongly *supported*. By contrast, **H2** was *rejected*. While learners were very supportive of the usefulness of mobile devices they were also aware of its negative impact and hence an increase in usage would make them aware of both the positive and negative aspects; this would explain the observed lack of correlation. A negative correlation between gaming access and mobile attitudes was also found which was not in our theoretical model; (future work).

5.7   Impact of Friends on IT Attitudes

H4 was *accepted*. This is also substantiated by our descriptive analysis of the high amount of social gaming. Gaming together with friends is likely enjoyable. According to [35], males are more inclined to participate in such gaming due to its instrumental, activity-based style of intimacy and friendship.

5.8   Attitude Towards IT Versus Attitude Towards CRD

H5 and **H6** were both *rejected*. Attitude towards mobile devices did not influence attitude towards CRD; surprisingly attitude towards gaming was found to influence attitude towards CRD *negatively*.

5.9   Final Framework

Our study yields evidence for gaming and mobile device usage varying by gender. Girls use mobile devices more while in contrast boys do more gaming. The following relationships were validated:

- Gender influences learners' mobile and gaming usage.
- Mobile usage influences gaming usage positively.
- Gaming usage influences gaming attitudes positively.
- Friends' attitudes towards gaming influences learners' attitude towards gaming positively.

Our final framework is presented in Fig. 3 (including Spearman R and t-test values). The initial theoretical framework (Fig. 1) was thus *not* validated. For gaming, the relationships from self-efficacy theory, social cognitive theory and the theory of reasoned action were found to be valid, but *not* the relationship between attitude towards the technology and attitude towards CRD.

The relationships from self-efficacy theory, social cognitive theory and the theory of reasoned action were found to be valid for gaming technology and are shown in the validated framework in Fig. 4 called the *IT usage attitude model*. However, the main relationship which we intended to validate was not validated; instead the reverse relationship was found. It seems that increased gaming and positive attitude towards gaming have a negative effect on attitude towards CRD. *Our study questions the commonplace argument that more IT usage by learners would increase their interest in CRD.* Recently, learners have

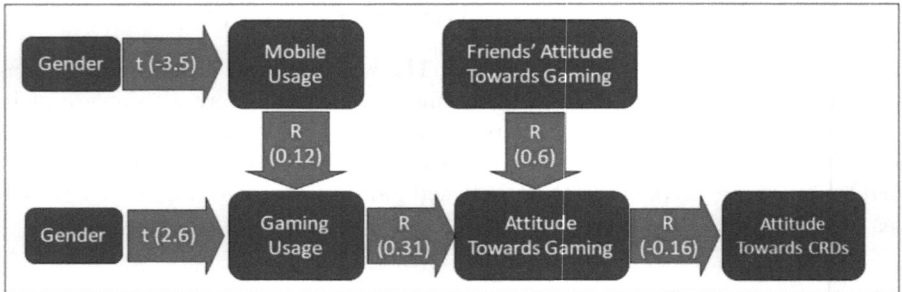

Fig. 3. Final validated framework

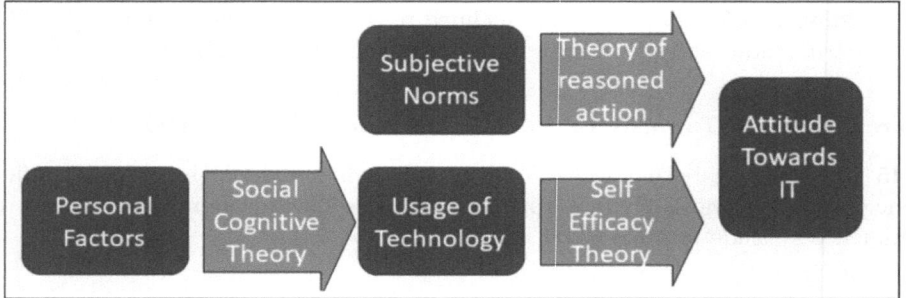

Fig. 4. IT usage attitude model

been exposed to ever more IT whereas interest in CRD has *not* increased.[3] Indeed our study provides support—end explanations—for the reverse argument and thus refutes the preumption that school learners more exposed to IT will be more likely interested in CRD. Our finding is similar to the one of [55] according to which positive attitude towards IT by girls does not translate into considering CRD—for contradictory findings on the impact of gaming on choosing to study CS see [40].

Our study re-confirmed that learners at school do not know enough about CRD at university and have misperceptions of computing salaries and job availability. Hence, interventions need to focus on computing career education. It has been confirmed that attitudes about CRD can be improved by exposing students to the work-life activities of IT professionals [55]. Students studying CRD have indicated that their parents, career counsellors and school teachers assisted them in chosing a career path more so than friends [11]; hence those advisors should be included in this CRD-related career education.

---

[3] Is this perhaps a *technological saturation* effect? For comparison, is there any correclation between nowadays ubiquitous availability of motor cars and student enrolment numbers in automotive engineering?

This leaves us with little advice for how to increase interest in CRD amongst female learners. Recent research shows that a 'sense of belonging' is a strong predictor of interest in science and technology amongst women. Therefore computing stereotypes that women do not identify with reduce their sense of belonging in computing [12]. For example, the perception that computing work is isolating does not match with the communal goals often valued by women. The literature has described four characteristics that students believe to be required for belonging into computer science: being singularly focused on computing, asocial, competitive, and male [27]. However, welcoming and supportive cues, especially from men, can increase women's sense of belonging and hence interest in computing in the face of those stereotypes [30]. Universities also play a role in dissuading women. For example, undergraduate assessment practices that emphasise competition over cooperation tend to advantage men over women [13]. Working against this practice some universities have started increasing the mentoring of women in undergraduate technical fields [50].

6 Conclusion

The lack of interest in CRD is a global issue. Of special concern are the misperceptions about these degrees among school learners and the lack of interest by girls. This study from the Western Cape of SA confirms school learner misperceptions of the entrance requirements and economic value of these degrees and the general lack of interest especially among girls in CRD. Our study also highlighted for parents and educators the nature of the increased mobile and gaming usage patterns of school learners in the Western Cape. From a practical perspective our study indicates that more career and degree information is needed in schools to correct those misperceptions.

It has been suggested that IT usage influences the attitude towards CRD. Given the increase in the number of smartphones in schools, and the exponential growth in gaming during the last five years, it might be expected that such IT usage will have a measurable influence on the future career decisions made by school learners. However, our study found no positive correlations between increased IT usage and attitude towards CRD. Conversely, for our surveyed sample, positive attitude towards gaming showed a negative correlation to attitude towards CRD. Given the increasing interest in gaming this could lead to a further drop in interest and graduates, which is cause for concern.

From a theoretical perspective, our study proposes the IT usage attitude model based on self-efficacy theory, social cognitive theory the theory of reasoned action and validates it for gaming technology. We found that school learners' usage of mobile and gaming technology varies significantly by gender. Furthermore, the subjective norm factor, friends' attitudes towards gaming, effects attitude towards gaming technologies.

In retrospect we identified several limitations to this exploratory study as well as areas for future research. The targeted sampling resulted in usage patterns and perspective not representative of the entire country, such that a broader understanding of IT usage in schools is still desirable. Hence we recommend a more in-depth study of the relationships between IT attitudes and attitude towards CRD. The reverse relationship for gaming is interesting but not yet understood and requires further study, too. It might be possible that different game types are differently influencial and that the preference for these game types varies between the genders.

Last but not least, this study only looked at mobile and gaming technologies, whereas IT in the school curriculum and social media should also be included in future studies. In terms of subjective norm variables, the attitude of teachers, parents and other positive role models should also be considered in future work.

References

1. ACM: Computing Curricula 2005. Technical report. ACM (2005)
2. Akoojee, S., Arends, F.: Intermediate-level ICT skills and development in South Africa: private provision from suited to national purpose. Educ. Inf. Technol. **14**(2), 189 (2009)
3. Anthony, E.: Computing education in academia: toward differentiating the disciplines. In: Proceedings 4th Conference on Information Technology Curriculum, pp. 1–8 (2003)
4. Asur, S., Huberman, B.A.: Predicting the future with social media. In: Proceedings International Conference on Web Intelligence and Intelligent Agent Technology, pp. 492–499. IEEE (2010)
5. Bandura, A.: On the Functional Properties of Perceived Self-Efficacy Revisited. SAGE, Newcastle upon Tyne (2012)
6. Bandura, A.: Self-efficacy: toward a unifying theory of behavioral change. Psychol. Rev. **84**(2), 191 (1977)
7. Bandura, A.: Social cognitive theory: an agentic perspective. Annu. Rev. Psychol. **52**(1), 1–26 (2001)
8. Blaikie, N.: Designing Social Research, 2nd edn. Polity, Cambridge (2009)
9. Bureau of Labor Statistics: Occupational Outlook Handbook. Governmental Communication, USA (2010)
10. Calitz, A.P., Evert, C., Cullen, M.: Promoting ICT careers using a South African ICT career portal. Afr. J. Inf. Syst. **2**(7), 40–61 (2015)
11. Calitz, A., Greyling, J., Cullen, M.: The influencers of scholars' ICT career choices. Educ. Res. Soc. Change **2**(1), 64–81 (2013)
12. Cheryan, S., Drury, B.J., Vichayapai, M.: Enduring influence of stereotypical computer science role models on women's academic aspirations. Psychol. Women Q. **37**(1), 72–79 (2013)
13. Cohoon, J.M.G., Aspray, W.: Women and Information Technology: Research on Underrepresentation. MIT Press, Cambridge (2006)
14. Compeau, D.R., Higgins, C.A.: Computer self-efficacy: development of a measure and initial test. MIS Q. **19**(2), 189–211 (1995)
15. Crothers, L.M., Hughes, T.L., Morine, K.A.: Theory and Cases in School-based Consultation. Routledge, Abingdon (2008)
16. Deloitte: South Africa's ICT Skills Shortage Starts in Matric. Technical report (2014)

17. Department Basic Education: Annual Schools Surveys: Report for Ordinary Schools 2010 and 2011. Governmental Communication, Republic of South Africa (2013)
18. Downey, J., McGaughey, R., Roach, D.: Attitudes and influences toward choosing a business major: the case of information systems. J. Inf. Tech. Educ. Res. **10**(1), 231–251 (2011)
19. Govender, I., Naidoo, E.: Perceived relevance of an introductory information systems course to prospective business students. S. Afr. Comput. J. **51**, 1–9 (2013). Old counting, without iss
20. Harris, L.: Mind the ICT skills gap – South Africa's ICT industry will face a serious skills crunch in the future unless it focuses on developing its young professionals. Technical report, ITWEB (2012)
21. Heeks, R.: Do information and communication technologies (ICTs) contribute to development? J. Int. Dev. **22**(5), 625–640 (2010)
22. Holmes, W.: Crisis communications and social media: advantages, disadvantages and best practices. In: Proceedings 33rd Annual CCI Research Symposium (2011)
23. IT News Africa: Study reveals African Mobile Phone Usage stats. Technical report (2015)
24. Jacobs, C., Sewry, D.A.: Learner inclinations to study computer science or information systems at tertiary level. S. Afr. Comput. J. **45**, 3–10 (2010)
25. Lagesen, V.A.: A cyberfeminist utopia? Perceptions of gender and computer science among Malaysian women computer science students and faculty. Sci. Technol. Hum. Values **33**(1), 5–27 (2008)
26. Lenhart, A., Kahne, J., Middaugh, E., MacGill, A.R., Evans, C., Vitak, J.: Teens, video games, and civics: teens' gaming experiences are diverse and include significant social interaction and civic engagement. Pew Internet & American Life Project, ERIC (2008)
27. Lewis, C.M., Anderson, R.E., Yasuhara, K.: I don't code all day: fitting in computer science when the stereotypes don't fit. In: Proceedings International Conference on Computer Education Research, pp. 23–32. ACM (2016)
28. Liebenberg, J., Mentz, E., Breed, B.: Pair programming and secondary school girls' enjoyment of programming and the subject information technology (IT). Comput. Sci. Educ. **22**(3), 219–236 (2012)
29. Madden, M., Lenhart, A., Duggan, M., Cortesi, S., Gasser, U.R.: Teens and technology 2013. Pew Internet & American Life Project, ERIC (2013)
30. Master, A., Cheryan, S., Meltzoff, A.N.: Computing whether she belongs: stereotypes undermine Girls' interest and sense of belonging in computer science. J. Educ. Psych. **108**(3), 424–437 (2016)
31. Michener, W.K.: Quantitatively evaluating restoration experiments: research design, statistical analysis, and data management considerations. Restor. Ecol. **5**(4), 324–337 (1997)
32. Mishra, S., Kavanaugh, L., Cellante, D.L.: Strategies to encourage females beginning in middle school and high school to participate in the computing field. Strategies **14**(1), 251–259 (2013)
33. Muller, H., Gumbo, M.T., Tholo, J.A.T., Sedupane, S.M.: Assessing second phase high school learners' attitudes towards technology in addressing the technological skills shortage in the South African context. Afr. Educ. Rev. **11**(1), 33–58 (2014)
34. Nelson, J.: Do you think Mobile Gaming will ever be taken seriously in the Minds of 'Hardcore' Gamers? Technical report, TouchArcade (2014)
35. Ogletree, S.M., Drake, R.: College students' video game participation and perceptions: gender differences and implications. Sex Roles **56**(7/8), 537–542 (2007)

36. Ovide, O.: Addressing the lack of women leading tech start-ups. Wall Street J. Blog. https://blogs.wsj.com/venturecapital/2010/08/27/addressing-the-lack-of-women-leading-tech-start-ups/
37. Parry, T.: IITPSA gender statistics. Private Communication (2017)
38. PayScale. http://www.payscale.com/research/ZA/
39. Porter, G., Hampshire, K., Milner, J., Munthali, A., Robson, E., Lannoy, A., Bango, A., Gunguluza, N., Mashiri, M., Tanle, A.: Mobile phones and education in Sub-Saharan Africa: from youth practice to public policy. J. Int. Dev. **28**(1), 22–39 (2016)
40. Prescott, J.: Gender Considerations and Influence in the Digital Media and Gaming Industry. IGI Global, Hershey (2014)
41. Pricewaterhouse-Coopers: Entertainment and Media Outlook 2015–2019: South Africa, Nigeria, Kenya. Technical report, PWC (2015)
42. Rogowsky, M.: The money's in mobile: supercell and the rise of gaming's new giants. Technical report, Forbes (2014)
43. Schofield, A.: JCSE ICT skills survey report. Technical report, JCSE (2014)
44. Scholtz, B., van Turha, T., Johnston, K.: Internet visibility and cyberbullying: a survey of cape town high school students. Afr. J. Inf. Commun. **15**, 93–104 (2015)
45. Seymour, L.F., Hart, M., Haralamous, P., Natha, T., Weng, C.W.: Inclination of scholars to major in information systems or computer science. In: SAICSIT 2004 Proceedings Annual Conference on South African Institute of Computer Scientists and Information Technologists, pp. 97–106 (2004)
46. Seymour, L.F., Serumola, T.: Events that lead university students to change their major to information systems: a retroductive south african case. S. Afr. Comput. J. **28**(1), 18–43 (2016)
47. Smyth, R., Gable, G., Pervan, G.A.: SWOT analysis of the IS academic discipline in Australia. In: Proceedings ACIS 2016, paper #39, Wollongong (2016)
48. Sorenson, M.C.: A qualitative study of 4-H state and field faculty use of social media to communicate with youth, volunteers, and stakeholders. Louisiana State University, Dissertation (2011)
49. Statistics South Africa: Statistical Release – Census 2011. Governmental Communication (2012)
50. Sullivan, P., Moore, K.: Time talk: on small changes that enact infrastructural mentoring for undergraduate women in technical fields. J. Tech. Writ. Commun. **43**(3), 333–354 (2013)
51. Swan, K., Hooft, M., Kratcoski, A., Unger, D.: Uses and effects of mobile computing devices in K8 classrooms. J. Res. Technol. Educ. **38**(1), 99–112 (2005)
52. The Entertainment Software Association Group: Sales, Demographic and Usage Data: Essential Facts about the Computer and Video Game Industry. Technical report (2015)
53. Vesa, M., Hamari, J., Harviainen, J.T., Warmelink, H.: Computer games and organization studies. Organ. Stud. **38**(2), 273–284 (2017)
54. Volman, M., van Eck, E.: Gender equity and information technology in education: the second decade. Rev. Educ. Res. **71**(4), 613–663 (2001)
55. Walstrom, K., Schambach, T.: Impacting student perceptions about careers in information systems. J. Inf. Technol. Educ. Res. **11**(1), 235–248 (2012)

Questioning the Value of Vodcasts in a Distance Learning Theoretical Computer Science Course

Colin Pilkington$^{(\boxtimes)}$

School of Computing, Univeriosity of South Africa, Pretoria, South Africa
`pilkicl@unisa.ac.za`

Abstract. Newer technologies have the potential to enhance how teaching and learning in higher education happen. The context of the provision of 12 supplementary vodcasts in a distance learning theoretical computer science course drove questions about whether these led to a statistically significant improvement in examination marks. Vodcast access numbers and examination marks for four semesters for which vodcasts were available, as well as for four preceding semesters, were analyzed. There was *no* statistically significant difference in marks that could convincingly be linked to the provision of the vodcasts. The results are discussed in relation to the concepts of communication, accessibility, and online use of learning materials.

Keywords: Theoretical computer science education · Distance learning · Vodcast · Video podcast

1 Introduction

Vodcasts are often used in higher education online courses [4] and have the potential to transform how teaching and learning happen [14,15]. Newer technologies do not just offer the possibility of a new way of broadcasting information [22], but also are believed to provide tools that can enhance a student's learning experience [23,30]. The assumption of more digitally involved students, who are able to access learning materials via a range of ICT devices in everyday life, has led to the increasing adoption of tools such as wikis and blogs [22,30,46], and the use of vodcasts has seen increasing interest since 2006 [26,39].

While technology should not determine pedagogy, teaching and learning practice should take up the affordances of technology where vodcasts are a possible tool of instruction [1,27]. Reasons for using vodcasts are often linked to attempts to use blended learning approaches. Although blended learning may be seen to mean the merging of face-to-face and web-based, online learning [15,47], it may also be seen as a mix of different pathways of delivering learning material, building on variation theory, and relying on diverse experiences to aid and encourage learning [18,35]. Of course, in distance learning educational environments where there are no traditional face-to-face lecturers, vodcasts can become, in a sense,

© Springer International Publishing AG 2017
J. Liebenberg and S. Gruner (Eds.): SACLA 2017, CCIS 730, pp. 83–98, 2017.
https://doi.org/10.1007/978-3-319-69670-6_6

the voice (and even face) of the lecturer and, thus, blend normally text-based material with online, audio-visual formats.

There are also views that the use of online technologies should be seen as an asset whose potential should be tapped in attempts to maintain a competitive edge [15,16,22]. Also, ICT approaches may be introduced as external forces pressuring universities to remake themselves in more technologically modern and efficient forms [16], allowing such approaches to become an integral part of the university experience [30]. Thus, academic staff may be required to develop for, and implement, online learning approaches [15], in which vodcasts are seen to play a role, and show the academics' attempts to adopt more technological platforms often *for largely non-academic reasons* [16].

Many computer science degree programmes include some work in theoretical computer science (or formal languages and automata theory), which introduce the thinking required to understand the abstractions of the foundations of computer science [2]. Students find these abstract concepts difficult [2,6], which can lead to poor understanding of the material, which is exacerbated in learning distance environments where there is little face-to-face practical help [19] and students have to try to follow the algorithms in text format. In an effort to support student learning and produce better summative results at the end of a semester, 18 vodcasts were prepared as additional material for a second-year course in theoretical computer science at an open distance learning institution. The study material that is usually provided for this course is a prescribed textbook and university-written study material—all text-based materials. Even where study material is available online via the university's learner management system (LMS), it is usually also distributed in printed format to support student learning (where students may not have access to the web-based LMS). In an attempt to help students follow some of the algorithmic procedures contained in the course, which are quite difficult to follow in a textbook, vodcasts were created to show the steps that should be taken and how to structure problem solutions.

The research question is, thus, as follows: *has the release of vodcasts led to any statistically significant improvement in students' examination scores?* This will be examined for both the overall examination score and the marks for a specific question by comparing the examination marks for the semesters prior to the vodcasts being released to those once the vodcasts were available.[1]

It has been noted that vodcasting is used in a variety of contexts, disciplines, and levels of study [16,25], often making it difficult to generalize about its effects on student performance [34]. It is hoped that this study of the use of vodcasts in a distance education, theoretical computer science course will add to the work already done and the generation of a more solid basis from which to use vodcasts in higher education, so as much detail (as called for by [25]) as possible is given to promote comparisons with other studies.

---

[1] Ethical clearance was obtained from the relevant university committees both to do the research and to gather university data.

2   Vodcasts in Education

A vodcast, or video podcast, is a video file that is made available on the Internet in digital format [30], and the increased availability of high-speed Internet access has allowed vodcasts to become more popular [25]. Similar to an audio podcast, which, strictly speaking, is distributed via user-subscribed syndication feeds, vodcast files can be downloaded in their entirety to personal computers or media devices and are usually not streamed [9,17]. However, there is no requirement as to what format should be used for such a downloadable file to be called a vodcast [39], and in fact, vodcasts do not even have to be video files, but can range from recordings of lectures to voiced-over slide show presentations (often termed 'enhanced vodcasts' [5,34]), as well as complex animations [22,39].

2.1   Theoretically Informed Creation of Vodcasts

Vodcast design and use should be guided by beliefs about learning, such as the theoretical bases provided by cognitive load theory (CLT) and the cognitive theory of multimedia learning (CTML) [1,15]. Guidelines for vodcast creation are based on the understanding that [1,29]:

- there are separate channels for visual and auditory content,
- each channel has limited capacity,
- active processing from selecting material, organizing it, and then integrating it uses cognitive resources.

As cognitive capacity is allocated to the initial processing tasks first, only what is left can be used for the task of integration, meaning that the first two processing levels should be supported as much as possible by reducing anything non-essential and offloading information from the visual to the auditory channel [1]. The following represents a broad outline of recommendations that are given in terms of CLT and CTML, as well as recognized best practice [1,24,29,41,54]:

1. Use voice, text, and pictures. Ensure legibility. Use pictures rather than words, where possible (remembering that simply adding pictures will not help). Use static images rather than animations.
2. Avoid any unnecessary information. Ensure an uncluttered look. Avoid background music.
3. Highlight the important material.
4. Do not read what is already on the screen. Use a conversational style rather than a formal one.
5. Have an image of the narrator on the screen.
6. Pace carefully, allowing pauses as necessary.
7. Optimal length is less than 10 min. Keep it short, and present material in meaningful, bite-sized chunks.

It has also been noted that the quality of the message is more important than the quality of the presentation, and so the level of effort put into the more technical side of creating a vodcast should be determined by how long it is going to be useful [41].

2.2   How Vodcasts are Used in Teaching

Several review articles of podcast and vodcast use have been published [16,17, 25,45], and various classification schemes have arisen. Teaching vodcasts can be one of three basic types:

Lecture-based or classroom vodcasts are simple recordings of lectures, whether presented by the usual, or a guest, lecturer [17,23,24]. This type has also been referred to as substitutional vodcasts, although the aim is often more to provide the lecture for review purposes rather than to replace the lecture entirely [16,18,30,42].

Supplementary vodcasts are used to supplement the material presented in lectures. Such vodcasts can be made up of extra explanations, worked examples, feedback and comments, course-related guidelines, summaries, or field guides [5,9,16,17,24].

Creative vodcasts are those prepared by students as part of their learning experience, often in collaborative forms of learning [16,17,24,30].

For the lecturer-provided vodcasts above, the vodcast can use existing material or be one created by the lecturer [17].

It has been suggested that students' awareness of vodcasts, and their varied uses, need to be promoted among students [50]. Hew [17] presents several ways in which vodcasts can be integrated in the learning process. Among these is the idea that vodcasts are released after an event, either as a copy of the event (which can be used for review) or as supplementary material (which adds relevant information to what was covered). In such cases, the vodcast is often used as revision in preparation for an assessment [9,18,28], and it has been shown that, when used in this manner, it leads to poorer results than would have been expected [54]. When vodcasts are used by students during the learning process, however, better results have been recorded, and it is surmised that such students had the metacognitive ability to realize early that their understanding was not what it ought to be [54]. Having students discuss the content of the vodcast encouraged students to use it earlier in the learning process [44].

Instead of vodcasts being used after an event, it is also possible that they be used before a teaching event in a preparatory fashion. Preparatory vodcasts are similar to lecture-based vodcasts, except that, when preparatory, they are more substitutional in flipped-classroom approaches, where the intent is that students view these before the face-to-face lecture and that lecture time is spent in a more applied and practical manner [15,39,42]. Such use of vodcasts, when integrated with questions, helped focus students and facilitated learning [39].

Students may access and watch vodcasts out of curiosity or because of their novelty [52], which supports arguments that student attitude towards vodcasts, along with perceived ease of use and usefulness, best determines the level to which they are used [9,20]. Viewing rates as high as 70% have been found [52], although cases well below this have also been recorded [50]. Along with attitudes towards vodcast use, it is also believed that lack of access to high-speed Internet, and its associated costs, can lead to lower levels of use [51]. Furthermore, as vodcasts become more commonly available, the novelty will wear off [52].

Although a major benefit of a vodcast is that it can be viewed on mobile devices, evidence suggests that mobile devices are used largely for entertainment purposes [48] and that learning material vodcasts are viewed sitting at a personal computer working, often with other learning materials at hand [16,17,28,30]. Thus, such vodcasts are rarely watched in a multitasking environment, while, for example, the student is travelling [17].

2.3   Benefits and Problems

The benefits of vodcasting for supporting learning are regularly mentioned, and these often relate to overcoming the hindrances of time and space [17] and are linked to the following properties.

Flexibility: the user has the ability to control the flow of the vodcast, pausing and replaying as necessary [11,34,45]. Students can also return to the vodcast as often as required [22].

Accessibility: vodcasts can be released to students quickly [22], can be downloaded and watched whenever it suits students, and are, thus, not bound by the confines of the lecture hall [11,21]. Vodcasts are also easy to use on various viewing devices and platforms [22,49].

Text-based formats are considered too static, and vodcasts are generally preferred over textbooks [11,24,26]. Although a textbook does allow a student to return to previously read material, textbooks often do not highlight what is important, which is something that vodcasts should be doing [42]. Also, for second-language learners and those with learning difficulties, vodcasts provide unique opportunities for students to work at their own pace [1,22]. Thus, vodcasts are believed to provide support for individual learner needs [9].

There are, however, also drawbacks to the use of vodcasts. They generally leave the student in a passive role, often with low levels of engagement and interaction [21,30], and require a high level of discipline and self-regulated learning ability on the part of the student [11,15,40]. For some vodcast types, the presenter also cannot see the student who is watching the vodcast and so cannot rely on non-verbal cues such as a nodding head or blank look to determine whether the content is being grasped or not [40]. In cases where students have chosen not to use vodcasts, such decisions are linked to [18,25,26]:

– more experienced students not believing that they needed the help or finding the pacing too slow,
– lack of knowledge of the availability of the vodcasts,
– not having the time to watch them.

Furthermore, technical difficulties related to the hardware required, file size, download speeds, and Internet connectivity availability and cost are often mentioned [22,26]. Poor sound quality has also been noted as a weakness [18].

Considering the acknowledged benefits of vodcasts and their recognized drawbacks, *what is the evidence for their effect on student learning?* This can be viewed in three ways: their effect on experience, performance, and cognition [16]. Generally, students have reported a very positive experience of the use of vodcasts where they are perceived to be useful and effective to improve under- standing and even to be motivating [4,18,25,34,44,48]. This is especially true when compared to textbooks [24], and rejection of vodcasts is rare [16]. How- ever, there are conflicting reports on whether vodcasts lead to better student learning and knowledge retention, usually measured as improved student perfor- mance in final examination scores [22,25]. Some studies have reported significant gains [25,39,49], whereas others have found no such gains [9,17,18,40,42,52]— particularly when lecture attendance was good [54]. Furthermore, where vod- casts were used in a substitutionary (rather than a supplementary) manner, no significant difference between the groups has been noted [43]. Where positive results have been found is in improved cognition in experimental-type situations with pre- and post-tests, where vodcast use does appear to lead to improved short- and medium-term knowledge retention [11,22,26,39,54]. One study [26] has noted that there is a positive correlation between time spent viewing vod- casts and improved performance. Apart from one study where vodcast use led to lower scores in some test questions [9], there do not appear to be any detrimental effects to the use of vodcasts [16]. As positive results may depend on the types of vodcasts used and may be more prevalent in specific knowledge areas [25], this study seeks to expand a little further on the body of knowledge on vodcast use, particularly in a distance learning scenario.

There have been cases where vodcasting was used to teach theoretical com- puter science [12,33]. However, in both cases, it was used in flipped-classroom environments where the vodcast material was used to prepare students for more practical face-to-face sessions and not as supplementary material to a textbook in a distance learning institution. Positive student experiences were recorded in both cases, and in one [33], a positive trend in examination scores was reported as well.

3   Vodcast Details and Data Gathering

Of the 18 vodcasts that were produced only 12 were used in the research, as they could be directly linked to specific examination questions (see Table 1). The vodcasts were all supplementary, worked example type vodcast, and were of two major varieties: vodcasts 1–8 were PowerPoint slides with a voiced-over commentary, whereas vodcasts 9–12 were handwritten in a PDF document using a DigiMemo (see Fig. 1), with the narrator explaining what was being done in each step of the algorithm (using a 'thinking out loud' approach)—an approach that has been referred to as 'pencasting' [12]. For these latter vodcasts, the final PDF document giving the full worked solution was also made available. Note that vodcasts 9–12 are longer than the recommended maximum of 10 min, but as one worked example was being presented in each vodcast and could be seen as one continuous chunk, it would not have made sense to split the explanation into two parts simply to keep to the time recommendation.

Table 1. Details of the vodcasts that were used in the research (where file size is given in MB and FA stands for finite automaton)

No.	Content	Format	MB	mm:ss
1	Recursive definition of a given language: ODDAB	mwv	5.4	3:35
2	Recursive definition of a given language: EVENnotAA	mwv	7.6	5:30
3	Mathematical induction $(11 + 15 + \ldots + (4n + 7) = 2n^2 + 9n, n > 0)$	mwv	8.1	6:36
4	Mathematical induction $(2n - 3 \leq 2^{(n-2)}, n \geq 5)$	mwv	7.0	5:56
5	Build a finite automaton that accepts a described language	mwv	7.0	4:54
6	Convert a transition graph to a regular expression	mwv	4.0	3:31
7	Pumping lemma with length: prove $\{a^n b^n a^m, n > 0 \ m > 0\}$ non-regular	mwv	4.1	2:53
8	Pumping lemma with length: prove $\{ab^{n+1}c^5 a^n, n > 0\}$ non-regular	mwv	4.9	3:47
9	Given FA_1 and FA_2, find $FA_1 + FA_2$ (the union language)	mp4	13.2	11:19
10	Given FA_1 and FA_2, find $FA_1 FA_2$ (the product language)	mp4	22.8	17:24
11	Given FA_1, find $FA_1{}^*$ (the closure language)	mp4	10.1	9:15
12	Given FA_1 and FA_2 that accept L_1 and L_2, respectively, find an FA that accepts $L_1 \cap L_2$ (the intersection language)	mp4	14.6	12:44

Fig. 1. Handwritten worked example using a DigiMemo in a PDF document

The vodcasts were recorded, following the guidelines on vodcast design given above, on my laptop computer using the freeware program CamStudio (as .avi), with the sound being captured using the built-in microphone. These were converted to .wmv, so that they could be edited in MS Movie Maker. Most of the vodcasts were then published in .wmv format, but some were converted and published in .mp4 format. There have been no requests for the .wmv-formatted vodcasts to be made available in an alternative format. These were then made available to students via the university's LMS; viewing was optional. Vodcasts 1–8 were uploaded in the second semester of 2014 (and all subsequent semesters), and vodcasts 9–12 were uploaded from the second semester of 2015; there are two semesters per year. Students were informed of the availability of the vodcasts via announcements sent from the LMS as well as by SMS (in 2015 semester 2).

The study participants were the students registered for the course under consideration (10 semesters from 2012_S1 to 2016_S2, where S1 and S2 refer to semesters 1 and 2, respectively), and it is expected that students would be similar from one semester to the next. The following data were gathered from the LMS and examination statistics:

1. The number of times each vodcast was accessed by students. These were recorded as overall totals for each vodcast (rather than which students accessed which vodcast) and analyzed as a percentage of the number of

students registered (as not all the data was available per student). This approach is in line with a recommendation that access to such material be used rather than simply its availability to students [32]. Also, this was not an experimental pre- and post-test design, so there was no intention to compare results for specific students on the basis of whether they viewed the vodcasts. These totals were only available from the second semester of 2014, as no vodcasts had previously been uploaded.

2. The overall pass rates for each semester from 2012 semester 1 to 2016 semester 2. The pass rates were calculated as a percentage of those who passed against those who wrote the examination (rather than those who were registered for the module or who were admitted to the examination).
3. The average percentage obtained for the course for the same 10 semesters.
4. The mark detail per question (the number of students in each mark range, the percentage passed, the average mark, and the standard deviation) for the same 10 semesters was recorded.

4   Results

4.1   Comparison of Final Overall Marks

An initial view of the final mark data for the 10 semesters (2012_S1 to 2016_S2) can be seen in Table 2 and Fig. 2. Plotted with the data in Fig. 2 is the percentage

Table 2. Final mark descriptive statistics for the 10 semesters under review (showing number of students, mean marks, median marks, and standard deviation)

Sem.	2012S1	2012S2	2013S1	2013S2	2014S1	2014S2	2015S1	2015S2	2016S1	2016S2
N	235	266	328	280	324	306	320	308	318	267
Mean	42.0	43.0	48.2	42.3	47.6	48.9	45.0	47.8	41.4	46.3
Median	40.0	42.5	50.0	40.0	50.0	50.0	44.0	50.0	40.0	45.5
SD	18.9	15.8	17.0	18.0	20.1	16.0	17.7	15.6	17.3	17.0

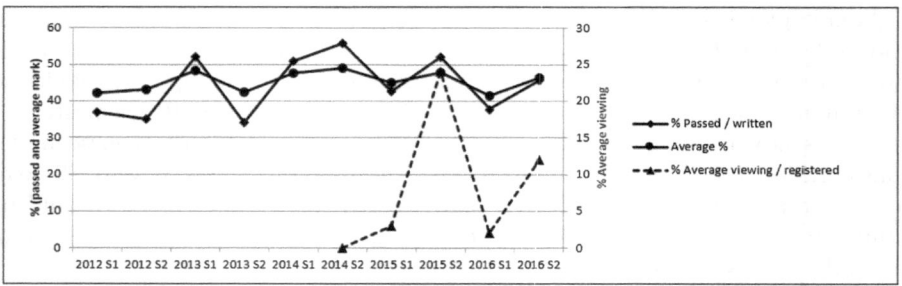

Fig. 2. Overall results: pass rate (%) and average mark (%) plotted against vodcast viewing (%)

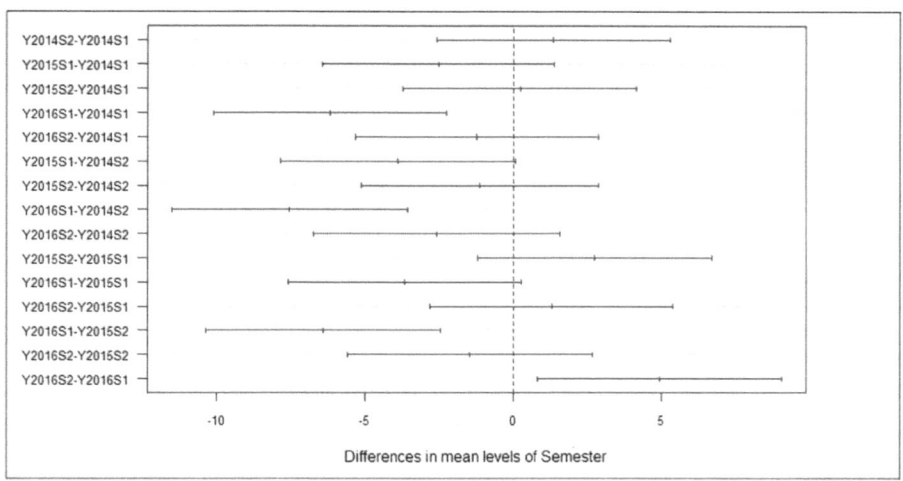

Fig. 3. Final mark Tukey HSD plot for 2014_S1 to 2016_S2 (at 95% family-wise confidence level)

of students who accessed the vodcasts, where this percentage was calculated as an average of the percentages of all the vodcasts that were used in the research; note that even though vodcasts were available from 2014_S2, there was a very low viewing percentage (0.2%) in this first semester as well as in 2016_S1 (2%). Not only does this show the very low uptake of students accessing the vodcasts (highest at 24%), but there appears to have been very little difference in the overall pass percentages or average mark obtained. The growth in the vodcast viewing average in 2015 S2 can be attributed to the use of both announcements and SMSs to inform students in an attempt to increase the number of students using the vodcasts in that semester.

However, it needs to be checked whether there is any significant difference statistically, and an ANOVA test was carried out based on the last six semester marks: from 2014_S1 (where there were no vodcasts available), through 2014_S2 and 2015_S1 (where vodcasts 1–8 were available), to 2016_S2 (when vodcasts 9–12 were also available). There was a statistically significant difference between groups as determined by one-way ANOVA ($F(5,1837) = 7.591$, $p < .0001$). The Tukey HSD post-hoc test was done to determine where the differences between groups occurred (see Fig. 3). It is clear that four statistically significant differences were found; however, in three of these cases (2016_S1–2014_S1, 2016_S1–2014_S2, and 2016_S1–2015_S2, $p \leq .0001$) both the percentage pass rate and average percentage obtained dropped, and in only 2016_S2–2016_S1 ($p = .008$) did the marks rise. A similar ANOVA for all 10 semesters from 2012_S1 to 2016_S2 also showed that there was a statistically significant difference between the semesters ($F(9,2942) = 8.055$, $p < .001$). A further Tukey HSD test showed 17 significant inter-semester differences; however, six of these showed differences between semesters where there were no vodcasts (that is, all prior to 2014_S2), and where there had been vodcasts available, in four of the 11 cases the marks were lower than before vodcasts were used. This may be an indication that the source of the difference is not related to the use of the vodcasts.

4.2   Comparison of Kleene's Theorem Exam Question Marks

To see the effect of vodcasts at the question level, the differences around the release of vodcasts 9–12 were checked, specifically, from 2015_S2. These vodcasts all relate to specific algorithms for the purpose of proving Kleene's Theorem, which are all very similar to one another and require little interpretation on the part of the student. These four vodcasts also all relate to the same question (Kleene's Theorem: proof algorithms) in the final examination. As the other vodcasts relate to how to approach questions and lay out the answers, their approaches are not as 'mechanismic' as those explained in vodcasts 9–12, and as there was a substantial increase in the numbers of students accessing the vodcasts in 2015_S2, it is believed that if there were to be a positive effect on student performance, then it would be visible in this particular question. Table 3 presents the basic descriptive statistics for this question for the 10 semesters under review.

ANOVA was used again to determine whether there was any significant difference in the means of the last six semesters: there was a statistically significant difference between groups as determined by one-way ANOVA ($F(5,1837) = 11.53$, $p < .001$). The Tukey HSD post-hoc test was done to determine where the differences between groups occurred (see Fig. 4).

Although statistically significant differences were noted in seven cases, two of these were prior to the vodcasts relating to this question being released, and of the remaining five, only two showed an improvement in marks once the vodcasts had been released. Considering the significant differences between 2015_S2 (the semester where the highest video access was recorded) and 2014_S2 and 2015_S1 (when the marks improved) and 2016_S1 (when the marks decreased again), it could be argued that there is a link between vodcast access and the marks obtained in this particular question, but this is not a consistent result and it is again possible that some other factor led to the differences in the means observed. It should also be noted that no significant differences were noted when considering specific algorithms from before and after the release of the vodcasts, making it even less likely that it is the vodcasts that are leading to these differences.

Table 3. Descriptive statistics for the 10 semesters under review (showing mean and standard deviation) for the Kleene's Theorem algorithm 10-mark exam question

Sem.	2012S1	2012S2	2013S1	2013S2	2014S1	2014S2	2015S1	2015S2	2016S1	2016S2
Mean	3.54	1.47	3.10	3.43	3.81	2.33	2.14	3.61	2.49	2.84
SD	4.03	2.61	4.11	4.08	4.25	3.16	3.14	4.21	3.18	3.62

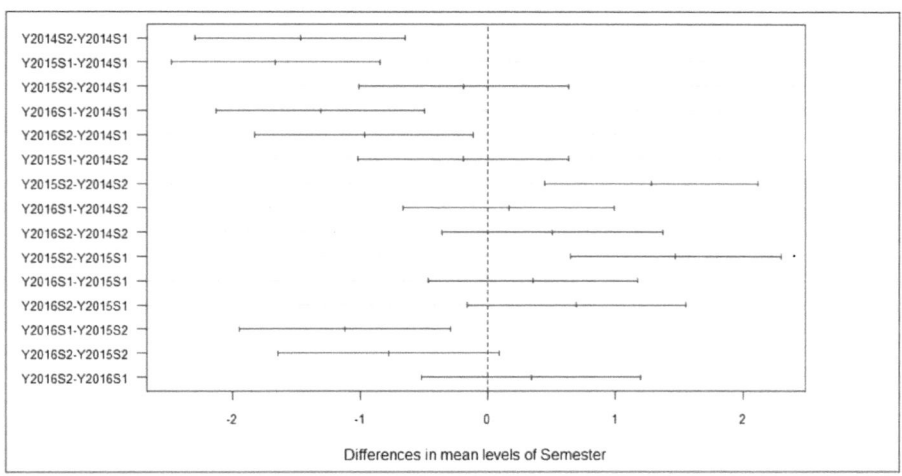

Fig. 4. Kleene theorem algorithm Tukey HSD plot for 2014_S1 to 2016_S2 (at 95% family-wise confidence level)

5   Discussion

It is too simplistic to argue that vodcasts have not been producing any measureable increase in student performance in final examinations and that their use and value should be questioned. It is true that this study produced results similar to others showing that the introduction of vodcasts did not lead to a significant difference, but a positive effect cannot be ruled out [16], and other factors also need to be considered. This would be in keeping with the conclusions drawn by authors of the notorious 'no significant difference' phenomenon [7,38] where the choice of media has little effect in learning or teaching effectiveness; rather, it is the multitude of factors that make up a learning experience that determine its effectiveness, which can often not be controlled in the research of such effectiveness.

Low rates of use of the vodcasts in this study lead to questions of communication and accessibility. Just because a distance learning university uses the Internet to distribute materials to its students does not mean it is accessible to all its students [46]. Even after the availability of the vodcasts was broadcast via LMS announcements and SMS, there were still students who were not aware of them (as noted by a student in a survey); this breakdown in communication will need to be examined further, as it is unknown whether the messages simply never reached the students or whether the students did not remember, or did not read, messages sent to them.

Related to this is the fact that emailed announcements are sent only to the email address established by the university and not to the primary email address of choice of a student. Students changing email addresses over the lifetime of their studies and not informing the university of this change clearly points to the value of such a university-provided address, and it is true that email can be forwarded from one email address to another, but this may well be beyond the technical ability of many students.

It may be the case that students become frustrated with technical problems accessing the university LMS and logging into its associated email and, thus, abandon it altogether. It is also possible that students may ignore the emailed announcements, as they are overwhelmed by the volume of email that is sent [13] or by the volume of work that a part-time, distance education student already has to handle [10]. A further reason could also be that the vodcasts are not compulsory learning material (even though they cover work that is), and so are ignored by students as of less significance.

However, even where there is access, this does not imply use. Although some level of self-regulated learning would be required for a student to succeed in a distance education qualification, low levels of accessing online materials and low levels of responses to online surveys may point to low levels of online engagement, which could lead to lower academic performance [48]. There is, furthermore, the possibility that other material in the examinations, and varying question difficulty, are introducing statistical noise that inhibits the positive effects of the vodcasts from showing up in the overall results [52]. Furthermore, questions need to be asked around how such low levels of online interaction relate to the concept of the digital native who has grown up in the digital world and uses technology with ease [36].

The 'digital native' versus 'digital immigrant' distinction has been criticized as being inaccurate and lacking evidence [3,31,53], and 'net-generation' students may not be as tech-savvy as is often believed [8]. Prensky [37] has since argued that that distinction is less relevant and has proposed focusing on 'digital wisdom'—gained both from, and in the use of, technology. An alternative continuum has been offered [53] based on the metaphor of 'place': digital 'residents' at the one end, who occupy the Web as a social 'space', and digital 'visitors', at the other, who remain anonymous, only using the Web as a tool when necessary. With this background, it is possible to understand why modern students have been referred to as 'digital socialites' [15]. However, the question remains open regarding the extent to which current students are 'digital learners' and whether their interactions online are directed at learning rather than establishing a 'digital identity' or simply seeking information.

It has been argued that it may well be worth continuing with vodcast production, even though there may be no discernible results [34], due to students' positive response to them and the fact that students find them motivating. However, then it may be worth exploring when these vodcasts are used, seeking to integrate them into the learning process earlier rather than at the end solely for revision purposes, as their use may simply be assisting in the development of good study habits [16]. It could possible to try strategies to make their viewing compulsory, such as asking specific questions in assignments that relate directly to content in the vodcasts. This may then also mean that the vodcasts are used during the learning process rather than just at the end of a semester. The challenge that remains is to get the 'digital socialite' to engage with the learning material as a 'digital learner'.

The current study has its limitations. As in other studies [52], this was not a fully experimental design, and so the influence of other factors (such as question difficulty) could not be ruled out. The low rate (with a maximum of 24%) at which the vodcasts were accessed also meant that the link between vodcast use and final/question scores was less easy to measure; further work needs to be done in this context on ways to raise the levels of interaction and engagement with learning materials that have been made available. Further research could also focus on the type of vodcasts used, and whether lecture-based vodcasts (for example) would have a more observable effect than supplementary ones in theoretical computer science courses.

6 Conclusion

E-learning is in a state of constant flux [46], and higher education's attempts to use online platforms to enhance teaching and learning have not been studied fully [54]. It must be remembered, though, that simply changing the mode of delivery of study material will not necessarily lead to improved learning results [15]. Although the value of vodcasts may have been overstated [28], studies have shown that there is some benefit, even if linked only to the student experience of learning, and that vodcasting is an appropriate instructional approach [4] and should be continued. This research has sought to test whether a link can be found between the availability of vodcasts in a distance learning, theoretical computer science course and the marks obtained for the course. Not only was there a very limited uptake of the vodcasts, but there did not appear to be any significant difference in either examination or question scores that could be convincingly linked to the vodcasts. It has been suggested that positive results may depend on the types of vodcasts used and the knowledge areas in which they are used [25], and further research may make this clearer.

The use of vodcasts in higher education teaching and learning needs to be carefully considered in its various contexts, realizing that while there are certainly educational affordances that are offered by such technologies, we also need to remember: *"The field of educational technology has suffered a surfeit of fools and poseurs claiming to be futurists and visionaries. For several decades now they have heralded the arrival of various toys and technologies as if they signal the advent of a Brave New World. Many of these prophets and visionaries have been selling us fool's gold"* [31].

References

1. Andersen, L.: Podcasting, cognitive theory, and really simple syndication: what is the potential impact when used together? J. Educ. Multimed. Hypermed. **20**(3), 219–234 (2011)

2. Armoni, M., Rodger, S., Vardi, M., Verma, R.: Panel proposal: automata theory: its relevance to computer science students and course contents. In: Proceedings of the 37th SIGCSE Technical Symposium on Computer Science Education, pp. 197–198, Houston (2006)
3. Bennett, S., Maton, K.: Beyond the 'digital natives' debate: towards a more nuanced understanding of students' technology experiences. J. Comput. Assist. Learn. 26(5), 321–331 (2010)
4. Bolliger, D.U., Supanakorn, S., Boggs, C.: Impact of podcasting on student motivation in the online learning environment. Comput. Educ. 55(2), 714–722 (2010)
5. Carvalho, A.A., Aguiar, C., Maciel, R.: A taxonomy of podcasts. In: Proceedings of the Association for Learning Technology Conference, ALT-C'2009, pp. 132–140, Manchester (2009)
6. Chesñevar, C.I., González, M.P., Maguitman, A.G.: Didactic strategies for promoting significant learning in formal languages and automata theory. In: ACM SIGCSE Bulletin, vol. 36, no. 3, pp. 7–11 (2004)
7. Conger, S.B.: If there is no significant difference, why should we care? J. Educ. Online 2(2), 1–4 (2005)
8. Coombes, B.: Generation Y: are they really digital natives or more like digital refugees? Synergy 7(1), 31–40 (2009)
9. Dupagne, M., Millette, D.M., Grinfeder, K.: Effectiveness of video podcast use as a revision tool. Journal. Mass Commun. Educ. 64(1), 54–70 (2009)
10. Fose, L., Mehl, M.: Plugging into students' digital DNA: five myths prohibiting proper podcasting pedagogy in the new classroom domain. MERLOT J. Online Learn. Teach. 3(3), 277–287 (2007)
11. Fößl, T., Ebner, M., Schön, S., Holzinger, A.: A field study of a video-supported seamless-learning-setting with elementary learners. Educ. Technol. Soc. 19(1), 321–336 (2016)
12. Gnaur, D., Hüttel, H.: How a flipped learning environment affects learning in a course on theoretical computer science. In: Popescu, E., Lau, R.W.H., Pata, K., Leung, H., Laanpere, M. (eds.) ICWL 2014. LNCS, vol. 8613, pp. 219–228. Springer, Cham (2014). https://doi.org/10.1007/978-3-319-09635-3_25
13. Hara, N., Kling, R.: Students' frustrations with a web-based distance education course. First Monday 4(12) (1999)
14. Harris, H., Park, S.: Educational usages of podcasting. Br. J. Educ. Technol. 39(3), 548–551 (2008)
15. Hayward, D.V., Montgomery, A.P., Dunn, W., Carbonaro, M.: Pedagogy, digital learning spaces, and design: lessons from blended learning. In: Proceedings of the E-Learn: World Conference on E-Learning in Corporate, Government, Healthcare, and Higher Education, pp. 114–119, Kona (2015)
16. Heilesen, S.B.: What is the academic efficacy of podcasting? Comput. Educ. 55(3), 1063–1068 (2010)
17. Hew, K.F.: Use of audio podcast in K-12 and higher education: a review of research topics and methodologies. Educ. Tech. Res. Dev. 57(3), 333–357 (2009)
18. Hill, J.L., Nelson, A.: New technology, new pedagogy? Employing video podcasts in learning and teaching about exotic ecosystems. Environ. Educ. Res. 17(3), 393–408 (2011)
19. Holland, S., Griffiths, R., Woodman, M.: Avoiding object misconceptions. In: ACM SIGCSE Bulletin, vol. 29, no. 1, pp. 131–134 (1997)

20. Jiménez-Castillo, D., Sánchez-Fernández, R., Marín-Carrillo, G.M.: Dream team or odd couple? Examining the combined use of lectures and podcasting in higher education. Innov. Educ. Teach. Int. **54**(5), 448–457 (2017). https://doi.org/10.1080/14703297.2016.1148622
21. Jones, K., Doleman, B., Lund, J.: Dialogue vodcasts: a qualitative assessment. Med. Educ. **47**(11), 1130–1131 (2013)
22. Kalludi, S., Punja, D., Rao, R., Dhar, M.: Is video podcast supplementation as a learning aid beneficial to dental students? J. Clin. Diagn. Res. **9**(12), 4–7 (2015)
23. Kao, I.: Using video podcast to enhance students' learning experience in engineering. In: Proceedings of the 115th Annual ASEE Conference and Exposition, pp. 1–10, Pittsburgh (2008)
24. Kay, R.H.: Developing a framework for creating effective instructional video podcasts. iJET **9**(1), 22–30 (2014)
25. Kay, R.H.: Exploring the use of video podcasts in education: a comprehensive review of the literature. Comput. Hum. Behav. **28**(3), 820–831 (2012)
26. Kay, R.H., Kletskin, I.: Evaluating the use of problem-based video podcasts to teach mathematics in higher education. Comput. Educ. **59**(2), 619–627 (2012)
27. Loomes, M., Shafarenko, A.: Teaching mathematical explanation through audiographic technology. Comput. Educ. **38**(1–3), 137–149 (2002)
28. Malan, D.J.: Podcasting computer science E-1. In: Proceedings of the 38th ACM Technical Symposium on Computer Science Education SIGCSE 2007, pp. 389–393, Covington (2007)
29. Mayer, R.E.: Cognitive theory of multimedia learning. In: Cambridge Handbook of Multimedia Learning, pp. 31–48. Cambridge University Press (2005)
30. McGarr, O.: A review of podcasting in higher education: its influence on the traditional lecture. Australas. J. Educ. Technol. **25**(3), 309–321 (2009)
31. McKenzie, J.: Digital nativism, digital delusions and digital deprivation. From Now On: Educ. Technol. J. **17**(2) (2007)
32. Miller, N.B.: Student access of supplemental multimedia and success in an online course. Am. J. Distance Educ. **27**(4), 242–252 (2013)
33. Morisse, K.: Inverted classroom: from experimental usage to curricular anchorage. In: Proceedings of the International Conference on e-Learning, ICEL 2015, pp. 218–226, Nassau (2015)
34. Nash, J.: Using video podcasts to teach procedural skills to undergraduate students. In: Proceedings of the 44th Conference of the Southern African Computer Lecturers' Association SACLA 2015, pp. 77–84, Johannesburg (2015)
35. Oliver, M., Trigwell, K.: Can 'blended learning' be redeemed? E-Learning **2**(1), 17–26 (2005)
36. Prensky, M.: Digital natives, digital immigrants. Horiz. **9**(5), 1–6 (2001)
37. Prensky, M.: Homo sapiens digital: from digital immigrants and digital natives to digital wisdom. Innov. J. Online Educ. **5**(3), 1–9 (2009)
38. Ramage, T.R.: The 'no significant difference' phenomenon: a literature review. J. Instr. Sci. Technol. **5**(1) (2002)
39. Raupach, T., Grefe, C., Brown, J., Meyer, K., Schuelper, N., Anders, S.: Moving knowledge acquisition from the lecture hall to the student home: a prospective intervention study. J. Med. Internet Res. **17**(9), e223 (2015)
40. Schreiber, B.E., Fukuta, J., Gordon, F.: Live lecture versus video podcast in undergraduate medical education: a randomised controlled trial. BMC Med. Educ. **10**(68), 1–6 (2010)
41. Sheridan-Ross, J., Gorra, A., Finlay, J.: Practical tips for creating podcasts in higher education. ACM SIGCSE Bull. **40**, 311 (2008)

42. Smith, G., Fidge, C.: On the efficacy of prerecorded lectures for teaching introductory programming. In: Proceedings of the 10th Australasian Computing Education Conference, ACE'2008, pp. 129–136, Wollongong (2008)
43. Solomon, D.J., Ferenchick, G.S., Laird-Fick, H.S., Kavanaugh, K.: A randomized trial comparing digital and live lecture formats. BMC Med. Educ. **4**, 27 (2004)
44. Tarver-Grover, S.: Predicting the adoption of video podcast in online health education: using a modified version of the technology acceptance model (health education technology adoption model HEDTAM). In: Proceedings of the E-Learn: World Conference on E-Learning in Corporate, Government, Healthcare, and Higher Education, pp. 1100–1104, Kona (2015)
45. Taslibeyaz, E., Aydemir, M., Karaman, S.: An analysis of research trends in articles on video usage in medical education. Educ. Inf. Technol. **22**(3), 873–881 (2017)
46. Thomson, R., Fichten, C.S., Havel, A., Budd, J., Asuncion, J.: Blending universal design, e-learning, and information and communication technologies. In: Universal Design in Higher Education: From Principles to Practice, 2nd edn., pp. 275–284. Harvard Education Press (2015)
47. Ting, K.Y.: Blended learning as a theoretical framework for the application of podcasting. Engl. Lang. Teach. **7**(5), 128–135 (2014)
48. Tolulope, A.E., Adenubi, O.S., Oluwole, F.C.: Evaluating undergraduates' attitude towards the use of podcast for learning selected educational technology concepts. Glob. Med. J. **13**(special issue), 1–10 (2015)
49. Vajoczki, S., Watt, S., Marquis, N., Holshausen, K.: Podcasts: are they an effective tool to enhance student learning? A case study from McMaster Univeristy, Hamilton, Canada. J. Educ. Multimed. Hypermed. **19**(3), 349–362 (2010)
50. van Heerden, M.E., Goosen, L.: Using vodcasts to teach programming in an ODL environment. Progressio **34**(3), 144–160 (2013)
51. van Rooyen, A.A.: Integrating MXit into a distance education accounting module. Progressio **32**(2), 52–64 (2010)
52. Walker, J.D., Cotner, S., Beermann, N.: Vodcasts and captures: using multimedia to improve student learning in introductory biology. J. Educ. Multimed. Hypermed. **20**(1), 97–111 (2011)
53. White, D.S., Le Cornu, A.: Visitors and residents: a new typology for online engagement. First Monday **16**(9) (2011)
54. Williams, A.E., Aguilar-Roca, N.M., O'Dowd, D.K.: Lecture capture podcasts: differential student use and performance in a large introductory course. Educ. Tech. Res. Dev. **64**(1), 1–12 (2016)

Regex Parser II: Teaching Regular Expression Fundamentals via Educational Gaming

Ariel Rosenfeld[1]([envelope]), Abejide Ade-Ibijola[2], and Sigrid Ewert[1]

[1] School of Computer Science and Applied Mathematics,
University of the Witwatersrand, Johannesburg, South Africa
ariel@rosenfeld.tech, sigrid.ewert@wits.ac.za
[2] Formal Structures Algorithms and Industrial Applications Research Cluster,
Department of Applied Information Systems, University of Johannesburg,
Auckland Park, South Africa
abejideai@uj.ac.za

Abstract. We present an educational game that is expected to improve the teaching of and stimulate the interest in regular expressions. The game generates regular expressions using pseudo-random numbers and predefined templates and asks the player to provide strings that are in the language corresponding to the given regular expression while a timer counts down. Points in the game are awarded on how many strings the player was able to enter before time runs out (The game, which requires the JRE, is available via http://bit.ly/RegexParserII).

Keywords: Computer science education · Regular expressions · Educational games

1 Introduction

Games have been used in education since ancient times. In India, during the Gupta Empire (320–550 CE), a predecessor to Chess was used to teach military strategy [12]. In the early 1800s, Friedrich Fröbel developed the early concept of the kindergarten which is a preschool education where children learn by playing [6]. In the 1980s, due to the invention of the personal computer, learning games such as *Oregon Trail* and *Reader Rabbit* became digital and were commercialised [10]. In modern times, the type of games that people play has changed due to the increased availability and performance of electronic devices. Video games have become increasingly popular, especially in the field of education [3]. There are various platforms for and genres of educational video games ranging from mobile applications intended for toddlers and preschoolers to video games with the purpose of teaching skills like negotiation and time management. Although the technology and game types have evolved from ancient times, the main objective of teaching various concepts through play has not changed.

As it stands, there are many methods to teaching regular expressions. These methods include formal textbooks such as [14] as well as a plethora of online

© Springer International Publishing AG 2017
J. Liebenberg and S. Gruner (Eds.): SACLA 2017, CCIS 730, pp. 99–112, 2017.
https://doi.org/10.1007/978-3-319-69670-6_7

resources such as [4,19]. As the theory behind regular expressions has not changed since its conception, the resources are all relevant despite their datedness. While there exist interactive tutorials and games for regular expressions, none make use of the challenge of a timer or of automatically generating regular expressions at each level of the game or tutorial. Including these features in this type of game would ensure that a player has an enjoyable experience (as outlined by [9]) while gaining knowledge and experience in interpreting regular expressions.

In this project, our aim was to create an educational game that is expected to improve the teaching of and stimulate the interest in regular expressions (*REs* for short, *RE* in singular). The game generates regular expressions using pseudorandom numbers and predefined templates and asks the players to provide strings that are in the language corresponding to the given RE while a timer runs down to zero. Points in the game are awarded on how many strings the player was able to enter before time runs out. The game has adjustable difficulty and RE alphabet (Σ). The RE generation algorithm has been implemented in Java, and the Java Matcher class [16] has been used for RE validation. A simple 2D interface was developed in Java.

This project is based on *Regex Parser I* [1], which is a Windows application developed to aid the comprehension of regular expressions. *Regex Parser I* asked the user to input an RE and a string and it would check whether the string was in the language corresponding to the RE. In this project we answered the following questions. How do we:

1. design an algorithm for the automatic generation of RE problems of varying difficulty;
2. develop a novel educational game that presents the RE problems generated in point 1 (above) as puzzles to be completed at game time, and are stretched across the difficulty levels of the game?

It is important to note that our project capitalises on the well documented background on educational games that suggests that such games have been proven to aid the comprehension and are used to support the teaching of many difficult subjects [23]; in this paper we will not discuss this fact again.

The remainder of this paper is structured as follows: Sect. 2 presents our findings for works and research related to this project, whereby we cite and analyse works in the fields of Computer Science education, regular expressions, educational games, and game design. Our research method is outlined in Sect. 3. Section 4 presents the results of our research, whereby the template structure and algorithm for regular expression generation are explicitly defined and examples of REs generated by the algorithm are given. The gameplay the player can expect is described and screenshots of the game's interface are displayed. Further topics relating to our research are discussed in Sect. 5, including methodological changes, regular expression generation, scoring design as well as possible future work. Finally, Sect. 6 concludes this paper.

2 Background and Related Work

2.1 Computer Science Education (Especially Automata Theory)

The discussion of Computer Science education with relation to our project is relevant, as this project aimed to create a game that assists with the learning of regular expressions: a topic of Automata Theory and in Computer Science in general.

Regex Parser I [1] was developed at the University of the Witwatersrand, Johannesburg, to assist with the learning of regular expressions (RE) in a third-year Formal Languages and Automata Theory course. The program asked students to enter an RE and a string. The program then checks whether the string is in the language corresponding to the RE. The interface and the RE-string validity were developed using the Microsoft.NET Framework. This program is the foundation of the project from a programming perspective. Its source code is used as an aide for this project.

The automated grading of context-free grammars (CFGs) is discussed in [21] in an independent study. The paper focuses on expanding on the algorithm introduced by [2] for CFG equivalence and ambiguity-checking which converts CFGs such that decidable versions of undecidable problems are produced. The paper then outlines the preconditions and problem constraints of the study as well as the algorithm of the tool that was used in the study. Tests were run with various alphabets and the grammars were of a complexity level appropriate for an introductory Computer Science theory course. The paper concludes that if a marker is only interested in the correctness of a student's work it would be straightforward to repeatedly run the tool for equivalence checking, however assigning credit (i.e.: how close the given solution is to the marker's solution) requires additional work.

This paper is relevant to this project for two reasons. First, it explores a method of automated marking. The game will automatically verify whether a given input string is in the language that the game generates in the form of a regular expression. Second, it discusses the usefulness of the tool used in Computer Science education. As previously stated, it is useful in the aspect of whether a solution is correct or incorrect, but not when a marker is assigning credit. The game will have a similar issue with regards to scoring. How do we assign scores to different input strings that are correct? We will address this later in this paper.

A discussion of the errors made by students when designing finite automata (FAs) can be found in [20]. The paper attempts to analyse and categorise the errors made by students when designing FAs through an exploratory case study. For the case study, Computer Science students from the University of South Africa were asked to complete five questions where the student had to draw the finite automaton according to the instruction (given in plain English). The questions were set such that the difficulty would increase with each question. To remove the factor of time pressure (such as in an exam situation), the questions were sent out during a less busy time of the year and students were given a month to complete them. The paper then makes comments on the responses received (the paper notes that in future work, respondents should be asked to explain their work to avoid misinterpreting the solutions). The general understanding of the study is that students have a conceptual misunderstanding of the theory. Mainly, students have difficulty with smaller concepts that lead to extra, unnecessary states. The paper comments that there may be a misconception with matching FAs with their corresponding regular expressions and visa versa.

The major contribution of this paper to the project is the concept of having FAs of varying difficulties. The paper shows that the questions that were deemed more difficult resulted in fewer correct answers from the respondents. This would lead one to recognise that REs, too, can be of varying difficulties. This would be implemented in the game and would thus increase the students' learning experience and takeaway.

2.2 Regular Expressions

Regular expressions and the theory behind them are paramount to this project. They are the main focus of the project and what we will be spending most of our time on. We note here that we were unable to find works on what makes a regular expression difficult (i.e., how can we have different game levels of difficulty). Therefore we looked at examples given in Automata Theory textbooks and used these as templates for REs of varying difficulties.

A method and a tool developed by Microsoft Research, called *Rex*,[1] is introduced in [24], for symbolically expressing and analysing regular expression constraints. The method translates REs into symbolic representations of finite automata (SFAs). In an SFA, moves are labelled by formulas representing sets of characters rather than individual characters. An SFA is translated into a set of (recursive) axioms that describe the acceptance condition for the strings accepted and build on the representation of strings as a list. This set of axioms is asserted to the tool for processing. The report revisits several classical algorithms for finite automata and describes the corresponding algorithms for SFAs. The report evaluates the performance of these algorithms based on *Rex*. The report then defines some special case SFAs. The performance of *Rex* was evaluated on a collection of sample REs. For each RE r, the SFA and special case SFAs were constructed for r using algorithms described in the report. The report then comments on the experiments concluding that *Rex* is fast and scalable and that in most cases, not all the special case SFAs need to be constructed for this application.

This report and tool contribute to this project in the sense that the theory in it could be implemented in the game. We could draw automata for the generated REs as a method of validating the player's input string and also as a method of helping the player understand the RE that is given.

As mentioned in Sect. 1, there are many online resources one could use to gain experience in and knowledge of regular expressions. We have found a number of

---

[1] http://rise4fun.com/Rex.

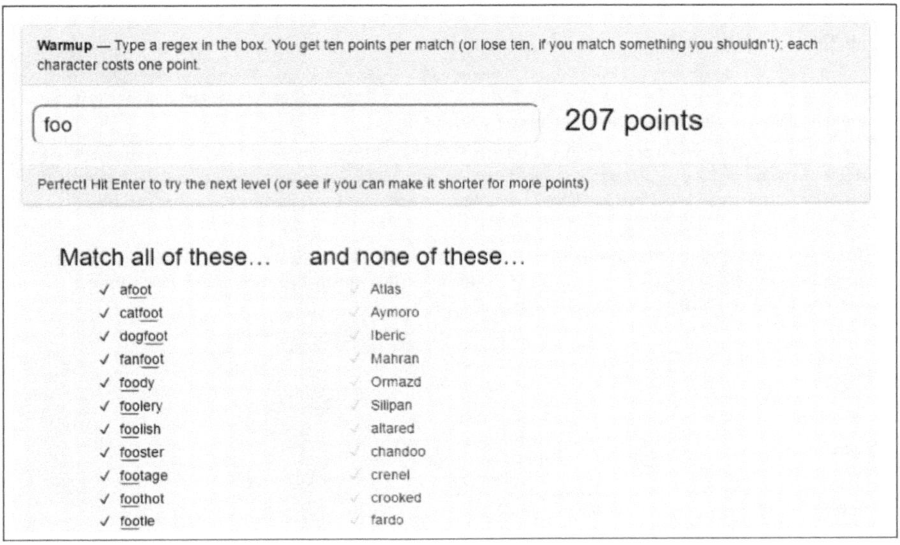

Fig. 1. Level one of the game [18]

RE-based games and interactive tutorials that we will expand on and analyse, namely [17–19].

In [18] we can find *"a game in which programmers attempt to solve a given programming problem using as few characters as possible, analogous to the number of golf shots it takes to reach the goal"* [5]. The player is given two sets of strings and he/she must enter an RE whose language accepts the first set but rejects the second set (see Fig. 1). The game intelligently scores the player based on the RE input. The shorter the RE, the higher the score given.

An interactive tutorial, with exercises which increase in difficulty, can be found in [19]. The user is given a short introduction and explanation of the concept that should be used to complete the exercise. For the exercise, as with [18], the user is given a number of strings and the RE that the user inputs must either accept or reject the string as indicated (see Fig. 2).

A game similar to crosswords, where the clues are REs and the answers on the grid must match both the horizontal and vertical clues, is described in [17] (see Fig. 3). There are several levels in the game, the difficulty of which go from simple one or two character clues to dictionary words to difficult repeating REs. The game makes use of multiple RE operations in its clues including ^, +, *, ., and |.

These games and tutorials are examples of what we want to achieve in our research, both in practical concept and in particular field of study. While these are still relevant and helpful to the user in learning about regular expressions, the game we developed has the added challenge of adjustable difficulty and time pressure.

Lesson 2: The Dot

In some card games, the Joker is a wildcard and can represent any card in the deck. With regular expressions, you are often matching pieces of text that you don't know the exact contents of, other than the fact that they share a common pattern or structure (eg. phone numbers or zip codes).

Similarly, there is the concept of a **wildcard**, which is represented by the **.** (dot) metacharacter, and can **match any single character** (letter, digit, whitespace, everything). You may notice that this actually overrides the matching of the period character, so in order to specifically match a period, you need to escape the dot by using a slash **\.** accordingly.

Below are a couple strings with varying characters but the same length. Try to write a single pattern that can match the first three strings, but not the last (to be skipped). You may find that you will have to escape the dot metacharacter to match the period in some of the lines.

Exercise 2: Matching With Wildcards

Task	Text	
Match	cat.	⊘
Match	896.	⊘
Match	?=+.	⊘
Skip	abc1	

\.|	Continue ›

Solve the above task to continue on to the next problem, or read the Solution.

Fig. 2. Lesson two of [19]

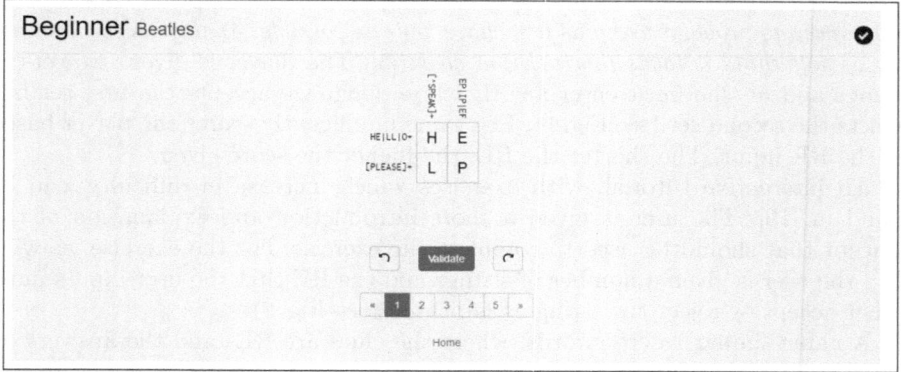

Fig. 3. Puzzle one, of level 'Beginner', of [17]

2.3   Educational Games

Educational games are a type of serious game with the aim of learning or teaching concepts through play. These games are not necessarily electronic. Board games and card games are often used for educational purposes, but due to the rise of digital technology in modern times, the popularity of electronic and digital games has increased, especially with respect to educational games. Educational games are often confused with serious games. 'Serious game' is the umbrella term for any game that is not purely designed for entertainment purposes. Serious games include games used by institutions like the military, politics, engineering, management, as well as education. Educational games have a great potential and the positive results and effects have been mentioned in news and television. We note here that while many recent publications on this topic exist [3,11,15,23], they mainly focus on the components of a game that lead to an effective educational game rather than the learning outcome of an educational game. Also, these publications are often literature searches or systematic literature reviews based on older publications on the topic of educational games. The review [23] recapitulates the learning effects of educational games in order to gain more insight into the conditions under which a game may be effective for learning. The review used three components to analyse the results: environment (e.g., content, game type, game elements), moderating variables (i.e., learner characteristics such as gender, age, socio-economic status), and learning outcomes (e.g., cognitive learning outcomes, attitude, enjoyment). With regards to environment, the review states that there is no explicit definition and thus there is no uniformity on what aspects are crucial to constitute an educational game. The review then presents its results and concludes that while educational games are an advantage to the learning of various subject matters, there are many factors and conditions under which educational games should be used. It is also noted that educational games cannot be the main source of learning in many cases. The review notes that the articles often focused on a single domain (e.g., history, mathematics, languages) and thus could not conclude whether educational games were more successful in a specific context than others. The review also notes that game elements such as feedback, interactivity, competition and background music aid in the effects of the game on the player. With regards to cognitive learning outcomes, the review finds that the majority of articles reported a better result for participants that played educational games on post-tests measuring their knowledge in comparison to participants playing no games. Results were also positive in science and mathematics [23].

This review serves as a major contributor to our project. Firstly, it provides ample confirmation that the game that we have developed will aid in the learning of regular expressions. Also, the findings in the review were regarded in the design of the game to improve the effectiveness of it.

2.4   Game Design

The exploration of game design for this project will aid in improving the overall success of the game. A well designed game will lead the player to a more enjoyable experience while playing this educational game which may lead to a greater educational gain.

The *Mechanics, Dynamics and Aesthetics* (MDA) framework for game [9] is a formal approach to understanding games; it attempts to bridge the gap between game design and development, game criticism and technical game research. It formalises the consumption of games by breaking them into their distinct components and establishing their design counterparts. From the designers' perspective, the mechanics give rise to dynamic system behaviour, which in turn leads to particular aesthetic experiences. Each game pursues multiple aesthetic goals, in varying degrees. From the players' perspective, aesthetics set the tone, which is borne out in observable dynamics and eventually, operable mechanics. Adjusting the mechanics of a game helps fine-tune the game's overall dynamics. By applying changes to the fundamental rules of play, players might remain competitive and interested for longer periods of time. The article recognises that most people idealise Artificial Intelligence components as black-box mechanisms that, in theory, can be injected into a variety of different projects with relative ease. Rather, game components cannot be evaluated in a vacuum, aside from their effects on a system and player experience. The article concludes by stating that simple changes in the aesthetic requirements of a game will introduce mechanical changes for its AI on many levels, sometimes requiring the development of entirely new systems for navigation, reasoning, and strategic problem solving.

Although the MDA framework was developed with video games such as *Counter Strike* and *The Sims* in mind, it can be applied to the game that we have developed. Following these guidelines in creating our game should ensure a successful educational and gaming experience for the players.

2.5   Motivation

Our research project is relevant to the field of Computer Science education as it serves to provide an educational game that is expected to improve the teaching of and stimulate the interest in regular expressions. Through the playing of this game, the player shall expect to gain a better understanding of regular expression theory as well as the skill of being able to interpret difficult regular expressions. When applied to a structured university course, the instructor shall expect to see an improvement in the students' understanding of regular expressions, as with any form of tutorial exercise, and ultimately an increase in students' marks. Aside from the impact that *Regex Parser II* makes on Computer Science education, the RE generation algorithm developed has many uses outside of the game, for instance, to be used in setting tests or exams.

3   Research Method

In order to answer the questions set out in Sect. 1, we have done the following:

- Find examples from Formal Languages textbooks and online resources and classify them by difficulty.
  - We found 30 examples: 12 easy, 12 medium and 6 hard.
  - We used examples from [8,13,14], as well as the online resources mentioned in Subsect. 2.1, namely [17–19].
  - Examples that provided an RE were preferred, however, other types of examples were used and reinterpreted for our purposes (for example, should the question be posed as a Formal Automata, we converted it to an RE and used it in our template).

- For the online resources, these were generally set out in such a way that the difficulty increases as you proceed, therefore classifying these examples was straightforward.
- For the textbooks, the authors often flagged questions or exercises that are particularly difficult. We took this into consideration when classifying examples.
- Convert the examples into templates.
    - The examples were analysed and converted into Java RE syntax.
- Design and develop the game engine.
    - This consists of the RE generator, using the RE-string validation from the Java Matcher class, and the game settings for RE alphabet, difficulty and timer duration.
    - This was done in Java.
- Design and develop the game interface.
    - The main interface includes a textbox that displays the generated RE to the player, a textbox that displays the remaining time, a textbox that displays all the previously entered answers, an interactive textbox where the player enters his/her answers and a menu bar for starting a new game and changing the settings of the game.
    - This was done in Java.
- Design and develop game dynamics.
    - This includes the scoring design and continuity between levels.

4   Results

4.1   Regular Expression Generation

Templates. For RE generation, examples were taken from the resources stated in Sect. 3, classified into easy, medium and hard using subjective criteria, and were converted into templates. The templates were structured as follows:

- The 'back quote'[2] symbol was used to represent the boundaries of the substrings within the template.
- For each substring:
    - If the substring was the letter 'c' followed by a single digit, it represented a placeholder for random characters, where the digit was the number of characters to be used.
    - Otherwise, the substring was used as is in the generation of the RE. These substrings include RE operations, parentheses and whole words.

Note that for our purposes we will be using the notation where + represents the Kleene-plus (for repetitions), and | represents alternation.

Algorithm. The algorithm for generating REs from a template is shown in Table 1. The algorithm ensures that each template is only used once in the session and that a given random character will not be chosen again until all the other random characters have been chosen.

Examples. Table 2 provides examples of templates and the REs generated from those templates for an alphabet of $\{a, b\}$.

---

[2] http://www.computerhope.com/jargon/b/backquot.htm.

Table 1. Algorithm for the generation of REs

1:	**Function** GENRE(difficulty,$\Sigma\{\}$)
2:	// *initialise variables:*
3:	pattern ←null; list{} ←all templates of given difficulty; available{} ← templates that have not been used in this session; Σ_{new} ← Σ.clone();
4:	rand ← [0,available.size()−1];
5:	REindex ← available[rand];
6:	RE ← list[REindex];
7:	available ← available.delete(rand);
8:	split[] ← RE.split(');
9:	**for all** s in split **do**
10:	**if** s is a placeholder for random characters **then**
11:	**for** j ← 0 to number of random characters **do**
12:	**if** $|\Sigma_{new}| = 0$ **then** Σ_{new} ← Σ.clone(); **end if**
13:	rand ←random$[0, \Sigma_{new}$.size()−1];
14:	pattern ←pattern+Σ_{new}[rand];
15:	Σ_{new} ← Σ_{new}.delete(rand);
16:	**end for**
17:	**else** pattern ←pattern+s;
18:	**end if**
19:	**end for**
20:	**return** pattern
21:	**end Function**

Table 2. Templates and REs generated for $\Sigma := \{a,b\}$

Easy	
['^'c3']'c2	[^aba]ba
c1'*|'c1'*	b*|a*
('c1'|'c2'|'c2')'c1'*	(a|ba|ab)b*
c3'{2,5}'c2	abb{2,5}aa
Medium	
\w+@\w+\.\w+	\w+@\w+\.\w+
('c1'*'c1'*)*'c3'('c1'|'c1')*	(b*a*)*abb(a|a)*
^[a-f]*$	^[a-f]*$
(c2'*|'c2'*|'c1'*'c1')+('c1'*'c1')*	(ba*|ab*|b*a)+(b*a)*
Hard	
\b('c3').*\1\b	\b(bab).*\1\b
[a-z][\.\?!]\s+[A-Z]	[a-z][\.\?!]\s+[A-Z]
['c2']('c1'|'c1')['c1'\s]\1['c1']+\1	[ab](a|b)[b\s]\1[a]+\1

4.2   Gameplay

The gameplay is as follows:

- The player starts the game and a level begins.
- An RE is generated for the chosen difficulty (easy by default) and displayed to the player.
- The player can now enter as many strings as he/she can within the allotted time (60 s by default).
- The score is based on the number of correct strings entered by the player.
  - Repeated answers are allowed and will count towards the score.
  - The empty string (λ) is allowed.
  - Each correct string is worth one point. Alternate scoring designs are addressed in Subsect. 5.2.
- Once the timer has run out, the score is displayed to the player and a new level begins.
- Once the player has completed three levels of the same difficulty, the difficulty is increased (from easy to medium or from medium to hard).

The player can change the settings for the difficulty, RE alphabet and timer duration at any time. If this is done, the current level is ended and a new one begins and all templates become available for RE generation. For a picture of the main screen of our game see Fig. 4.

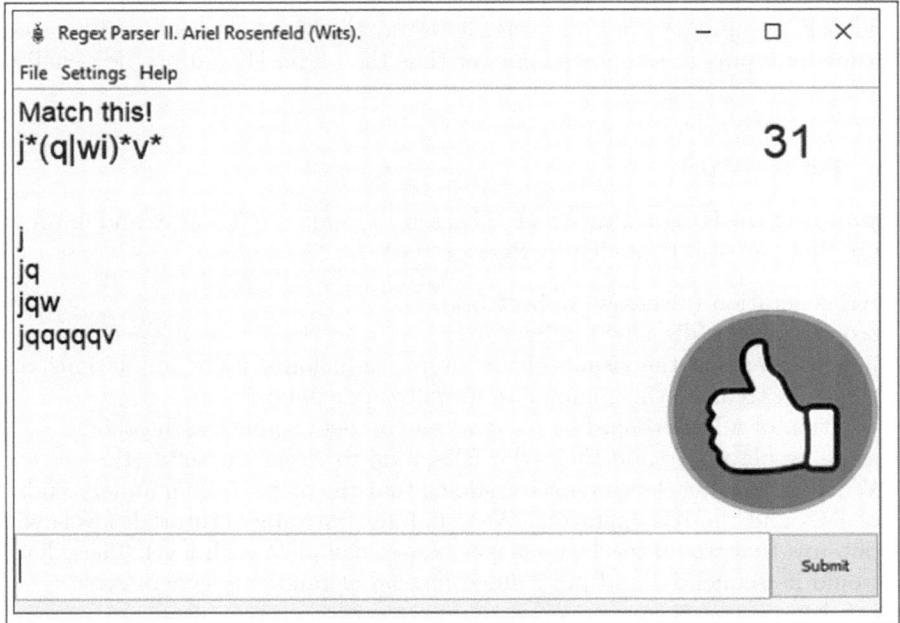

Fig. 4. Main screen of Regex Parser II

5   Discussion

5.1   Regular Expression Generation

Originally we intended that the generation of REs would include randomly select-ing RE operations (e.g. alternation, concatenation, Kleene operators, and the like). We realised that this may cause generated REs to be invalid and not read-able by the Java Pattern class. We thus opted to exclude the random selection of RE operations. In retrospect, it is possible to allow the random selection of Kleene operators (Kleene star, Kleene question mark,[3] Kleene plus) as this would not break the validity of any generated RE.

5.2   Scoring

We designed the scoring to be such that each correct string entered is worth one point. Thus the score at the end of the level is the number of correct strings entered. When designing the scoring, we came across two features that could be added to the scoring design. The first is string distance measurement or substring verification between an answer that was just entered versus answers that were previously entered for the groups of the RE that are under the Kleene star.[4] For example, if the RE is a*b, ab should receive a higher score than aab as a is a substring of and shorter than aa (a and aa are the groups of the string that are under the Kleene star.). This presented many challenges and was therefore disregarded. The second is fractional points awarded for entering previously accepted correct answers, however, the player is penalised by using up time by typing in a repeated answer thus the player should not be penalised twice.

5.3   Future Work

Expanding on Regex Parser II. There are a number of features and improve-ments that can still be added to *Regex Parser II*:

- RE generation (discussed in Sect. 5.1);
- scoring design (discussed in Sect. 5.2);
- improvements to the visuals of the interfaces (colours, fonts, animations, and the like), to make the game more visually appealing;
- addition of a leaderboard so players can compete against each other;
- allowing players to add their own REs, akin to *Regex Parser I* [1].
- When developing the game, we assumed that the player has an understanding of REs and the RE operators. We could also introduce 'tutorial' levels with pop-ups that would teach the player how to complete each level. These levels would be sequential and pre-defined instead of randomly generated.

---

[3] RE group must be repeated 0 or 1 time(s).
[4] http://www.regular-expressions.info/brackets.html.

– Further developing the RE generation algorithm to produce REs the structure of which is randomised, rather than just the symbols and operations.

Future Work (Not in the Scope of this Paper).

– As the focus of this research was on the RE generation algorithm and the game developed, we did not conduct a study on the effects of *Regex Parser II* on students' marks and enjoyment. We hope to evaluate this research and its impact on teaching in the future.
– A challenge we faced when executing this research is the classification of REs into difficulties. We opted to use our own subjective criteria as there is, to our knowledge, no research on the difficulty of REs. A study on the perceptual difficulty and difficulty by measurements of string distance (and perhaps a hybrid) of REs should be done.

6   Conclusion

This paper has described in some detail how *Regex Parser II* was designed and implemented, the related works done previously, the method followed to complete the research, the results of the completed project, and how the research can be improved upon and expanded. Through this paper, we have shown that this research is relevant to the fields of Computer Science education and educational gaming. We designed a developed an algorithm which allowed us to automatically generate regular expressions that were used in the novel game created to teach Computer Science students about regular expressions. The game was made to be configured and suited to each player that may play it through varying difficulties, Σ alphabet and other aspects of the gameplay itself. We believe the way forward with this research is conducting studies on the impact that *Regex Parser II* has on the teaching of regular expressions, and on how one would be able to objectively classify the difficulty of regular expressions. We would also like to expand on the RE generation algorithm to make it more robust, with the hopes of achieving an algorithm which generates truly random regular expressions. With this in mind, we call on fellow scientists for collaboration to test our game in an educational setting and to improve upon what this research has established.

References

1. Ade-Ibijola, A.: Regex Parser I: Unpublished Teaching Aid. Personal communication (2014)
2. Axelsson, R., Heljanko, K., Lange, M.: Analyzing context-free grammars using an incremental SAT solver. In: Aceto, L., Damgård, I., Goldberg, L.A., Halldórsson, M.M., Ingólfsdóttir, A., Walukiewicz, I. (eds.) ICALP 2008. LNCS, vol. 5126, pp. 410–422. Springer, Heidelberg (2008). doi:10.1007/978-3-540-70583-3_34
3. Battistella, P.E., von Wangenheim, C.G.: Games for teaching computing in higher education – a systematic review. IEEE Technol. Eng. Educ. **9**(1), 8–30 (2016)
4. Codeproject: the 30 minute regex tutorial. Technical report (2004)

5. Explain xkcd. http://www.explainxkcd.com/wiki/index.php/1313:_Regex_Golf
6. Fröbel, F.: Autobiography of Friedrich Froebel. C.W. Bardeen, Syracuse (1889)
7. Glendenning, P., Xu, J.: Method and apparatus for compiling regular expressions. U.S. Patent US8,726,253B2, Micron Technology Inc. (2014)
8. Hopcroft, J.E., Motwani, R., Ullman, J.D.: Introduction to Automata Theory, Languages, and Computation. Pearson, London (1979)
9. Hunicke, R., LeBlanc, M., Zubek, R.: MDA: a formal approach to game design and game research. In: Proceedings AAAI Workshop on Challenges in Game AI (2004)
10. Institute of play: history of games & learning. http://www.instituteofplay.org/
11. Kirriemuir, J., McFarlane, A.: Literature review in games and learning. Futurelab Series Report 8 (2004)
12. Kulke, H., Rothermund, D.: A History of India. Psychology Press, Hove (2004)
13. Linz, P.: An Introduction to Formal Languages and Automata. Jones & Bartlett, Burlington (2011)
14. Martin, J.C.: Introduction to Languages and the Theory of Computation, 4th edn. McGraw-Hill, New York City (2011)
15. Mitchell, A., Savill-Smith, C.: The Use of Computer and Video Games for Learning: A Review of the Literature. Learning and Skills Development Agency, London (2004)
16. Oracle: Matcher. Technical report (2002)
17. Regex Crossword (2013). https://regexcrossword.com/
18. Regex Golf. http://regex.alf.nu/
19. RegexOne (2015). http://regexone.com/
20. Sanders, I., Pilkington, C., van Staden, W.: Errors made by students when designing finite automata. In: SACLA 2015 Proceedings Southern African Computer Lecturer's Association, Johannesburg (2015)
21. Sorrell, J.: Towards automated grading of student context-free grammars. Technical report, Optimist Industries (2015)
22. van Eck, R.: Digital game-based learning: it's not just the digital natives who are restless. Educause Rev. **41**(2), 16 (2006)
23. Vandercruysse, S., Vandewaetere, M., Clarebout, G.: Game-based learning: a review on the effectiveness of educational games. In: Handbook of Research on Serious Games as Educational, Business, and Research Tools, pp. 628–647. IGI Global (2012)
24. Veanes, M., de Halleux, P., Tillmann, N.: Rex: symbolic regular expression explorer. Technical report, MSR-TR-2009-137, Microsoft Research (2009)
25. Wilkinson, C.K.: Chessmen and Chess. Metropolitan Museum of Art, New York (1943)

Keeping ICT in Education Community Engagement Relevant: Infinite Possibilities?

Leila Goosen[1,3(✉)] and Ronell van der Merwe[2,4]

[1] Department of Science and Technology Education,
Muckleneuk Campus Pretoria, Pretoria, South Africa
[2] School of Computing, Florida Campus Johannesburg,
Johannesburg, South Africa
[3] University of South Africa (UNISA), Pretoria, South Africa
goosel@unisa.ac.za
[4] University of South Africa (UNISA), Johannesburg, South Africa
vdmerwer@unisa.ac.za

Abstract. The motivation for the investigation reported on in this paper relates to objectives with regard to implementation progress made towards the achievement of the policy goal of the White Paper on e-Education, the continued validity and relevance of associated Information and Communication Technologies (ICT) concepts, and filling gaps identified in literature. Our study locates ICT, e-education and e-schools within applicable conceptual and theoretical frameworks. Drawing on the latest research, we present strategic objectives to structure implementation of the policy goal. We discuss findings from a sample of 43 respondents from South African schools in order to assess to what extent they can already be characterized as e-schools.

Keywords: Community engagement · Computer lecturers

1 Introduction

According to [21], Information and Communication Technologies (ICT) are increasingly playing a significantly meaningful role at local, national and global levels, where the use of these emerging technologies is affecting everyday life. ICT generally also affects government policies, as well as worldwide commercial and economic growth [23]. According to the South African Deputy Minister of Basic Education, there is *"a deep grasp of the urgency, particularly for developing countries, to bridge the digital divide, to achieve developmental goals and improve people's lives"* [20]. The Deputy Minister further drew attention to policy enabling the improvement and support for integrating Information and Communication Technologies (ICT) into teaching and learning: these are key catalysts in the process of transforming education systems and public sector schools to equip their students with 21st century skills [20]. Also in South Africa (as in many other countries) ICT had a revolutionary effect on the development of the curriculum and the delivery of school practice [23]. However, many policy makers

© Springer International Publishing AG 2017
J. Liebenberg and S. Gruner (Eds.): SACLA 2017, CCIS 730, pp. 113–127, 2017.
https://doi.org/10.1007/978-3-319-69670-6_8

tend to understand ICT as being limited to only computers, satellite and internet technologies [5]. In that context it was also pointed out that more 'traditional' technologies, such as radio and television, also form part of the ICT that can be used for supporting pedagogical curriculum delivery to improve effective and efficient teaching and learning practices [1]. Although attention was called to the fact that policy requirements are local and are laid down by the government of the particular country for which curricula are developed [12], it was also emphasized that an e-education policy specifically plays a key role in the effectiveness of educational reform [5]. South Africa in particular has already developed its own structured and focused e-education policy, including strategies for using ICT to transform school teaching and learning [20]. Although this is currently the lone e-education policy in South Africa [1], the integration of e-education into teaching has ascended on the educational agenda in South Africa [24]. The South African government responded with an e-education policy goal according to which all South African students *"in the General and Further Education and Training bands will be ICT capable (that is, use ICT confidently and creatively to help develop the skills and knowledge they need to achieve personal goals and to be full participants in the global community) by 2013"* [3]. Three objectives have been identified in this regard, with the motivation of the project discussed in this paper being to make a significant and substantial contribution towards obtaining these:

1. To what extent had the e-education policy goal been achieved, since its 'due date' (2013) has come and gone?
2. More than ten years have passed since the publication of the aforementioned White Paper on e-education. With regard to the issue of working in an area of fast-moving change it was stated that the validity and continued relevance of assumptions and claims made in such an 'old' document in the field of ICT and e-education may be time-contingent and in need of investigation [10].
3. A 'continuing paucity' of the associated research base was mentioned [10].

Therefore, this paper aims at creating a platform for computer lecturers to become involved in discussions with the objective of filling the gaps in knowledge identified in the literature.

2 Background

A number of terminological concepts relevant to this paper are often being confused, are used interchangeably and/or no clear distinction is drawn between certain terms in existing research. It is therefore necessary not only to clarify what these definitions refer to in education in general, but also to specify the way in which they shall be understood in this paper. The aforementioned e-education policy document abstractly and technically defined *ICT* as *"the convergence of information technology and communication technology"* [3]. Such ICT is *"the combination of networks, hardware and software, as well as the means of communication, collaboration and engagement that enable the processing, management and exchange of data, information and knowledge"*. ICT as a resource for reorganizing curriculum integration is also embraced [3]. According to [24], *e-education* is about *"integrating technology into one's lessons to support and enhance learning"*, echoed by [18] whith objectives to enhance students' learning experiences. In accord with the e-education policy of [3], e-education was seen as being about more than just developing computer literacy and learning the skills needed to use different ICT [24]. The explanation further highlights tools and communication aspects, envisioning ICT as communication and collaboration tools for educators and students as well as for management, that contribute to national development. In [3] *e-schools* are characterized as institutions that have: students who utilize ICT to enhance learning; qualified and competent leaders using ICT for planning, management and administration; qualified and competent educators using ICT to enhance teaching and learning; access to ICT resources supporting curriculum delivery; connections to ICT infrastructure; connections to their communities. According to [1], the years since the advent of a new democracy in South Africa (1994) has seen the development of dramatic changes throughout the education and training system as part of the democratization process. The government intends these changes to redress past inequalities and to provide access to new learning opportunities [3]. Hence, [16,20] consider it imperative to understand the contribution that advances in e-education could make towards demonstrating the unflinching commitment of the South African Government to education transformation. According to [14], the introduction of ICT into especially the higher education community has necessitated new approaches, such as the creation and implementation of supple platforms and tools, being adopted as an alternate system towards improving the quality of teaching and learning. However, [19] warned that the implementation of educational technology policies *"is a highly contested domain within the South African Higher Education landscape"*. Although the abstract of [19] indicated that an analysis of the South African government's e-policy and the impact thereof on higher education would be provided, that indication did not materialize. This project will therefore aim to provide an analysis of the progress being made on the implementation of the e-education policy in South Africa. The government acknowledged a massive investment required [3], due to the magnitude of the task of implementing the e-education policy goal. Along with [1], the government called for a long-term implementation *"strategy that will provide a framework for specific priorities and actions"*, set out in a multi-year programme. The government thus established the following strategic objectives for using ICT to turn schools into centers of quality learning and teaching for the 21st century [3].

2.1   ICT Professional Development for Management, Teaching and Learning

Educators, to be able to adequately respond to changing workplace requirements, must develop the necessary skills to maximize the usefulness of computers for education purposes [2]. According to [3] it is therefore of utmost importance that increased access to ICT for teaching and learning as well as the provision of software must go hand-in-hand with adequate professional development of educators and the actual implementation of e-education. If educators do not provide e-education, and/or are not trained to effectively handle the challenges of having ICT in their classrooms, it is highly unlikely that any significant improvements will be obtained. In [22] it was stated that educators *"viewed professional development as a broad concept that encompasses their practice, the community and the teaching profession within a global context"*. Hence, all leaders, educators and administrators in schools should have access to the knowledge, skills and support needed for creating opportunities to integrate e-education into the curriculum [20]. However, [22] warned that the process of implementing educational change to improve the quality of professional practice can be difficult. Higher Education Institutions (HEI) are therefore encouraged to improve educator training as well as their participation in other education events concerning education pedagogy for application in various educational contexts.

2.2 Electronic Content Resource Development and Distribution

According to [15], the school curriculum should be supported by ensuring that a comprehensive set of effective, engaging and sustained software, electronic content, resources, tools and information across all grade levels and learning areas in several South African languages are freely and electronically accessible for re-use and adaptation. According to [12] such contents should be developed by societies and organizations which have a vested interest in the particular discipline. This should allow students, educators, administrators and content developers to develop a view on providing and structuring learning resources, resulting in an attempt to contribute effectively to such resources [17].

2.3 Access to ICT Infrastructure

All students, educators, leaders and administrators also need access to educational technology infrastructure. According to [3], one of the major challenges for the success of ICT involves institutions being able to allow educators and students to have increased and regular access to reliable educational technology infrastructure that is *"specifically suited to Africa"*. Additionally, *accountability* mechanisms have to be put in place for maintaining such infrastructure [5].

2.4 Connectivity

According to [20], a great contribution towards improving the quality of teaching and learning needs to be made by expanding the access of all educators and students to Internet connectivity in both primary and secondary schools.

2.5 Community Engagement

In line with the specific theme of this paper, [15] mentioned that e-schools should work in partnership with families and the wider community in ensuring shared knowledge about ICT and creating extended opportunities for community member e-education and development through ICT. On a note closer to the computer lecturers audience of our conference (SACLA'20017), [26] conjectured that ICT professionals ought to 'pay back' for the opportunities they were given, for the benefits they received, and for general support from society and industry. According to [26], computer lecturers have an obligation to 'pay forward' to ensure that there are replacements for us, to ensure that those being educated for the future have the ability and opportunity to thrive, and to ensure that the computing professions set an example in providing opportunities for all who have the ability to contribute. In particular, volunteering can help to foster such a sense of community [26]. In [4] those sentiments were echoed officially, whereby community engagement has been a concept with which the South African higher education system has grappled for more than a decade. According to [4], community engagement is one of three 'core functions' of institutions for tertiary education, along with research and teaching. Accordingly, community engagement has a direct relationship to academic programmes and research, and therefore forms part of a learning that engages students in community work as an official part of their academic programmes.

2.6   Research and Development

Research, evaluation and collaboration represent the best ways for learning and understanding how to improve practice. To this end, the South African government aims to bring together educators, scientists and the ICT industry in action-oriented research, to evaluate and develop leading-edge applications for e-education. Accordingly, [21] implored research and development communities, as specifically represented by higher education institutions, to support education departments by sharing the e-education knowledge and research produced at South African institutions for tertiary education. The government emphasized that this can be achieved by continuously assessing current practices, and by exploring new technologies, methodologies and techniques that reliably support educators and administrators in e-education and e-administration [3]. Research on e-education should thereby not only be linked to general research on learning, but also to practice. Since the education rofession has an obligation to play an important role in generating ideas, testing prototypes and implementing strategies, they—in collaboration with the government and other relevant stakeholders—will need to formulate a research agenda on e-education [3]. The objectives set out by [3] provide a strategic framework within which different governmental departments, provincial education departments, business and industry, non-profit organizations, higher education institutions, general and further education and training institutions, local communities and other stakeholders can collaborate to respond to the challenge of harnessing emerging ICT. This could be achieved by ensuring that institutions are supported in meeting the needs and interests of students and communities during the implementation of new ICT.

3   Method

A *mixed research method* was chosen for our study in which both *qualitative and quantitative* modes of inquiry or data collection are combined [13], though only our quantitative data are discussed in this paper. Research of 'descriptive design' offers a review of a current phenomenon—in our case: particular schools—by assessing the features of current circumstances. Quantitative data were mainly collected using a our own purpose-specific questionnaire; such kind of surveys are regularly done in educational research. Our questionnaires were distributed at a *community engagement seminar* held at our campus; the sample for this study consisted of all attendants who agreed to take part in the survey. Whether these respondents are sufficiently 'representative' is discussed below. Since no untoward pressure was applicable regarding participation, we may reasonably assume the trustworthiness of yielded responses. A case study research design considers a restricted system (the 'case') that uses numerous sources of data located in its context [25]. In our project, each case is represented by a particular school, with a collection of persons limited by time and location. Each case is selected for use as an example of a particular instance. The focus was on several entities (schools), making this a multi-site study [13]. Research sites in this paper mainly consisted of schools from a specific education district in the Gauteng province of South Africa.

4   Findings

4.1   Demographic Details

The composition of our respondents (Table 1) with regard to gender closely matches the spread obtained by [24] where the survey sample consisted of 48.5% men and 51.5% women. However, that neither of these two samples are strictly representative of the general population of educators in South Africa, as these do not reflect the fact that there are many more female educators in South African schools than male. Whereas only 43% of the sample population of [24] lived in *township* areas,[1] almost three quarters our respondents represented such areas (Table 2). Just more than a quarter of the schools represented in this study were from urban areas, comparable to those in [24] which had almost a third (31%) of

Table 1. Gender of e-schools' community engagement respondents

Gender	Respondents	Percentages
Female	23	53.5%
Male	20	46.5%

---

[1] The term 'township' has different meanings in different Enlish-speaking countries. The specifically South African notion of 'township' is reasonably well explained in https://en.wikipedia.org/wiki/Township_(South_Africa).

Table 2. Location of the surveyed schools

Area	Schools	Percentages
Township	31	72%
Urban	12	28%

Table 3. Types of schools represented by the respondents

Type	Schools	Percentages
Primary	33	77%
Secondary	7	16%
(other)	3	7%

Table 4. Respondents' age

Age in years	Respondents	Percentages
20–29	2	5%
30–39	6	14%
40–49	21	49%
50–59	11	26%
60+	3	7%

respondents from urban areas. For the district, on which the seminar reported on in this paper focused, no respondents from rural area schools were represented in the sample, compared to [24] with a rural area percentage of more than a quarter (26%).

The sample of [24] had 59% 'General Education and Training' (GET) educators and 41% 'Further Education and Training' (FET) educators. Comparably, more than three quarters of the respondents of our survey were from primary schools, and almost a fifth from secondary schools (Table 3). Respondents indicating 'other' included a representative from the district, as well as one each from a special school and a combined school.

Just less than half of our respondents fell in the 40–49 years of age category, followed by just over a quarter in the 50–59 years group (Table 4). For comparison: the 30–39 years and 60+ categories accounted for 14% and 7% of the respondents in [24].

More than a third of our respondents had more than 20 years of teaching experience (Table 5). The percentages with regard to the number of years of teaching experience for the intervals between 11–15 years of teaching experience and 0–5 years were almost the same as those in [24]; the categories 6–10 years and 16–20 years were comparably evenly distributed.

Table 5. Respondents' teaching experience

Experience in years	Respondents	Percentages
0–5	6	14%
6–10	5	12%
11–15	11	26%
16–20	5	12%
20+	15	36%

Table 6. Numbers of computers at the surveyed schools

Computers	Schools	Percentages
0	2	5%
1–10	3	7%
11–20	5	12%
20+	33	77%

Table 7. Frequency of ICT-integrated lessons

Frequency of lessons with ICT	Schools	Percentages
More than once per month	19	44%
About once per month	7	16%
Less than once per month	9	21%
Never	8	19%

In what could arguably be considered as illustrative of the situation across the Gauteng province, more than three quarters of the schools represented by these respondents have more than 20 computers at their disposal (Table 6).

Findings for the respondents in this paper compared favourably to the findings reported by [24] with regard to ICT-integrated lessons taking place more than once a month (48.5%) and about once a month (13.5%): see Table 7. Although the percentage for 'less than once per month' in the current study was considerably higher than the 9.2% of [24], fewer respondents in our study never used ICT-integrated lessons, compared to the more than a quarter (28.8%) reported by [24]. Thereby, the frequency of ICT-integrated lessons seems to be independent of the number of computers a specific school possesses (Table 8).

In line with the findings summarized in Tables 6, 7 and 8, more than half of our respondents reported that computer laboratories are being used more than once a month (Table 9). Although more than a quarter of the respondents indicated that computer laboratories are being used less than once a month, incidences where computer laboratories are being used about once a month or never are significantly lower.

Table 8. Relation between computers and frequency of ICT-integrated lessons

ICT-integrated lessons	0 Comp.	1–10 Comp.	11–20 Comp.	21+ Comp.
More than once per month	1 (13%)	0	1 (13%)	6 (75%)
About once per month	0	1 (11%)	1 (11%)	7 (78%)
Less than once per month	0	1 (14%)	1 (14%)	5 (71%)
Never	1 (5%)	1 (5%)	2 (11%)	15 (79%)

Table 9. Frequency of computer laboratory use

Frequency of computer lab-use	Schools	Percentages
More than once per month	23	53%
About once per month	3	7%
Less than once per month	12	28%
Never	5	12%

Table 10. Respondents' professional roles

Role	Respondents	Percentages
Principal	10	23%
Deputy principal	5	12%
Head of department	9	21%
Educator	16	37%
Administrator	1	2%
(other)	2	5%

The largest segment (more than a third) of the respondents were educators, with principals and heads of departments making up almost a quarter each (Table 10). Of the two respondents who selected 'other', one specified herself as an ICT coordinator. Although the other person did not explicitly select the 'provincial official' answer option, she indicated informally that she was indeed from the district office.

4.2   About the Surveyed Institutions Being e-Schools

W.r.t. the surveyed institutions *having students who utilise ICT to enhance learning* [6], just less than three quarters of the respondents (71%) 'agreed' or 'strongly agreed' that students at their institutions responded to ICT-integrated lessons by helping each other, compared to 91% of the respondents in [24] agreeing with a similar statement. Almost two thirds of the respondents agreed or strongly agreed that students at their institutions responded to ICT-integrated lessons by producing work that is more creative, with two fifths of them agreeing with this statement. More than half of the respondents (55%) agreed that students at their institutions responded to ICT-integrated lessons by working together, compared to 88% of the respondents in [24] agreeing with a similar statement. In [24], 94% of the respondents agreed with a statement relating to students at their institutions responding to ICT-integrated lessons by becoming actively involved—in our study almost three quarters of the respondents (74%) agreed or strongly agreed with this statement. Respondents' opinions regarding student activities at their institutions changing towards increasingly working on group projects show that almost two thirds of them (60%) agreed or strongly agreed—very close to the 61% in [24] agreeing with a similar statement. Although the largest segment of respondents agreed that student activities at their institutions were changing towards increasingly presenting their work to the class, two fifths (40%) of the respondents either disagreed or strongly disagreed. In line with the envisaged progress of e-learning outlined by the government [3], more than half of the respondents agreed that students at their institutions are learning *about* ICT (exploring what can be done with ICT), that students at these institutions are learning *with* ICT (using it to supplement normal processes or resources), and that students at these institutions are learning *through* the use of ICT (using it to support new ways of teaching and learning). With regard to achievement of the e-education policy goal, 50% of our respondents agreed or strongly agreed, versus 50% disagreeing or strongly disagreeing that their institutions had ICT-capable students.

W.r.t. these institutions *having qualified and competent leaders who use ICT for planning, management and administration* [6], more than half of the respondents agreed that every leader had the means to obtain a personal computer for personal use, administration and preparation of lessons; institutional leaders have access to in-service training on how to integrate ICT into management and administration; all institutional leaders integrate ICT into management and administration; the Department of Education uses ICT seamlessly in planning, management, communication, monitoring and evaluation; provincial leaders are trained in educational technology integration to offer support to institutions. For two items, about on-going support to leaders being provided at different levels of the system and about every leader having access to basic training in the use of ICT, the percentage of respondents who agreed versus disagreed with each of these statements were equal. Finally, although 46% of the respondents agreed or strongly agreed with the item relating to a set of case studies and examples being available to leaders on how to integrate ICT into management, teaching and learning, the largest segment of the respondents disagreed with this statement.

W.r.t. the surveyed institutions *having qualified and competent educators who use ICT to enhance teaching and learning* [6], more than half of all respondents agreed that these institutions had a dedicated educator to manage the facility and to champion the use of ICT in these institutions, while almost two thirds of the respondents (64%) agreed or strongly agreed that every educator has access to basic training in the use of ICT, and that educators have access to in-service training on how to integrate ICT into teaching and learning (59%). More than half of the respondents disagreed that technology incentives for institutions and educators to use ICT are installed through the 'Most Improved Schools Awards' programme and other schemes, that all educators integrate ICT into the curriculum, and that all educators are ICT-capable. Almost half of the respondents disagreed that every educator has the means to obtain a personal computer for personal use, administration and preparation of lessons, while just slightly less disagreed that educators have access to ICT technical support training, that all educators are trained in basic ICT integration into teaching and learning, and that a set of case studies and examples is available to educators on how to integrate ICT into management, teaching and learning.

W.r.t. these institutions *having access to ICT resources that support curriculum delivery* [8], more than half of the respondents agreed that these institutions are using educational content that was developed according to set national norms and standards, and that these institutions have access to educational content on the educational portal 'Thutong', while exactly 50% of the respondents agreed that 'Thutong' provides access to resources in all learning areas in the GET phase and all subjects in the FET phase. Just less than half of the respondents agreed that their institutions use educational software of high quality, while slightly less agreed that these institutions have access to an updated database of evaluated content resources and are able to select content for their usage. Two-thirds of the respondents disagreed that educators are producing digital content of high quality and making it available to other educators, while the same number of respondents disagreed that their institutions have access to digital libraries. Just more than half of all respondents disagreed that their institutions use the educational portal to communicate, collaborate and access content, while just less than half of the respondents disagreed that the province is collaborating and pooling ICT resources where appropriate. Although the largest segment of respondents disagreed that their institutions use the educational portal for teaching and learning in an outcome-based education fashion, almost the same number of respondents agreed with this statement.

W.r.t. the surveyed institutions *having connections to ICT infrastructure* [8], more than half of our respondents agreed that their institutions use electronic means to communicate with provincial offices, that these institutions have a computer and software for administrative purposes, and that they have legal software and also use it. Just less than half of the respondents agreed that their institutions have access to a networked computer facility for teaching and learning that is safe, effective, designed to facilitate ICT integration into teaching and learning, and is in working condition, that these institutions have access to a networked computer facility for teaching and learning, and that their ICT facilities are safe. Although just over two fifths of the respondents agreed that facilities are being used effectively integrate ICT into teaching and learning, and that ICT facilities are safe, effective, suited for integration into teaching and learning, and also in good working condition, almost the same numbers of respondents disagreed with those two statements. Almost two-thirds of the respondents disagreed that their institutions have access to an 'e-rate', i.e.: a discounted connectivity rate specifically for GET and FET institutions according to [3]. Although just more than two fifths of the respondents disagreed that their institutions are connected to the educational network, that networks are safe and that information security is monitored, again almost the numbers of respondents agreed with those two statements.

W.r.t. these institutions *connecting with their communities* [8], the majority of respondents disagreed; almost two-thirds of all respondents disagreed that their institutions serve as a venue for business advisory services or training for community-based small computer and repair businesses, that local small-medium- and micro enterprises (SMME) have been developed and trained to provide technical support to those institutions, or that SMME provide technical support to them. More than half of all respondents disagreed that their communities are integrally involved in these institutions or that communities have access to the schools' computer equipment and services in the after-hours. Finally, just less than half of all respondents indicated that community involvement supports their institutions to sustain their ICT facilities.

5   Conclusions and Outlook

Pre-university schools world-wide are exposed to rapid 'digitalization' [11,27], whereby schools in developing countries have (understandably) not yet achieved the same level of ICT usage as in developed ones. The motivation of the project described in this paper is therefore to make a significant and substantial contribution towards achieving *three national objectives*, which have been expressed in the South African e-education policy goal, as well as a number of associated matters. A summary of our findings in this regard includes the following points:

Concerning the *first objective*, i.e.: achievement of the e-Education policy goal, there was no consistent agreement among our respondents about their institutions having students who are ICT-capable.

W.r.t. the *second objective*, i.e.: validity and continued relevance of some of the assumptions and claims made in the White Paper on e-Education, we can provide the following results: The majority of our respondents agreed with 90% of the survey statements relating to students using ICT to enhance their learning; for five of those statements 50% or more respondents agreed. The implications of these results for teaching and learning at tertiary level are positive: computer lecturers can expect increasingly ICT-capable students in their classes, and it is therefore recommended that computer lecturers consider this in their teaching at tertiary level. More than half of the respondents agreed with 5/8 survey statements about their institutions having qualified and competent leaders that can use ICT for planning, management and administration purposes. For two statements, concerning on-going support to leaders being provided at different levels of the system, and every leader having access to basic training in the use of e-Learning, the percentages of respondents who agreed versus disagreed with each of those survey statements were the same. Finally, although a total of 46% of the respondents agreed or strongly agreed with the item relating to a set of case studies and examples being available to leaders on how to integrate ICT into management, teaching and e-Learning, the largest sub-group of our respondents disagreed with that statement. For eight of the survey statements concerning these institutions having ICT infrastructure and connectivity, the majority of respondents agreed, with three of those representing more than half of the respondents. Although the majority of the respondents therefore only disagreed with only three of those survey statements, one of the latter, relating to these institutions having access to an 'e-rate', represent almost two thirds of all respondents.

W.r.t. the *third objective*, the publication of this paper is justified in contributing some new and original insights to our field of study. In similar and closely related contexts, [6] portrayed students wanting to be exposed to ICT to keep their education relevant, [7] reported on a limited data set relating to a community engagement project around ICT in education for growing innovative e-schools in the 21st century, and [9] presented results relating to students, educators and leaders at e-schools in South Africa. Alongside [8], the merit of our study presented in this paper, and its relevance to the SACLA'2017 conference, is also justified w.r.t. the frequent changes of ICT facilities and usage in the education environment which educators have to contend with again and again. Although members of the audience attending the presentation of this paper at SACLA'2017 seemed to be impressed, they did not respond significantly in terms of substantive comments. Some conference participants, however, asked whether we would be planning any follow-up evaluations, in light of the interactions we have had with educators during the previous 18 months. It was finally suggested that we might conduct a similar survey with our first-year students at university, to gauge their experiences during their pre-university school years in terms of the implementation of the governmental policy: Do the university students have the same impression of what is or was happening to them? Obtaining their answers might put things into a new light, possibly in contrast the answers provided by the respondents of the survey described in this paper.

Acknowledgments. Thanks to all respondents who took part in our survey. The research for this paper received funding from the *Women in Research Support Programme* (WiR-SP) of the University of South Africa.

References

1. Blignaut, S., Els, C.: Not yet where we want to be: South Africa's participation in SITES 2006. US-China Educ. Rev. **7**(2), 55–66 (2010)
2. Dagada, R.: Educator competence in integrating computers for teaching and learning within the framework of the gauteng online project. Educ. Change **8**(2), 103–133 (2004)
3. Department of Education: White Paper on e-Education: Transforming Learning and Teaching through Information and Communication Technologies (ICTs). Republic of South Africa: Government Gazette no. 26734, pp. 3–46 (2004)

4. Department of Higher Education and Training: White Paper for Post-School Education and Training: Building an Expanded, Effective and Integrated Post-School System. Republic of South Africa: Governmental Communication (2013)

5. Evoh, C.: Policy networks and the transformation of secondary education through ICTs in Africa: the prospects and challenges of the NEPAD e-schools Initiative. Int. J. Educ. Dev. ICT **3**(1), 64–84 (2007)

6. Goosen, L.: Educational technologies for growing innovative e-schools in the 21st century: a community engagement project. In: Proceedings of South Africa International Conference on Educational Technologies (SAICET), Pretoria, pp. 49–61 (2015)

7. Goosen, L.: We don't need no education? Yes, they do want e-learning in basic and higher education! UNISA (2016)

8. Goosen, L., Breedt, M.: Educating in the changing environment of computer applications technology. In: Proceedings of the Annual Conference of the Southern African Computing Lecturers (SACLA), Bloemfontein, p. 57 (2012)

9. Goosen, L., van der Merwe, R.: E-learners, teachers and managers at e-schools in South Africa. In: Proceedings of the 10th International Conference On E-Learning (ICEL), Nassau, pp. 127–134 (2015)

10. Heeks, R.: Do information and communication technologies (ICTs) contribute to development? J. Int. Dev. **22**(5), 625–640 (2010)

11. Huang, R., Kinshuk, P., Jon, K. (eds.): ICT in Education in Global Context: Comparative Reports of Innovations in K-12 Education. LNET. Springer, Heidelberg (2016). https://doi.org/10.1007/978-3-662-47956-8

12. Marshall, L.: Developing a computer science curriculum in the South African context. In: Proceedings of the Computer Science Education Research Conference, CSERC 2011, pp. 9–19 (2011)

13. McMillan, J.H., Schumacher, S.: Research in Education: Evidence-Based Inquiry. Pearson, London (2010)

14. Mouyabi, J.S.M.: Higher education in the wake of new ICT: reaping benefits or creating more problems through e-learning? S. Afr. J. High. Educ. **25**(6), 1178–1189 (2011)

15. Mpehle, Z.: Is gauteng on-line or off-line? an empirical study of the gauteng department of education. J. Public Adm. **46**(1), 709–719 (2011)

16. Park, T., van der Merwe, A.: The transformative role of ICT in higher education: a case study of the alignment of educational technology utilization with the vision of Stellenbosch university. S. Afr. J. High. Educ. **23**(2), 356–372 (2009)

17. Rogier, E., van der Veer, G.C.: Adult CS learning: flexible learning from each other. In: Proceedings of the 3rd Computer Science Education Research Conference, CSERC 2013, (2013)

18. Russell, I., Coleman, S., Markov, Z.: A contextualized project-based approach for improving student engagement and learning in AI courses. In: Proceedings of 2nd Computer Science Education Research Conference, CSERC 2012, pp. 9–15 (2012)

19. Sesemane, M.J.: E-policy and higher education: from formulation to implementation. S. Afr. J. High. Educ. **21**(6), 643–654 (2007)

20. Surty, E.: Address at the Handover of the 150 Telkom Foundation Schools Project to the Western Cape Education Department by Mr. Enver Surty, Deputy Minister of Basic Education, Ruyterwacht Pre-School, Cape Town, 29 July. Governmental Communication, Republic of South Africa (2010)

21. Terzoli, A., Dalvit, L., Murray, S., Mini, B., Zhao, X.: Producing and sharing ICT-based knowledge through english and African languages at a South African university. S. Afr. J. High. Educ. **19**(7: special iss.), 1486–1498 (2005)

22. Thomen, C.: Education practitioners' understanding of professional development and associated competencies. S. Afr. J. High. Educ. **19**(4), 813–821 (2005)
23. Whelan, R.: Use of ICT in education in the South Pacific: findings of the pacific eLearning observatory. Distance Educ. **29**(1), 53–70 (2008)
24. Wilson-Strydom, M., Thomson, J., Hodgkinson-Williams, C.: Understanding ICT integration in South African classrooms. Perspect. Educ. **23**(4), 71–85 (2005)
25. Yin, R.K.: Case Study Research, Design and Methods, 5th edn. Sage, Thousand Oaks (2014)
26. Young, F.H.: Paying back and paying forward. In: Proceedings of the 46th ACM Technical Symposium on Computer Science Education, SIGCSE 2015, p. 2. ACM (2015)
27. Zhang, J., Yang, J., Chang, M., Chang, T. (eds.): ICT in Education in Global Context: The Best Practices in K-12 Schools. LNET. Springer, Heidelberg (2016). https://doi.org/10.1007/978-981-10-0373-8

Augmenting a Data Warehousing Curriculum with Emerging Big Data Technologies

Eduan Kotzé[✉]

Department of Computer Science and Informatics, University of the Free State,
Bloemfontein, South Africa
kotzeje@ufs.ac.za

Abstract. The demand for graduates with big data and data warehousing skills far exceeds the supply of students graduating with these skills. This paper addresses this problem by means of a pilot study in which big data topics were integrated into a classical data warehouse course at postgraduate level. Courses like this could be helpful in supporting hands-on learning experience with big data warehousing.

Keywords: Data warehousing · Big data · Hadoop · Hive · Curriculum

1 Introduction

Traditionally, data warehouses were designed to store and process structured data. Structured data tend to be organized in entities such as XML documents or relational database tables. With the advent of 'big data', modern data warehouses need to accommodate new datasets, which could include webpages, spreadsheets, social media and sensor logs, to name a few. These new large datasets are not structured by nature, but tend to be more semi-structured, such as a spreadsheet, or unstructured with clickstreams or log files. Industry has emphasized the importance of incorporating these new types of datasets into existing data warehouses using emerging big data technologies [3,21]. The integration and expansion of data warehousing platforms with big data technologies have received some attention in literature [3,42,47]. However, this integration and expansion also offers many other challenges. One such challenge is to ensure that universities educate enough graduates with sufficient big data and data warehousing skills. Recent research indicate that the current demand for graduates with expertise in big data and analytics far exceeds the supply of such graduates [37]. In [43,52] it was argued that inadequate staffing or skills, in terms of big data, are hampering organizations to diversify the platform types of their data warehouses. It is therefore imperative that data warehousing and/or business intelligence graduates must acquire specific skills, such as big data warehousing skills, to be ready for industry [2]. Additionally, [55] expressed the need for more big data education and suggest that universities should augment existing data warehousing and business intelligence concepts with big data, big data topics and learning objectives. In South Africa, the number of universities offering courses

J. Liebenberg and S. Gruner (Eds.): SACLA 2017, CCIS 730, pp. 128–143, 2017.
https://doi.org/10.1007/978-3-319-69670-6_9

incorporating big data is limited [28,36]. Some work has been done to address the challenge by incorporating big data technologies with existing database curricula [33,34,44]. However, more research is needed in terms of incorporating big data technologies with data warehousing curricula to prepare the next generation workforce [27]. The purpose of this paper is to explore the feasibility of augmenting an existing data warehouse module with emerging big data technologies. The main contributions of this paper are to:

- identify important big data concepts and technologies relevant to data warehousing, and provide guidelines on when to use them;
- provide guidelines in terms of integrating big data technologies into an existing data warehousing course module;
- describe the learning resources that can be used to enable 'hands-on' experience.

A pilot study was done to assess the feasibility of integrating big data technologies into a data warehousing module. The results of the exploratory study are also presented.

2 Method

This research is of an exploratory nature and conducted a pilot study. The objective was to assess the feasibility of incorporating emerging big data technologies into an existing Data Warehousing postgraduate module. The module is part of a Bachelor's Honours degree in Computer Information Systems at a university in South Africa.[1] The Honours degree was at the time of writing this paper presented as either full-time or part-time with combined contact sessions after office-hours. The data warehousing module has a study credit weight of 120 h per semester and was presented over a 12-week period. In total 13 students enrolled for the module. In the final week, the students were asked to complete a student satisfaction survey. Data analysis was conducted on the survey and the success of incorporating big data into the module was assessed.

3 Data Warehousing

A data warehouse is an example of a data-driven decision support system that stores and analyses structured data in a relational format [39]. In [20] a data warehouse is defined as a collection of data that, in support of management's decision making-process, is subject-oriented, integrated, time variant and non-volatile. On the other hand, [24] defines a data warehouse as a collection of tightly

---

[1] For readers from outside South Africa: the South African 'honours' degree is an extension of the classical 'B.Sc.' degree which enables a student to commence with Master-studies thereafter. While already considered 'postgraduate' in South Africa, the 'honours' degree in South Africa is reasonably well comparable to the final study-year in the (longer) U.S.American 'B.Sc.' curriculum.

integrated data marts that are based on a dimensional model. In order to clearly understand the conceptual difference between a data warehouse and traditional operational information systems, a clear distinction needs to be made. Operational systems are online transaction processing (OLTP) systems that are used for the day-to-day operations of an organization. In other words, OLTP systems are designed to store and put data into the database. Data warehousing, on the other hand, are specially designed decision-support systems that focus on providing strategic information from a database [38]. Designing the data model and loading this data model with structured data from source systems are considered the two main issues in developing a data warehouse. In terms of designing a data model, the two most popular database design approaches are the traditional approach using entity-relationship modelling [20], and the alternative approach, known as 'dimensional modelling' [24]. The traditional approach makes use of entity-relationship diagrams and normalization techniques and are often used for the database design in OLTP systems [20]. The dimensional modelling approach makes use of star schemas to represent a multidimensional model and consists of fact and dimension tables [24]. This approach is very appropriate for decision support systems where efficiency in querying and loading of data are important [7]. Once a data model has been designed, the model must be loaded with data from operational information systems. To perform this function optimally, a data warehouse needs a stable and flexible technology platform to store and process the data. Relational database management systems (RDBMS) with Structured Query Language (SQL)-based analytics are considered the most popular platforms for this function. These relational systems are used to manage the process of extracting data and integrating it into either the normalized schemas or dimensional models. This process is often referred to as 'Extraction, Transformation and Loading', or ETL, and is an integral part of the data staging component of a data warehouse [23]. The main functions of ETL generally consist of a set of activities and tasks that will be performed to ensure that the data is cleaned, changed, combined, converted, deduplicated and prepared for storage and analysis in the data warehouse [38]. It is not uncommon to see data warehousing project teams spending as much as 70–80% of development time and effort on ETL [53]. For the purpose of this study, dimensional modelling was chosen as the preferred data model design paradigm since the entity-relationship model often have difficult navigation paths, is hard for users to understand [38] and is not well-suited for high volume queries [26]. A successful data warehouse project also requires a data warehouse architecture. Its purpose is to seamlessly integrate the dimensional model and data staging that will allow user application tools (such as an Online Analytical Programming reporting tool) to query the underlying data model. Since the module employed dimensional modelling as design paradigm, the Data-Mart Bus or 'Bus' architecture [24] approach was adopted for the proposed module. In this approach, the data warehouse will be developed adopting a bottom-up approach that begins at selecting a core business process, analyzing the requirements, designing a dimensional model, and finally, loading the model using ETL functions.

4   Big Data

'Big data' generally refers to a huge collection of data that is either semi- or unstructured, growing at a rapid speed, too large for traditional RDBMS, and hence requires a new type of database technology [41,50]. Moreover, [13] describes big data as *"high-volume, high-velocity and high-variety information assets that demand innovation information processing technologies to enhance insight and decision making"*. The purpose of the following section is to provide a brief overview into the reasons behind this new requirement. This will be followed by a brief discussion on key emerging information processing technologies used in modern day big data systems.

4.1   Issues with Relational Database Systems and Big Data

One of the most important goals of big data systems is to provide a stable and scalable environment for storing, analyzing and mining big datasets [18]. This provides interesting challenges for any data storage system since structured data is generally repeatedly queried using an RDBMS, while semi-structured and unstructured data are generally handled on an ad hoc basis, or sometimes once a month as a batch [18]. Also, a flexible infrastructure that is scalable for queries and easy to use should be put in place. However, it is widely acknowledged that traditional RDBMS and SQL are not suited or flexible enough to cope with the variety and size of big data sets in terms of data storage and querying [22,29]. The reason for this limitation is that a relational database is based on pre-defined schemas, which means that storing the data needs to adhere to ACID (atomicity, consistency, isolation, and durability) compliancy in terms of transaction management [6,29]. In other words, an RDBMS is best suited to query and update small chunks of structured data [54]. Also, an RDBMS can only scale with very expensive hardware and not with commodity hardware, such as a cluster [18]. Both these factors (heterogeneity and scalability) mean that traditional RDBMSs are inadequate to manage the growing data volumes often associated with more modern day applications such as sensor networks and e-commerce platforms [18,29]. In an effort to overcome these limitations, distributed file systems [14,17], NoSQL databases [6], MapReduce [11], YARN [51] and Hadoop [54] have emerged as information processing technologies for storage and management of big data sets. Each of these technologies will now be introduced briefly.

4.2   Distributed File Systems

The file system forms the basis of data storage within the data management framework of a big data system [18]. Google was one of the first large internet companies to design and develop a distributed file system, called Google File System or GFS [14]. Other distributed file systems include Microsoft's Cosmos and Facebook's Haystack [9]. At the time of writing of this paper, the Hadoop

Distributed File System (HDFS) is seen as the most popular open source deriv-
ative of GFS and is known for its ability to run on a cluster of inexpensive
personal computers [15, 42].

4.3   NoSQL

Databases, together with distributed file systems, are used for data storage in
a big data system [18]. NoSQL (Not Only SQL) database systems are currently
viewed as the preferred database technology to process large volumes of data that
are in varied formats [45]. NoSQL databases can broadly be classified into four
categories: key-values, column-oriented, document databases and graph databases [18]. Key-value databases store the data in schema-less or key-value pairs.
Column-oriented databases store and process data by columns instead of rows,
as is the case with relational database systems. Document databases store documents that contain a certain format, for example XML or JSON. Finally, graph
databases store graph-like data that consists of nodes and edges.

4.4   MapReduce and YARN

MapReduce was originally developed by Google to assist with web searches on
web documents [11]. These days, MapReduce is used as a processing model
to manage and process big data sets [45, 50, 54]. It is important to note that
MapReduce is not a programming language and was developed to be used by
programmers, not by business users [45]. Instead, MapReduce is a programming
model that enables parallelism and distributed computing on a computer cluster [18]. This programming model uses the distributed file system to store and
process the data, making it an ideal environment to batch process big datasets
with MapReduce applications. These applications can be developed using either
C, Java, Ruby and Python [52, 54]. YARN, or *Yet Another Resource Negotiator*,
is considered an improvement on the original MapReduce programming model.
YARN offers greater stability, higher efficiency and enables a large number of
different frameworks to share a cluster by providing a layer between the HDFS
and MapReduce [51].

4.5   Apache Hadoop

Apache Hadoop is a big data framework that consists of several open-source software components and is modeled on Google's MapReduce [30, 42]. This framework provides scalable and distributed computing on clusters of inexpensive
servers. The main purpose of this framework is to provide an ecosystem [12]
to process, store and analyze big volumes of structured, semi-structured and
unstructured data [45, 50]. A typical Hadoop ecosystem will include a host of
other technologies as well, such as a NoSQL database (HBase), a data warehousing system (Hive), a platform to manipulate the data (Pig), a tool to efficiently
transfer bulk data (Sqoop) and MapReduce/YARN as a resource management

tool. This set of technologies complement each other and should not be viewed as separate components. For example, MapReduce and HDFS run on the same cluster nodes, which therefore allows tasks to be scheduled on the nodes in which data are already stored.

4.6   Apache Hive

Developing MapReduce programmes can be quite challenging and as the processing becomes more complex, the complexity is reflected in the code. For this reason, developers commonly consider a higher-level programming model and language [54]. Since the MapReduce framework is an ideal environment to process batch data, it has also become a popular platform for data warehouses to process their batch-data during the extraction, transformation and loading (ETL) process. Apache Hive is considered the preferred data warehouse architecture for a Hadoop system since it provides a higher-level programming model and language. HiveQL, a SQL-like declarative language, was developed for Hive to allow programmers to work on a higher abstraction level instead of writing low level code in C or Java [46]. Programmers with SQL knowledge can now easily learn to write Hive applications, which are essentially SQL-based applications [5]. When a Hive application is executed, HiveQL is used to query and analyze the large datasets stored in the HDFS. Hadoop also internally converts the Hive queries to MapReduce tasks without the programmer's intervention [46]. This allows for high-level programming while maintaining full support for *map* and *reduce*, and presenting a familiar SQL abstraction. Hive also makes provision for ad hoc queries, aggregation and the ETL of a variety of data formats [31]. It should be noted that Hive is not a NoSQL database and does not provide large-scale record-level updates, inserts or deletes. Instead, Hive is suitable for a data

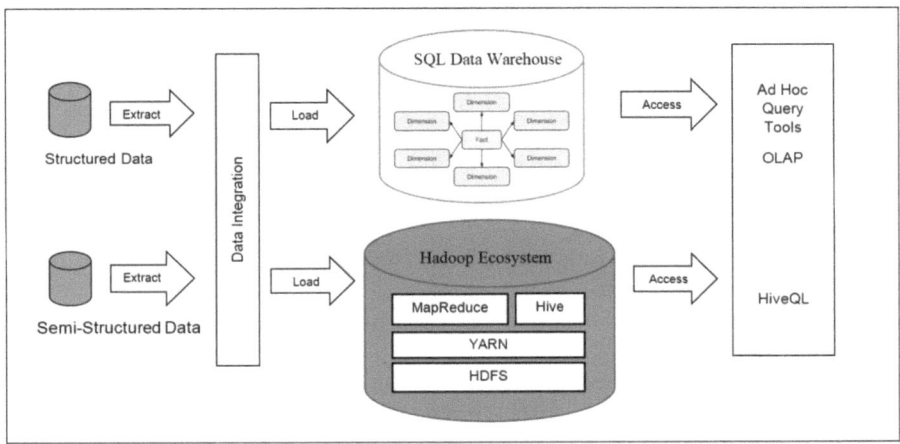

Fig. 1. Big data warehouse architecture, as adopted from [24]

warehouse application where data is relatively static, does not change rapidly and fast response times are not required [5]. If a large-scale OLTP is required, a NoSQL database, such as HBase or Cassandra, should be considered. For the purpose of this study, Hadoop with HDFS was selected as the big data ecosystem. This ecosystem allowed for the use of Hive (a big data warehouse), instead of a NoSQL database. The decision was based on the notion that the students are proficient in SQL and, by using HiveQL, it would reduce the learning curve. Figure 1 illustrates how the traditional data warehousing architecture and the big data ecosystem was working together. Due to time constraints, the data warehouse was not linked with the Hadoop Hive ecosystem for querying purposes. Sqoop, a tool designed to facilitate bulk data transfers between Apache Hadoop and a relational database, was also not implemented.

5   Learning Objectives

One of the most important aspects in developing computing curricula is to organize the theoretical concepts and application of technology into well-defined learning units. In [8] the need to integrate the concept of 'learning by doing' using 'hands-on' projects was emphasized, since big data analytics requires trail-and-error and experimentation. To introduce the postgraduate students to this, the following *core learning units* are proposed:

Data Warehousing: In this learning unit, the students shall be introduced to data warehousing principles, dimensional modelling and ETL. Students shall construct a data warehousing solution using structured data and an RDBMS such as Microsoft SQL Server 2014.

Big Data: In this unit, students shall be introduced to fundamental concepts regarding big data, which includes data variety, volume and velocity. Students shall use a big data warehouse to load and query semi-structured data. Big data technologies shall include Apache Hadoop, HDFS, MapReduce and Hive.

In terms of core *learning outcomes* (LO_i), students should be able to:

- distinguish/understand data warehousing development methodologies (LO_1);
- understand the difference between OLTP and OLAP, multidimensional modeling, star schema, data marts, normalization of dimensions and snowflaked schemas (LO_2);
- utilize and evaluate techniques and methods for extracting, transformation and loading data (LO_3);
- understand the fundamentals of big data and how it related to data warehousing (LO_4);
- apply knowledge by creating a data warehouse, loading it with semi-structured data and processing it with big data technologies (MapReduce) to meet business requirements (LO_5).

5.1   ACM/IEEE Curriculum Conformance

Although the IT 2008 Curriculum Guideline [32], the IS 2010 Curriculum Guideline [48] and the IEEE and ACM Computing Curricula 2013 Guideline [1] do not have a dedicated section for *Data Warehousing* on its own, relevant components could still be mapped to. The Data Warehousing module described in this paper maps to two components listed in the Information Management (IM) section of the IEEE and ACM Computing Curricula 2013 Guideline [1]. Specific areas covered include: IM/Data Modeling and IM/Distributed Databases. The module also maps to the IM/Data Modeling section of the IT 2008 Curriculum Guideline [32]. Finally, the module also maps to the 2010.2 Data and Information Management section of the IS 2010 Curriculum Guideline [48].

5.2   Data Warehousing Learning Resources

Enabling students to work on an actual data warehouse is seen as one of the best ways to introduce them to data warehousing [25]. A SQL Server 2014 instance was set up on an internal departmental server and the students were granted access. Students were also provided with a developer copy of SQL Server 2014, which they could install on their personal computers for practice purposes. All students had access to Microsoft Development Network (MSDN), where they could download manuals and white papers on building a data warehouse using Microsoft SQL Server Integration Services (SSIS). They were provided with sample datasets, including the popular 'AdventureWorks2014' database.

5.3   Big Data Learning Resources

A popular approach in teaching big data is to provide the students with a MapReduce framework to be used for hands-on, in-class exercises or project assignments [54]. It should be noted that a MapReduce framework would typically need a computer cluster. This could be challenging for some institutions that do not have a computer cluster available for teaching purposes. One approach is to use a pseudo-distributed installation of Hadoop, which allows a single node to perform operations using Hadoop MapReduce and HDFS [15]. This approach requires significant installation and configuring effort that could be challenging to a student who has not yet been exposed to such technology. Hence, a virtual machine (VM) with the following required components were configured:

1. Linux operating system (Linux Mint 17.2);
2. Hadoop 2.7.2 pseudo-distributed mode (including HDFS and YARN);
3. Java 7 and Eclipse;
4. Hive 1.2.1;
5. sample datasets.

It was decided not to use commercially available VMs, such as Hortonworks [16], Cloudera [10], MapR [35] and IBM [19], but rather follow a hands-on approach using open-source software to keep abreast of new developments. Finally, an in-house computer cluster described in [4] was configured for Hadoop 2.7.2 and Hive 1.2.1 to be used during the big data practical assessment. In order to facilitate the process of gaining hands-on experience on Apache Hadoop, students could re-use a MapReduce template (WordCount.java) that performed a basic word count. Large text files were then created by combining several text files from Project Gutenberg [40]. The students used these text files while gaining hands-on experience with HDFS and MapReduce, and completing their practical assignment at the same time. To give the students a better understanding of Hadoop Hive, they were provided with the Lahman Baseball Sample Database[2] to work with. This database is available in a comma-delimited version on the internet. The Batting and Master data sources were used to populate a schema in Hive. Students were then asked to develop HiveQL queries such as:

- show the number of runs per country;
- indicate which player scored the most runs;
- list the top 10 players.

5.4   Achieving Learning Objectives

In order to achieve the learning objectives explained earlier, the module was divided into two terms. In the first term, the focus was on managing structured or relational data. Students were introduced to big data in the second term. In order to lay a solid foundation, the first few lectures introduced the Kimball life cycle methodology and theoretical components of designing a data warehouse, which included project management, business requirements gathering and information packages. The objective of these lectures was to cover the architectural component of data warehouse design and implementation. Thereafter, students were introduced to the infrastructure component of data warehousing, which included hardware such as computer clusters.

The theory and design of data warehouse appliances were also covered as this is fast becoming a popular platform to implement a data warehouse. These topics concluded the lecture-based learning model for the first term, as the teaching mode then switched to a work session-based learning model for the remainder of the first term. The reason for switching from a lecture-based learning model to a work session-based learning model was to allow the students to gain hands-on experience. The work sessions commenced with a theoretical introduction of dimension modelling and ETL concepts. It is well known that these two phases cover approximately up to 70% of a data warehousing project. The students find dimensional modelling very challenging since their undergraduate database module solely focused on normalization of data. For a data warehouse project, they need to 'unlearn' some of those concepts and denormalize a data model. Students were provided with two case studies on dimensional modelling and some sample data. In-class discussions were held where the students evaluated all possible theoretical solutions to the case studies. In [49] it was suggested that students should also learn to integrate technologies, applications and data and

---

[2] http://www.seanlahman.com/baseball-archive/statistics/.

observe how disparate parts are brought together. To introduce the students to this concept, it was demonstrated to them how several ETL components and functions in SSIS and Visual Studio 2014 could be useful in solving the technological application of the case studies.

The students were also required to complete three practical assignments in class to demonstrate their understanding of ETL. Guidance was provided only during these work sessions, while the students familiarized themselves with ETL using SSIS and Visual Studio 2014. Once the work sessions were completed, the students had to submit a practical data warehouse project consisting of three milestones that had to be completed within a period of 6 weeks. In order to simulate a real-world data warehouse project, students were provided with two sample databases and a spreadsheet together with business questions. The first milestone was the information package, where they had to indicate the dimensions, categories, hierarchies and measures, as well as the table name and column of each attribute or measure. In case of semi- or fully additive measures, the calculation had to be provided. The second milestone was the dimensional model, and the final milestone was an ETL solution that they had to develop using SSIS and Visual Studio 2014. No guidance was provided to the students during any of the milestones. It was envisaged that the experience gained during the work sessions should have provided the students with sufficient knowledge to complete the project. The author is of the opinion that the use of the SQL Server 2014, as well as the described in-class activities during the work sessions, played a critical part in consolidating their understanding of data warehouse methodologies (LO$_1$), dimensional modeling (LO$_2$), ETL techniques and methods (LO$_3$) and gaining hands-on experience with designing and implementing a data warehouse (LO$_5$).

In the second term of the module, the students were introduced to semi- and unstructured data. The students found this to be an uneasy paradigm shift since their undergraduate training solely focused on relational or structured data. Big data technologies such as Apache Hadoop, MapReduce, HDFS and Hive were introduced and demonstrated. The demonstration of these technologies gradually exposed the students to the idea that modern data warehouses should also accommodate semi-or unstructured data. A pre-configured VM with a pseudo-distributed installation of Apache Hadoop was made available to the students. Furthermore, a guideline outlining the installation and configuration was also made available to accommodate students who wanted to know more (for example, how to reconfigure the VM themselves). The students were provided with sample data, which included several text documents.

Part of the practical assignment was the loading of semi-structured data into a Hive data warehouse and developing several HiveQL queries. The practical assignment also required the students to replicate their findings on the departmental Hadoop cluster. As a result, the students used the VM as an experimental sandbox to gain hands-on experience. Once the students were ready, they were assigned a two-hour period on the departmental cluster to complete the assignment. The rationale behind this approach was to provide the students with not only Hadoop hands-on experience, but also with real-world experience of working on a computer cluster.

The author is of the opinion that the use of the VM with Hadoop (including HDFS and YARN) and Hive, as well as the described in-class activities, have been valuable components with regard to increasing the students' understanding of big data (LO$_4$), as well as allowing them to gain hands-on experience with big data warehousing technologies (LO$_5$). Furthermore, by being provided with a ready-made VM, the students could focus on implementing and experimenting with MapReduce and Hive applications instead of tedious installation and configuration tasks.

6 Assessment and Results

The first and the second term were each concluded with a formal two-hour theoretical assessment. The marks for the two tests were aggregated into a single theoretical mark, (see Fig. 2). The average mark was 51%, which indicated that the students had trouble in mastering all the necessary theoretical concepts as outlined in the learning outcomes.

As mentioned above, the students had to submit practical assignments. On average, the students scored 51.0% for the Hadoop Assignment, 72.4% for the Dimensional Model assignment, and 50.6% for the ETL assignment (Fig. 3).

Both the Hadoop and ETL assignments required an element of development, while the dimensional model assignment focused on database design. More than 61% ($n = 8$) of the students scored above 80% for the dimensional model, which indicated that they conceptually mastered data warehouse design. However, almost 53% ($n = 7$) of the class scored below 50% for the Hadoop assignment and the ETL assignment. In order to understand the apparent low marks for the Hadoop assignment, the students were requested to complete a student satisfaction survey. The questions made use of a Likert scale (from $1 =$ 'strongly

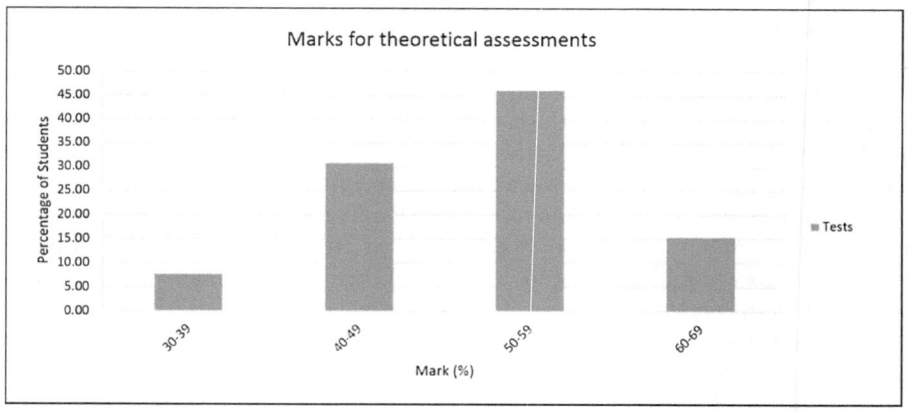

Fig. 2. Marks for theoretical assignments

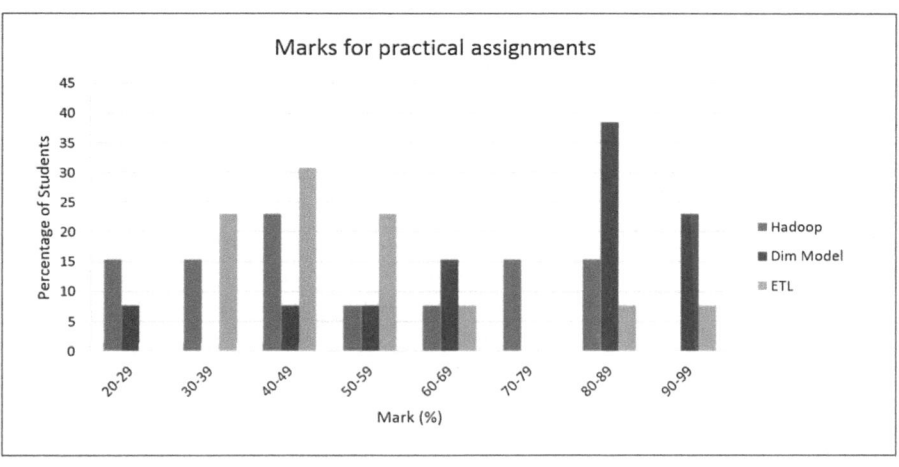

Fig. 3. Marks for practical assignments

agree' to 5 = 'strongly disagree') and collected data on demographics, lecturer and content engagement. Since this was a small class, all 13 students completed the survey. Some of the results were not promising; the scores are presented next.

With regard to the question *"The workload in this module was manageable"*, only 15.4% of the students agreed (strongly or agree) that the workload was manageable. The reasons provided include *"improper time management on my part"*, *"the module pace was too fast"*, *"course content is too much"*, and *"the hadoop content was very foreign and took long time to understand"*. It is possible that the Hadoop environment, which is Linux-based, could be quite challenging to the students who were mostly working in a Microsoft Windows environment during their undergraduate training. Furthermore, the Hive interface is command line-based and a very foreign environment to students who completed most of their software development using the Visual Studio Integrated Development Environment (IDE). Students also indicated, during the two-hour session on the computer cluster that they did not practice enough on the sandbox VM that was provided to them, and hence ran out of time to replicate their findings. The survey also indicated that the students were uncomfortable with the operating system configuration differences of the VM and cluster. For example, students had to login into the cluster, resulting in a directory structure that was different between the two environments. They also had to deal with file permissions, which is not necessary within a Microsoft Windows environment. This finding was confirmed when investigating the log files on the cluster. It is therefore suggested that students should have extensive experience with a scripting language (for example Bash) if they want to explore big data technologies such as Hadoop, HDFS and Hive. This is primarily because interacting with Hadoop, HDFS and Hive is done by executing command line instructions within a Linux Operating System environment. An argument could be made that an application using a Graphical User Interface (GUI) should instead be provided to the students for HiveQL interaction. However, the author was of the opinion that it would not provide the students with a real-world computer cluster environment, which predominately use command-line scripts for configuration and management.

With regard to the question *"The learning material helped me learn"*, 76.9% indicated they were unsure if the learning material helped them to learn, while only 23.1% agreed that it helped them. A more positive result was found towards the question *"are you satisfied with the digital resources available for this module"*, where 46.2% agreed while 46.2% were unsure. In the open-ended questions, where students were asked *"what aspects of this module helped you learn"*, the students gave positive feedback, e.g.: *"the in-class labs helped"*, *"the technologies used and provided examples"*, *"practical work tasks and assignments"*, *"assignments"*, and *"extra reading material provided"*. This finding supports the need to integrate the concept of 'learning by doing' using hands-on projects, since big data learning requires trial-and-error [8].

7   Conclusion

Modern data warehouses require the processing of structured, semi- or unstructured data warehouses. Recent research has reported the need for graduates with expertise in both big data and analytics or data warehousing. It has also been reported that universities should consider enriching existing data warehousing and business intelligence modules with big data concepts. The aim of this paper was to evaluate the feasibility of integrating emerging big data technologies into an existing data warehousing postgraduate module. A pilot study approach was followed and the module was evaluated at the end of the semester. Students indicated that they experienced difficultly in familiarizing themselves with the new big data computing environment and a different programming framework at the same time. However, even though the students struggled with the cognitive challenge, 10 of the 13 students passed the module successfully. Future research could focus on methods to reduce the cognitive load and follow a different pedagogical approach. This could include using only a virtual-machine pseudo-distributed installation of Apache Hadoop/Hive and abandoning the exposure to a real-world cluster. From the pilot study findings it also emerged that the students were not necessarily familiar with a scripting language. In the future, the module could be amended to make use of a GUI for Hadoop and Hive instead of the command-line. It is also recommended that the module be amended to include the usage of Sqoop to facilitate the transferring of bulk data between the Apache Hadoop ecosystem and the SQL Server data warehouse. Furthermore, provision could be made to create a link between the Hadoop Hive database and the SQL Server data warehouse. This would allow the students to query both environments in an integrated manner and build business intelligence artefacts such as dashboards or scorecards. The study has shown that it is important to expose data warehousing postgraduate students to emerging big data technologies.

The paper contributes by increasing our understanding of teaching big data and data warehousing. Firstly, the paper provided a brief overview of current data warehousing practices and emerging big data information processing technologies relevant to data warehousing. Secondly, the paper presented a practical approach to incorporate these technologies in a data warehousing module for postgraduate students by means of virtual machines, sample project and course resources. Thirdly, findings from the pilot study suggest that care should be taken to avoid cognitive overload of combining the theoretical concepts of data warehousing and big data within the same academic module.

Acknowledgements. Thanks to *Ian van der Linde* for having configured the cluster with Apache Hadoop and Apache Hive. Thanks also to the anonymous reviewers for their valuable comments and suggestions that have improved the quality of this paper.

References

1. ACM, IEEE: computer science curricula 2013: curriculum guidelines for undergraduate degree programs in computer science. Technical Report, ACM (2013)
2. Arnott, D., Dodson, G.: Decision support systems failure. Decis. Support Syst. **4**, 763–790 (2008)
3. Awadallah, A., Graham, D.: Hadoop and the data warehouse: when to use which. Technical Report, Cloudera (2012)
4. Botma, E., Kotzé, E.: Feasibility of a low-cost computing cluster in comparison to a high-performance computing cluster: a developing country perspective. In: Proceedings CONF-IRM 2016, p. 44, Cape Town (2016)
5. Capriolo, E., Wampler, D., Rutherglen, J.: Programing Hive. O'Reilly, Sebastopol (2012)
6. Cattell, R.: Scalable SQL and NoSQL data stores. ACM SIGMOD Rec. **39**(4), 12 (2011)
7. Chaudhuri, S., Dayal, U.: An overview of data warehousing and OLAP technology. ACM SIGMOD Rec. **26**(1), 65–74 (1997)
8. Chen, H., Chiang, R., Storey, V.: Business intelligence and analytics: from big data to big impact. MIS Q. **36**(4), 1165–1188 (2012)
9. Chen, M., Mao, S., Zhang, Y., Leung, V.C.: Big Data: Related Technologies, Challenges and Future Prospect. Springer, Heidelberg (2014)
10. Cloudera: Download QuickStarts for CDH 5.8. Technical Report, Cloudera (2017)
11. Dean, J., Ghemawat, S.: MapReduce: simplified data processing on large clusters. Commun. ACM **51**(1), 107 (2008)
12. Eckerson, W.: Big data analytics: profiling the use of analytical platforms in user organizations. Technical Report, TDWI (2011)
13. Gartner Consult: What is big data? – gartner IT glossary – Big Data. Technical Report (2016)
14. Ghemawat, S., Gobioff, H., Leung, S.T.: The Google file system. ACM SIGOPS Operating Syst. Rev. **37**(5), 29–43 (2003)
15. Hadoop: Hadoop documentation and open source release. Technical Report, Apache (2017)
16. Hortonworks: Hadoop sandbox – hortonworks. Technical Report (2017)

17. Howard, J.H., Kazar, M.L., Menees, S.G., Nichols, D.A., Satyanarayanan, M., Sidebotham, R.N., West, M.J.: Scale and performance in a distributed file system. ACM Trans. Comput. Syst. **6**(1), 51–81 (1988)
18. Hu, H., Wen, Y., Chua, T.S., Li, X.: Toward scalable systems for big data analytics: a technology tutorial. IEEE Access **2**, 652–687 (2014)
19. IBM: Hadoop Dev – Try it. https://developer.ibm.com/hadoop/try-it/
20. Inmon, W.H.: Building the Data Warehouse, 4th edn. Wiley, Hoboken (2005)
21. Intel: extract, transform, and load big data with apache Hadoop. Technical Report, Intel (2013)
22. Jukić, N., Sharma, A., Nestorov, S., Jukić, B.: Augmenting data warehouses with big data. Inf. Syst. Manag. **32**, 200–209 (2015)
23. Kimball, R., Caserta, J.: The Data Warehouse ETL Toolkit: Practical Techniques for Extracting,Cleaning, Conforming, and Delivering Data. Wiley, Hoboken (2004)
24. Kimball, R., Ross, M.: The Data Warehouse Toolkit: the Complete Guide to Dimensional Modelling, 2nd edn. Wiley, Hoboken (2002)
25. Kimball, R., Ross, M., Thornthwaite, W., Mundy, J., Becker, B.: The Data Warehouse Lifecycle Toolkit: Practical Techniques for Building Data Warehouse and Business Intelligence Systems, 2nd edn. Wiley, Hoboken (2008)
26. Kimball, R., Strehlo, K.: Why decision support fails and how to fix it. Datamation **40**(11), 40 (1994)
27. Kotzé, E.: A Survey of Data Scientists in South Africa. This volume: CCIS 730, (2017)
28. Kotzé, E.: An overview of big data and data science education at South African universities. Suid-Afrikaanse Tydskrif vir Natuurwetenskap en Tegnologie **35**(1) (2016). https://doi.org/10.4102/satnt.v35i1.1387
29. Krishnan, K.: Data Warehousing in the Age of Big Data. Elsevier/Morgan Kaufman, Amsterdam/Burlington (2013)
30. Lee, K.H., Lee, Y.J., Choi, H., Chung, Y.D., Moon, B.: Parallel data processing with MapReduce: a survey. ACM SIGMOD Record **40**(4), 11 (2011)
31. Loshin, D.: Big Data Analytics: From Strategic Planning to Enterprise Integration with Tools, Techniques, NoSQL, and Graph. Elsevier/Morgan Kaufman, Amsterdam/Burlington (2013)
32. Lunt, B.M., Ekstrom, J.J., Gorka, S., Hislop, G., Kamali, R., Lawson, E., le Blanc, R., Miller, J., Reichgelt, H.: Information technology 2008 curriculum guidelines for undergraduate degree programs in information technology. Technical Report, ACM and IEEE Computer Society (2008)
33. Lutu, P.: Big data and nosql databases: new opportunities for database systems curricula. In: SACLA 2015 Proceedings 44th Annual Southern African Computer Lecturers Association, pp. 204–209, Johannesburg (2015)
34. Mackey, A.L.: Incorporating big data technology into computing curriculum: conference tutorial. J. Comput. Sci. Coll. **31**(5), 38–39 (2016)
35. MapR: MapR Sandbox for Hadoop and MapR. https://www.mapr.com/products/mapr-sandbox-hadoop
36. Marshall, L., Eloff, J.H.P.: Towards an interdisciplinary master's degree programme in big data and data science: a south african perspective. CCIS **642**, 131–139 (2016)
37. Mills, R.J., Chudoba, K.M., Olsen, D.H.: IS programs responding to industry demands for data scientists: a comparison between 2011 and 2016. J. Inf. Syst. Edu. **27**(2), 131–141 (2016)
38. Ponniah, P.: Data Warehousing Fundamentals for IT Professionals, 2nd edn. Wiley, Hoboken (2010)

39. Power, D.J.: Understanding data-driven decision support systems. Inf. Syst. Manag. **25**, 149–154 (2008)
40. Project Gutenberg: Free eBooks by Project Gutenberg. https://www.gutenberg.org/
41. Provost, F., Fawcett, T.: Data science and its relationship to big data and data-driven decision making. Data Sci. Big Data **1**(1), 51–59 (2013)
42. Rahman, N., Iverson, S.: Big data business intelligence in bank risk analysis. Int. J. Bus. Intell. Res. **6**(2), 55–77 (2015)
43. Russom, P.: Evolving data warehouse architectures. Technical Report (2014)
44. Silva, Y.N., Dietrich, S.W., Reed, J.M., Tsosie, L.M.: Integrating big data into the computing curricula. In: SIGCSE 2014 Proceedings 45th ACM Technical Symposium on Computer Science Education, pp. 139–144 (2014)
45. Sharda, R., Delen, D., Turban, E.: Business Intelligence and Analytics: Systems for Decision Support, 10th edn. Pearson Education, Harlow (2014)
46. Thusoo, A., Sarma, J.S., Jain, N., Shao, Z., Chakka, P., Zhang, N., Antony, S., Liu, H., Murthy, R.: Hive – A petabyte scale data warehouse using Hadoop. In: Proceedings of International Conference on Data Engineering, pp. 996–1005 (2010)
47. Thusoo, A., Shao, Z., Anthony, S., Borthakur, D., Jain, N., Sen-Sarma, J., Murthy, R., Liu, H.: Artefact data warehousing and analytics infrastructure at facebook. In: Proceedings SIGMOD Conference, pp. 1013–1020, ACM (2010)
48. Topi, H., Valacich, J.S., Wright, R.T., Kaiser, K., Nunamaker, J.F., Sipior, J.C., de Vreede, G.J.: IS 2010: curriculum guidelines for undergraduate degree programs in information systems. Commun. Assoc. Inf. Syst. **26**(1), 359–428 (2010)
49. Trauth, E.M., Farwell, D.W., Lee, D.: The IS expectation gap: industry expectations versus academic preparation. MIS Q. **17**, 293 (1993)
50. Vaisman, A., Zimanyi, E.: Data Warehouse Systems Design and Implementation. Springer, Heidelberg (2014)
51. Vavilapalli, V.K., Seth, S., Saha, B., Curino, C., O'Malley, O., Radia, S., Reed, B., Baldeschwieler, E., Murthy, A.C., Douglas, C., Agarwal, S., Konar, M., Evans, R., Graves, T., Lowe, J., Shah, H.: Apache Hadoop YARN. In: SOCC 2013 Proceedings 4th Annual Symposium on Cloud Computing, pp. 1–16, ACM (2013)
52. Watson, H.J.: Tutorial: big data analytics: concepts, technologies, and applications. Commun. Assoc. Inf. Syst. **34**(1), 64 (2014)
53. Watson, H.J., Wixom, B.H.: The current state of business intelligence. IEEE Comput. **40**, 96–99 (2007)
54. White, T.: Hadoop: The Definitive Guide, 4th edn. O'Reilly, Sebastopol (2015)
55. Wixom, B.H., Ariyachandra, T., Douglas, D., Goul, M., Gupta, B., Iyer, L., Kulkarni, U., Mooney, B.J.G., Phillips-Wren, G., Turetken, O.: The current state of business intelligence in academia: the arrival of big data. Commun. Assoc. Info. Syst. **34**, 1–13 (2014)

Educational Cooperation with the ICT Industry

Information Systems as Creative Products: What Are Industry's Expectations?

Anine Kruger, Machdel Matthee[(⊠)], and Marita Turpin

Department of Informatics, University of Pretoria, Pretoria, South Africa
`machdel.matthee@up.ac.za`

Abstract. This paper presents the first step in exploring the match between IT managers' expectations of functional creativity within an information system (IS), and the functional creativity of the information systems developed by final-year undergraduate IS students. The Creative Product Assessment Model (CPAM) is used as a means to elicit the expectations that IT managers in various IT industry sectors have of functional creativity within information systems. Final-year IS student projects are evaluated for functional creativity by the same IT managers. Though IT managers value functional creativity in an information system, there are other creativity aspects considered even more valuable. These include the skills to design, as well as the end-user experience.

Keywords: Information systems · Functional creativity · Innovation · CPAM · IS students

1 Introduction

The Information Systems (IS) field requires the finding of practical solutions to complex problems. Creativity skills are generally acknowledged as an important prerequisite for solving complex problems [12,43]. In fact, creativity is included as a foundational skill in the IS2010 curriculum [39]. However, [36] pointed out a lack of research on creativity in IS, while [36] provides a valuable research framework which delineates the non-trivial matter of operationalizing the concept 'creativity'. Yet, despite this, creativity in IS still seems to be under-researched. As a result there is a limited formal knowledge base to draw from, and the term 'creativity' is often applied loosely and subjectively without a clear understanding of what is meant by it, and, from an educational context, what is required or expected from the students. In a previous study we investigated what is being done at eleven South African universities to develop the creative ability of undergraduate IS students [40]. We found that domain knowledge is regarded as more important for developing creative ability than creativity techniques, and that the nature of the problems presented to students is regarded as important for developing creative ability. Lecturers mentioned the importance of the problems being 'real-world' and authentic. The final-year 'capstone' project was singled out by respondents as a way to expose students to such problems, since it provides an

© Springer International Publishing AG 2017
J. Liebenberg and S. Gruner (Eds.): SACLA 2017, CCIS 730, pp. 147–161, 2017.
https://doi.org/10.1007/978-3-319-69670-6_10

ideal environment for solving real-life problems in a creative way [40]. All the universities that participated in our study provided such learning environments for IS students in one or the other way. The assumption is therefore that South African IS students are well prepared to be creative problem solvers when entering industry and hence universities produce graduates that can meet industry's needs. This paper focuses on investigating the plausibility of that assumption. We therefore need to ask the question: Do universities really deliver students with creativity skills that meet the expectations of the industry? In our attempt to address this question, we approached the problem in two phases:

1. Determining industry's expectations about creativity in the IT workplace;
2. Determining whether IS graduates meet these expectations.

Although the focus of this paper is on Phase 1, Phase 2 requires the evaluation of creativity. Industry members will have to use an evaluation framework according to which they evaluate aspects of creativity of IS students. We therefore decided to select a creativity evaluation framework as starting point for Phase 1.

This led us to the non-trivial matter of evaluating creativity. 'Creativity' can be defined in terms of 'the four P', namely: *Person, Process, Product* and *Press* (environment) [33]. Stated in terms of the four P, creativity is the *"interaction among aptitude, process, and environment by which an individual or group produces a perceptible product that is both novel and useful as defined within a social context"* [30]. How can this multi-faceted phenomenon be evaluated? Various instruments and approaches are known for assessing creativity. In an attempt to make sense of this assortment of approaches, [16] classified *criteria* for creativity evaluation into ten categories: tests of divergent thinking, attitude and interest inventories, personality inventories, biographical inventories, teacher nominations, peer nominations, supervisor rating, eminence, self-reported creative activities, as well as achievements and judgement of products. Most of the creativity evaluation mechanisms involve judgment from other people such as experts or peers [16]. In IS, most of the research on the evaluation of creativity focuses on the judgment of the creative product. In [25,36] the only creativity evaluation studies identified were those focusing on ideas or systems (both IS 'products'). There are good reasons for evaluating the creative product rather than the other aspects of creativity. The product is something tangible which reveals something of the creativity of its creator (who is more elusive to evaluate): *"Unambiguously creative products are constructed by unambiguously creative persons"* [20]. Some researchers believe that the evaluation of creative products should be the starting point for research on creativity and that this is perhaps the best way of evaluating creativity [4,18,22]. Still, research on the assessment of the creative IS product is scarce. There are *"surprisingly few studies aimed at assessing the creativity of products in the sense of tangible, scientific or technological products"* [10]. During 1998–2011, only 6% of creativity and IS-related research focused on the creative product [25]. The tangible nature of the product, as well as the lack of research on its evaluation in IS, influenced our decision to choose a framework focused on the evaluation of the creativity of the information systems (an IS product). The CPAM [4] was chosen for this purpose. Similar to [20] we believe that the creativity of the product reflects the creative ability of those that developed it.

Thus, the aim of our study was to explore the expectations of IT industry experts regarding the creativity of information systems. Representatives of four sectors within the IT industry were interviewed and the findings are presented. The Banking, Insurance, Telecoms and Software Development sectors were included in our study. In Phase 2 (*not* in the scope of this paper), IT industry experts shall evaluate the functional creativity of final year IS students' information systems. We believe that the findings will bring universities closer to a meaningful integration of creativity enhancing skills into the IS curriculum.

2   Literature

2.1   Functional Creativity and Information Systems

While the concept of 'novelty' is a necessary element of creativity in the domain of arts and æsthetic products, it does not necessarily play the same role when creativity is examined in a technological setting or when an engineering product is scrutinized [4,10]: here the creative idea or product must also be useful [2,8,10]. Accordingly, [26] defined creativity as the ability to produce both novel and useful ideas, or ideas that can be implemented practically to solve a meaningful and unique problem. In [9] the notions of 'novel' and 'useful' creativity were combined to into the notion of 'functional creativity'. Functional creativity of a product can be assessed when the observer or user of the product is aware of the initial problem so that the product can be appreciated as a functional creative solution, in other words: its usefulness can be assessed. This notion of functional creativity is applied to information systems as follows. An information system is a functional creative product that originates from a creative idea developed in response to a problem. The idea develops through a process of design and development in a specific social context, which is then expressed in a product that is both original and useful to industry. Products such as information systems, that perform tasks or solve problems, relate to a type of useful creativity or creativity with a goal [6,17]. If an information system can meet a customer's need, then it is not important whether the system is a completely novel solution to a problem or an existing system applied in a new manner [7]. In IS, novelty is concerned with the imaginative recombination of known elements [8]. The design and development of a functional creative product relies on the application of existing knowledge and skills in new ways to accomplish goals [37].

2.2   Evaluation Frameworks for Functional Creativity in IS Products

Related work on the evaluation of products in IS covers more than only the evaluation of information systems. Examples include studies that analysed the impact of creativity support systems or decision support systems on individual creativity [13, 24], the assessment of creative IS ideas [11, 21], and even the assessment of novel and useful software UML designs [14]. Some studies focus solely on one aspect of the creative information system, such as 'elegance' [23, 35] or 'usability' [15].

One of the first frameworks to measure creativity in information systems was introduced in [7]. The framework measured the 'utility' and 'novelty' criteria of a software product on a scale of 'low', 'medium', and 'high'. This framework was used by a panel of judges to assess six innovative software products to determine the products' contribution towards novelty (e.g., new technology, algorithms, etc.), economy (to increase return on investment, customer retention, retaining a market niche, etc.) [7]. The framework provides a starting point to look for creativity components in a software product. However, the focus of this framework is more on the creative product *after* its implementation, (whereas our eventual goal is to evaluate students' information systems design *before* the implementation phase). Over time a few rating scales have been developed to assess creativity in products, for example: the *Creative Product Inventory* scale [38], the *Consensual Assessment Technique* (CAT) [3], and the *Creative Solution Diagnosis Scale* (CSDS) to systematically assess the functional creative elements in engineering products [10].

The Creative Product Assessment Model (CPAM) is a comprehensive theoretical model used for the assessment of product creativity in general. This model is based on thirty years of empirical research and describes creativity in terms of 'novelty', 'resolution', and 'style' [4]. It can be considered as state-of-the-art [32]. Table 1 provides an overview of its component. The first component, novelty, refers to the newness of a product [4]. However, the idea or product does not have to be completely new to be novel. Sometimes a small change in an existing product can give a new fresh look and desirability to a product which in return also boosts its value. According to the framework, novelty includes the 'surprising' aspect as well as the 'original' aspect. A product might be highly original, but if the associated surprisingness is too high it might upset some users. The balance between these two factors needs to be considered for novelty to be appreciated by the end-user. 'Resolution' refers to the functionality of the working product [4]. Resolution contains the following four aspects: 'logical', 'useful', 'valuable', and 'understandable'. A new coffee machine (for example) can be very novel or original in design, but there is no value or usefulness in such a product if the end-user cannot grasp how to operate it. Usually a product originates from a problem that needs a solution. If the product's solution is not 'logical' or 'understandable' to the end-user, it affects the overall creativity and appreciation of the product. For example, a coffee machine *is expected to follow certain conventions* on how it receives instructions and dispenses coffee; otherwise it will not seem like a logical solution.[1] Also, if the user needs to read extensive user manuals before using the product, then it affects the product's understandability. The 'style' of a product refers to how well the product is presented. It speaks

---

[1] In other words, the coffee machine must have the status of a *device* [41].

Table 1. The CPAM according to [4,5]

Component	Description	Aspects
Novelty	The degree of newness in the product in terms of the number and extent of new materials, new concepts and new processes included	Surprising: The product presents unaccepted or unanticipated information to the user, listener or viewer Original: The product is unusual or infrequently seen in a universe of products made by people with similar experience and training
Resolution	The degree to which the product fits or meets the problematic situation	Logical: The product or solution follows the acceptable and understood rules of the discipline Useful: The product has practical applications Valuable: The product is judged worthy because it fills a financial, physical, social or psychological need Understandable: The product is communicated in a communicative, self-disclosing way, which is 'user-friendly'
Style	The degree to which the product combines unlike elements into a refined, developed and coherent whole, statement or unit, (how well the solution is presented to the world)	Organic: The product has a harmonious sense of wholeness or completeness to it. All the parts 'work well' together Well-crafted: The product has been worked and reworked with care to develop it to its highest possible level for this point in time, (quality) Elegant: The product shows a solution that is expressed in a refined, understated way, (simplicity)

about the product's 'personality' in relation to other products of its type, rather than how stylish it is. Style is understood in terms of three aspects: 'organic', 'well-crafted', and 'elegant'. The organic aspect of a product is concerned with the balance and harmony that exist among all its different parts [4]. It is all about a 'natural flow' in its appearance. The well-crafted aspect of the product is about how well each part of the product has been polished to yield a fine, finished result. The elegance of a product is all about the pleasure it brings to its end-user. It might be the colour of the product, how well it is packaged, or even just the ambience that a product creates while being beheld or used. Table 2 indicates how the CPAM compares with other creativity assessment techniques, in terms of focus as well as adoption measured in #citations (Google Scholar, at the time when we wrote this paper). A number of rating scales and theoretical models are thus available to assess the creativity of a product, of which some are appropriate for technical products. Of these, the CPAM is the most widely adopted. Hence we selected the CPAM for the purpose of our paper.

Table 2. Summary of various creativity assessment techniques

Method	[Ref.]	#Cit	Criteria	Comments
Creative product inventory	[38]	17	Generation, Reformulation, Originality, Relevance, Hedonics, Complexity, Condensation	Used for chemical products
Consen-sual assessment technique (CAT)	[1]	3	Not based on any theory of creativity. Judges use an agreed-upon consensual rating to assess products	Used for any creative product (e.g. poems, paintings, stories). No objective measurement; products are compared to each other
Creative product assessment model (CPAM)	[4]	179	Novelty, Resolution, Style	Products can be judged by expert and non-expert judges on a 55-item, bipolar adjective Liker scale. Tool collects information about consumer perception of existing products or services, or product concepts or protoyptes. Judges rate a product in three dimensions with high accuracy and consistence among various products of same type
Creative solution diagnosis scale (CSDS)	[10]	86	Relevance, Effectiveness, Novelty, Elegance, Genesis	Products can be judged by expert and non-expert judges on a 30-item, 5-point Likert scale to indicate the degree to which the CSDS item applies to the given product. Developed to measure functional creativity in engineered products, such as mouse traps, hands-free mobile phone holders etc.

3   Method

This section considers the approach used to answer the main research question of this paper, namely: *What are the expectations of industry experts regarding the functional creativity of information systems?* To this end, data were collected by semi-structured and open-ended interviews with several IT industry experts. The answers obtained from the experts were then interpreted [29, 42].

3.1   Data Collection

Data were collected through interviews with IT industry experts selected from four IT industry sectors where IS graduates often start their careers, namely:

Banking, Insurance, Software Development, and Telecommunications. It was initially presumed that the IT sectors may differ in their approaches to software development as well as their perceptions of creativity. Hence, a few prominent companies in each sector were identified, and their IT managers (or equivalent roles) were contacted for participation in our study. A requirement for inclusion was that the IT manager must have had more than five years experience as a manager in the specific sector and more than 10 years experience in the IS development field. It was assumed that the chosen IT managers would generally have expert knowledge in systems design and development for their industry sector, as well as knowledge about the company's viewpoints and strategies of software development. Of the experts who were contacted (and who met our requirements), nine were available to participate: four people from the Software Development sector, three from the Insurance field, one from Banking, and one from Telecommunications.[2] Apparently these industrial sectors were not equally represented, and the number of participants was too small to make sector-specific conclusions. Hence, the sectors that the respondents belonged to were captured during data collection and carried through to the data analysis merely for the sake of curiosity (to see from which sector a comment originated). Semi-structured face-to-face interviews were conducted with each IT manager separately, to individually determine their perceptions and expectations of 'functional creativity' in information systems. The CPAM was provided to each of the interviewees before the interviews commenced, so that they could familiarise themselves with our method. Interview questions were based on the CPAM. All interviews were recorded and transcribed, whereby the following questions stood in the centre of our interest:

Q1. *Is it important for an information system to be functionally creative? (motivate your answer).*

Q2. *How important or valid is each of the CPAM component in your industry? For example: how would you rate or weigh novelty, resolution and style according to their relevance for you?*

Q3. *Is there anything that you would like to add or omit from the CPAM?*

Q4. *Are there other elements in an information system (software product) that you value higher than functional creativity?*

3.2   Data Analysis

The interview answers were analysed and coded to identify any themes that relate to the perceived importance of functional creativity in information systems (Q1), comments and critique on the components of the CPAM (Q2), and other aspects [28] considered more important than functional creativity of information systems (Q3–4). Though the CPAM guided the analysis, themes unrelated to the CPAM also emerged. The written summaries were sent afterwards to each interviewee to confirm that they all agreed with the analysis [19]. This double-check helped to 'contain' any possible interpretation bias on our side.

---

[2] The mandatory procedures of gaining consent and protecting identities were followed; 'ethical clearance' was obtained.

4   Findings

The expectations about functional creativity in information systems have been discussed as described above. Below follows a discussion of the responses for each interview question. The sectors that respondents belong to are indicated for interest sake only. Apart from Q2, sector specific analyses are not provided (due to the small number of respondents per sector).

W.r.t. Q1, all IT managers responded that creativity is important in information systems development. One respondent mentioned the increasing importance of a creative system: *"Traditionally it was not important. The problem now is that you have customers who are beyond the point where they are not completely 'stupid' about the systems they want. They know what there is and they know what they can use. They do not want a system that is the same as a competitor's. They want something unique to their environment. Thus, creativity in IS today is very important".* However, another IT manager (Insurance sector) mentioned that sometimes creativity in information systems can have negative implications. This respondent mentioned that although a creative information systems is useful for innovation and moving forward, it can also be 'scary' if it becomes the standard in the organisation, and management starts to expect that same level of functional creativity in all information systems. The respondent believed that such a standard becomes very difficult to maintain because functional creativity is *"special and rare".* These remarks from IT managers come close to some of the concerns already expressed in 1992.

W.r.t. Q2, Table 3 provides a summary of the responses. During the interviews the respondents were requested to prioritise each CPAM component (novelty, resolution and style) as '1' (highest), '2', or '3' (lowest) for their specific industrial sector. A percentage weight was provided by each respondent for a finer indication of each CPAM component's importance (whith novelty% + resolution% + style% = 100%).

The responses are shown per industry sector. While there are not enough data to attach value to the variations between industrial sectors, it is interesting to note the consensus between the respondents of the different sectors: all respondents agreed that 'resolution' has the overwhelmingly highest priority (up to 80%), with 'style' second and 'novelty' third. This result corresponds with previous research findings [2,4,8,9] concerning the importance of the 'resolution' component in functional creativity which is generally most highly valued among IT industry experts; novelty is the lowest when IS-related products are considered.

The interviewed managers all agreed that on most IS projects the systems are already in place and that analysts basically follow a recipe to create the next version of a systems solution. It is only on rare occasions that there is the opportunity to be really creative and to invent a truly novel product. Most interviewees agreed that whenever they have the opportunity to bring novelty into the product, their first priority is to be original rather than surprising. This result also corresponds with previous research [4], which stated that the

Table 3. Importance rating assigned to CPAM components by interviewees

Component	Industry			
	Telecom	Banking	Insurance	Software eng.
	#Respond			
	1	1	3	4
Novelty: Priority importance	**3** (low) 5%	**3** (low) 5%	**3** (low) 5%–10%	**3** (low) 5%–10%
Resolution: Priority importance	**1** (high) 80%	**1** (high) 80%	**1** (high) 75%–80%	**1** (high) 50%–80%
Style: Priority importance	**2** (medium) 15%	**2** (medium) 15%	**2** (medium) 15%	**2** (medium) 15%–45%

surprising aspect of any creative product can shock people or consumers of the product so that they would rather avoid it.[3]

The 'style' element (15%) was rated by all respondents as second-most important, after resolution. Style was said to relate to the end-user in a visual way and to create the 'first impression' about an information system. A respondent from the Software Development sector stated that style is what sells an information system to the end-users even if they discover at a later stage that the resolution of the system is lacking or not of the required standard.

W.r.t. Q3, there were no IT managers who indicated that they wanted to remove components from the CPAM. In fact, when the interviewees were presented with the CPAM, some of them were pleasantly surprised and found this model very interesting because they had not considered creativity in this way before. Interviewees from the Software Development sector, however, mentioned some aspects which might be considered for further elaboration of the CPAM:

User Experience (for both customer and developer): Two interviewees mentioned that user experience should be added to CPAM as a sub-component of 'style', or should perhaps even constitute a new category.

System Integration: One interviewee mentioned that CPAM lacks a component for 'integration'. The interviewee said that if a newly developed information systems cannot be integrated with other existing information systems, then it impacts functional creativity negatively.

Affordability and profitability: If a system is not affordable for the client and profitable for the organisation, then being creative is irrelevant.

W.r.t. Q4, all interviewed IT managers indicated that while functional creativity in an information systems is important, there are other aspects or elements that are of equivalent value or even more valuable than a functional creative system. These other aspects range from the people and development process

---

[3] In *engineering*, mandatory *standards* also play an important role.

Table 4. Important aspects other than functional creativity in IS

Aspect	Examples
Process	Agile methodology, system maintenance, change management, requirements gathering process
People	Attitude of IS personnel, expert skills and knowledge, overcoming obstacles, resourcefulness, people who value quality
Environment (press)	Management style should foster creativity, freedom and trust, contemplation, time to think

aspects of the system to the environment in which the creative information systems are being developed. They can be categorised into the remaining of the aforementioned '4P' of creativity. Table 4 summarises our findings.

W.r.t. People, the aspect that the Banking sector interviewee valued most important was the IS employees who need to develop the system. These employees needed to value quality: if a person values quality within everything they do, then it will transcend to everything else they do; including developing creative information systems. He added that 'happy' employees create more creative end products. Respondents from the Insurance sector appeared to share the views of the IT manager from the Banking sector. They mentioned the importance of attitude of IS personnel. IS personnel's lack of positive attitude towards any challenge or project can stand in the way of developing a creative system. The interviewee from the Telecommunications sector mentioned the importance of staff's resourcefulness: *"How well does an engineer or software developer use what he has at hand to solve a problem? For me, that is resourcefulness — how he uses the building blocks around him to the best of his ability. This leads to the elegance of a solution. You can either build a solution that costs a lot because you buy new components, or you can deliver a solution that makes use of what you already have within the company and deliver it under budget. This is a more elegant and efficient solution".* Interviewees from the Software Development sector mentioned the importance of the skills and knowledge (both business and technical) of employees. It was stated that the challenge is not the development of a good creative information system, but rather the recruitment of good quality IS experts: if the right experts can be recruited, they will design the creative system that is required. All in all, the following people-related traits were mentioned as important: people who value quality, attitude of IS personnel, resourcefulness of staff, as well as skills and knowledge.

W.r.t. Process, the importance of the system development process was mentioned by interviewees from the Insurance as well as the Software Development sectors. There are different 'philosophies' and methodologies in information system development. These include 'Waterfall' (such as embedded in the SDLC) and 'Agile' (such as Extreme Programming and Scrum). According to the interviewees' responses the choice in process, SDLC versus Agile, plays a large role in the system being successful or unsuccessful. An iterative design process or agile methodology contributes towards a successful creative system that serves the needs and expectations of the end-user(s). An interviewee from the Software Development sector was of the opinion that a company cannot develop creative systems by continuously using the SDLC process. According to an interviewee from the Insurance sector, change management processes should not be neglected. Change management helps to prepare the system users for the novelty aspect of the system. If this process is not in place, then system users easily reject the entire information system. The crucial role played by the requirements gathering process was also mentioned by a respondent from the Insurance sector. He said that the problem statement for the product development should be thoroughly and regularly interrogated rather than only accepted to perform the required actions; see [31] for comparison. Without critical questions there cannot be a good quality creative product. The respondent further stated that IS personnel quickly develop a 'tunnel vision' about product development because they do not ask enough questions. This finding re-confirms the established knowledge [27, 31, 34] that the *requirements engineering* phase is the *most critical* phase in any Systems Development Life Cycle (SDLC). An interviewee from the Insurance sector added the aspect of system maintenance: If a creative information system is not well maintained after implementation, it becomes useless; then creativity means nothing and has no value for the organisation. All in all, the following process aspects were mentioned to be important: the system development process, change management, requirements gathering, and system maintenance.

W.r.t. Environment (Press), the importance of the environment for the development of a creative product was only mentioned by one respondent from the Software Development sector. It was stated that if a manager cannot help the IS employees to be productive and give them the trust and freedom they need, systems analysts will not perform as expected, and trust is broken. Several interviewees (from the Software Development, Telecommunications and Banking sectors) mentioned the importance of reflecting and taking time to think about the problem to be solved. According to them it is important to have sufficient time in order to be creative: it is difficult to invent a creative product or solution if there is no time to deeply think or 'day-dream' about it; see [33] for comparison. The consensus view in our interviews is that a good functional creative product implies that ample time was invested in understanding the problem and planning the system.

5   Discussion

We determined that IT industry respondents from all sectors valued functional creativity in information systems, which re-confirmed previously established findings [8]. All respondents valued *novelty*, *resolution* and *style* as the three main components of functional creativity, with *resolution* being the most important. We found that with tight budgets and timeframes the *resolution* aspect (logical, useful, valuable, understandable) becomes the most important one in the functional creativity framework. *Style* and *novelty* are accordingly regarded as 'nice to-have'. These results correlate with the literature [8,9].

A surprising finding of our study was that the IT industry respondents valued other creativity factors or elements more so than a functionally creative information system. Before the data was collected it was presumed that of the '4P' of creativity (Person, Process, Product, Press) the functionally creative information systems were the most valuable. The product, being the tangible result of the creative process, is what brings the financial boost or competitive advantage. However, the other three 'P' were found to be just as important, and in some cases even more important, than the functionally creative product itself, according to industry. For example, people aspects that were mentioned to be important are the traits of the person who develops the product, such as a love for quality and a can-do attitude, understanding the problem statement and reflecting on this problem before embarking on inventing the product solution. Process aspects mentioned as important included the user experience of the final creative information system, the maintenance and management aspects of the information systems when introducing it to the end-user, as well as the process and methodology used to develop the creative information system. A process aspect emphasised by respondents from all IT sectors was their belief that an agile methodology allows for better creativity in a creative information system than the typical SDLC methodology.

6   Conclusion

Creativity is a multi-dimensional concept which makes its operationalization and measurement a non-trivial matter. As IS educators we believe that the topic is important enough to be investigated, and started a number of such investigations. The topic of creativity in IS is in need of more research. The aim of this study was to investigate the expectations of industry regarding the functional creativity of an information system as IS product. The literature review recapitulated the evaluation of creativity in IS. Functional creativity was introduced as an appropriate measure, since it focuses on a tangible artefact and considers novelty as well as usefulness. In addition, functional creativity also alludes to the creativity of the developer behind the information system. In a search for instruments to evaluate functional creativity in IS, the CPAM [4] was selected with its components of *novelty, resolution* and *style*.

Information was retrieved from interviews with nine professionals out of four different sectors in the IT industry to understand their expectations of functional creativity in information systems, and also to determine whether there were other aspects more important to them than functional creativity. Our participants from the IT industry valued good, functional creative information systems according to the CPAM [4]. However, the interviews also revealed that it is not good enough to only examine functional creativity when a typical information system product is analysed. The respondents explained that they are keenly appreciative of behind-the-scenes work. Industry professionals want to know more about the traits of the person who went through a process within a specific setting to develop the functional creative information systems. This shows that it is difficult to isolate the creative product from a creative process and from the creative person who developed it. It was also mentioned that user-experience and product maintenance after implementation affect the value and appreciation of a creative product.

A *limitation* of our study is that it involved only nine representatives from the IT industry. No doubt the involvement of more experts would have revealed more or even contradictory statements. One implication of this study (w.r.t. its subsequent phase) is the limitation of evaluating information systems (in our case: 'capstone projects' of IS students) that are not actually implemented. From the interviews it is clear that the evaluation of the functional creativity of such systems is incomplete and will not give the full picture. This might imply adding aspects from existing creativity assessment instruments focusing on the creativity of individuals and/or groups. *Future work* should thus include representation from more IT industry sectors, as well as more representatives (IT managers) from each IT sector, to ascertain whether there are differences in their expectations and perceptions of functional creativity in IS.

Our paper makes a number of contributions. It contributes towards an understanding of what the IT industry expects and appreciates about functional creativity in information systems; as far as we know this is the first time that the CPAM is used in an interpretive study in IS. Our paper makes suggestions for expanding the CPAM to increase its relevance for application in IS. We also contribute to a better understanding for both industry and universities of what is expected from IS students. This may subsequently lead to further fostering creativity in IS education. Consequently we *recommended* that universities that offer IS degrees evaluate their curricula to ensure that efforts are made to prepare the students better for the reality of the IT industry. If it is true that the IT industry appreciates other factors more than the creative end-product, then universities should help students to cultivate the necessary qualities. For example: is there enough focus within the curriculum to develop the necessary design skills to help students better understand the user experience of a creative information system? Does the information systems design methodology allow for sufficient life cycle iterations to help the students to focus early on potential errors and to incorporate new ideas with client feedback [31]? Perhaps more companies need to get involved on a practical level with student projects at academic institutions to help students to prepare themselves better for their future careers as inventors of creative system solutions. Despite the shortcomings in evaluating the creativity of systems that are not yet in their implementation phase (e.g., IS student's 'capstone projects'), we believe that the subsequent phase of our study will reveal whether some of our suggestions for teaching are already practiced by IS educators. This, and additional suggestions, will hopefully transpire from the evaluation by industry members of the creativity of the capstone projects of IS students; (forthcoming).

References

1. Amabile, T.M.: A Consensual Assessment Technique. J. Pers. Soc. Psychol. **43**, 997–1013 (1982)
2. Amabile, T.M.: Creativity in Context: Update to the Social Psychology of Creativity. Westview Press, Boulder (1996)
3. Amabile, T.M.: The social psychology of creativity: a componential conceptualization. J. Pers. Soc. Psychol. **45**(2), 357 (1983)
4. Besemer, S.P.: Creating Products in the Age of Design: How to Improve Your New Product Ideas! New Forum Press, Stillwater (2006)
5. Besemer, S.P., O'Quin, K.: Analyzing creative products: refinement and test of a judging instrument. J. Creative Behav. **20**(2), 115–126 (1986)
6. Burghardt, M.D.: Introduction to the Engineering Profession, 2nd edn. Addison-Wesley, Boston (1995)
7. Couger, J.D., Dengate, G.: Measurement of creativity of IS products. In: Proceedings 25th Hawaii International Conference on System Sciences, pp. 288–298. IEEE (1992)
8. Couger, J.D., Higgins, L.F., McIntyre, S.C.: (Un)structured creativity in information systems organizations. MIS Q. **17**, 375–397 (1993)
9. Cropley, D., Cropley, A.: Elements of a Universal Aesthetic of creativity. Psychol. Aesthetics, Creativity Arts **2**(3), 155 (2008)
10. Cropley, D.H., Kaufman, J.C., Cropley, A.J.: Measuring creativity for innovation management. J. Technol. Manag. Innov. **6**(3), 13–30 (2011)
11. Dean, D.L., Hender, J.M., Rodgers, T.L., Santanen, E.: Identifying good ideas: constructs and scales for idea evaluation. J. Assoc. Inf. Syst. **7**(10), 646–699 (2006)
12. Dohan, S., Stapleton, L., Stack, A.: Problem solving skills in information systems development curricula. In: AISHE Proceedings All Ireland Society for Higher Education Conference (2000)
13. Elam, J.J., Mead, M.: Can software influence creativity? Inf. Syst. Res. **1**(1), 1–22 (1990)
14. Gomes, P., Seco, N., Pereira, F.C., Paiva, P., Carreiro, P., Ferreira, J.L., Bento, C.: The importance of retrieval in creative design analogies. Knowl.-Based Syst. **19**(7), 480–488 (2006)
15. Han, S.H., Yun, M.H., Kim, K.J., Kwahk, J.: Evaluation of product usability: development and validation of usability dimensions and design elements based on empirical models. Int. J. Ind. Ergon. **26**(4), 477–488 (2000)
16. Hocevar, D.: Measurement of creativity: review and critique. J. Pers. Assess. **45**(5), 450–464 (1981)
17. Horenstein, M.N.: Design Concepts for Engineers, 2nd edn. Prentice-Hall, Upper Saddle River (2002)
18. Horn, D., Salvendy, G.: Measuring consumer perception of product creativity: impact on satisfaction and purchasability. Hum. Factors Ergon. Manufact. Serv. Industr. **19**(3), 223–240 (2009)
19. Hycner, R.H.: Some guidelines for the phenomenological analysis of interview data. Hum. Stud. **8**(3), 279–303 (1985)
20. Kozbelt, A., Beghetto, R.A., Runco, M.A.: Theories of creativity. In: Cambridge Handbook of Creativity, vol. 20, p. 47 (2010)
21. Lobert, B.M., Dologite, D.G.: Measuring creativity of information system ideas: an exploratory investigation. In: Proceedings 27th Hawaii International Conference on System Sciences (1994)

22. MacKinnon, D.W.: Some critical issues for future research in creativity. In: Isaksen, S.G. (ed.) Frontiers of Creativity Research: Beyond the Basics, pp. 120–130. Bearly Limited, Buffalo (1987)
23. Madni, A.M.: Elegant systems design: creative fusion of simplicity and power. Syst. Eng. **15**(3), 347–354 (2012)
24. Massetti, B.: An empirical examination of the value of creativity support systems on idea generation. MIS Q. **20**, 83–97 (1996)
25. Müller, S.D., Ulrich, F.: Creativity and information systems in a hypercompetitive environment: a literature review. Comm. Assoc. Inf. Syst. **32**, 175–201 (2013)
26. Mumford, M.D., Gustafson, S.B.: Creativity syndrome: integration, application, and innovation. Psychol. Bull. **103**(1), 27 (1988)
27. Nguyen, L., Shanks, G.: A framework for understanding creativity in requirements engineering. Inf. Softw. Technol. **51**(3), 655–662 (2009)
28. Oates, B.J.: Researching Information Systems and Computing. SAGE, Thousand Oaks (2005)
29. Orlikowski, W.J., Baroudi, J.J.: Studying information technology in organizations: research approaches and assumptions. Inf. Syst. Res. **2**(1), 1–28 (1991)
30. Plucker, J.A., Beghetto, R.A., Dow, G.T.: Why isn't creativity more important to educational psychologists? Potentials, pitfalls, and future directions in creativity research. Educ. Psychol. **39**(2), 83–96 (2004)
31. Pohl, K.: Requirements Engineering Fundamentals Principles and Techniques. Springer, Heidelberg (2010)
32. Pritzker, S.R., Runco, M.A.: The creative product assessment model. In: Encyclopedia of Creativity, vol. 2. Academic Press, Cambridge (1999)
33. Rhodes, M.: An analysis of creativity. Phi Delta Kappa **42**(7), 305–310 (1961)
34. Robertson, J.: Requirements analysts must also be inventors. IEEE Softw. **22**(1), 48 (2005)
35. Salado, A., Nilchiani, R.: Using Maslow's hierarchy of needs to define elegance in system architecture. Proc. Comput. Sci. **16**, 927–936 (2013)
36. Seidel, S., Müller-Wienbergen, F., Becker, J.: The concept of creativity in the information systems discipline: past, present, and prospects. Comm. Assoc. Inf. Syst. **27**(1), 217–242 (2010)
37. Seltzer, K., Bentley, T.: The Creative Age: Knowledge and Skills for the New Economy. Demos, New York City (1999)
38. Taylor, I.A.: An emerging view of creative actions. In: Taylor, I.A., Getzels, J.W. (eds.) Perspectives in Creativity, Transaction, pp. 297–325. New Brunswick (1975)
39. Topi, H., Valacich, J.S., Wright, R.T., Kaiser, K.M., Nunamaker Jr., J.F., Sipior, J.C., de Vreede, J.G.: IS 2010: curriculum guidelines for undergraduate degree programs in information systems. Comm. Assoc. Inf. Syst. **26**(1), 18 (2010)
40. Turpin, M., Matthee, M., Kruger, A.: The teaching of creativity in information systems programmes at South African higher education institutions. Afr. J. Res. Math. Sci. Technol. Educ. **19**(3), 278–288 (2015)
41. Vincenti, W.: What Engineers Know and How they Know it: Analytical Studies from Aeronautical History. John Hopkins University Press, Baltimore (1990)
42. Walsham, G.: Interpreting Information Systems in Organizations. Wiley, Hoboken (1993)
43. Wolmarans, N., Collier-Reed, B.I.: Problem-solving discourse models: informing an introductory engineering course. Afr. J. Res. Math. Sci. Technol. Educ. **14**(2), 28–41 (2010)

A Critical Systems Perspective on Project-Based Learning: Guidelines for Using Industry Data for BI Student Projects

Roelien Goede(✉)

North-West University, Potchefstroom, South Africa
roelien.goede@nwu.ac.za

Abstract. Higher education institutions provide the IT industry with professionals who are good problem solvers and self-directed life-long learners with a firm theoretical knowledge base. To achieve this, project-based learning is included in the curriculum. Data and requirements are supplied by an industry partner to 4th-year students in business intelligence. The aim of the collaboration is to develop the students to the best of their ability, while also generating benefits to the faculty members and the industry partner. The goal of this paper is to develop guidelines for using industry data for business intelligence students. The guidelines shall support the interests of the students in terms of real-world experience, faculty members in terms academic results, and industry partners in terms of information protection.

Keywords: Business intelligence · Critical systems thinking · Computer science education · Project-based learning · Industry involvement

Preliminary Remark

Large amounts of detailed interview responses, which cannot all be reproduced in this paper due to page space limits, were collected in this study. Readers interested in obtaining further details may contact the author via the address of above.

1   Introduction

The aim of this paper, which continues a sequence of previous work [14,15,27, 33,35], is to create guidelines for projects to be used by *business intelligence* (BI) students, based on a critical reflection on project-based learning, and perceptions of stakeholders. The reflection is guided by *critical systems thinking*, which is a strand of *systems thinking* to understand a problem in context of a larger perspective [1]. *Critical* systems thinking is 'critical' in the sense that systems thinking is supported by the metaphysical assumptions stemming from the School of Frankfurt's critical social theory [19]. It aims to identify underlying assumptions that guide actions in order to understand different perspectives of

© Springer International Publishing AG 2017
J. Liebenberg and S. Gruner (Eds.): SACLA 2017, CCIS 730, pp. 162–174, 2017.
https://doi.org/10.1007/978-3-319-69670-6_11

stakeholders in order to achieve emancipation [34]. It is used in this paper to identify the different perspectives on project-based learning, and to understand different perspectives of various stakeholders regarding the use of industry data in BI student projects. The development of a set of guidelines for student projects in BI strives to achieve the emancipative goals of critical systems thinking.

4th-year education ('honours' level) in South Africa aims at developing students into 'life-long learners' and at providing students with 'integrated knowledge' of their chosen discipline.[1] These aims promote self-directed learning which is supported by project-based learning where students take control of their learning environment by completing a realistic project to learn the concepts required to complete the project [16].

BI is a growing field in the information technology (IT) industry worldwide and also in South Africa. BI is a generic term referring to the strategic use of information available to organisations in support of business [22]. The information available in an organisation is typically stored in a data warehouse [17].

Project-based learning for 4th-year IT students requires well-designed and sufficiently realistic projects. However, various stakeholders, including the faculty members, industry partners, and students have different expectations of the role of the industry to provide data and specifications for those projects. The above-mentioned guidelines are developed after a critical review of project-based learning and BI (including data warehousing) literature. Some initial guidelines are then enhanced on the basis of experiences from a current student project in BI where unanticipated difficulties occurred. Further reflection improves the guidelines derived from literature. An interpretive study critically reflected on the assumptions and expectations of industry and faculty members regarding the use of industry data for student projects. The results informed the further development of those guidelines.

The remainder of this paper is structured as follows. A short overview of critical systems thinking is given in Sect. 2. Basic principles of project-based learning are briefly recapitulated in Sect. 3. The required educational outcomes for 4th-year BI students are sketched in Sect. 4. Previous experiences are sketched Sect. 5. New findings from a stakeholder survey are outlined in Sect. 6, followed by reflections in Sect. 7 and conclusions in Sect. 8.

2   Background: Critical Systems Thinking

Critical systems thinking is used in this paper to reflect on different stakeholder perspectives of project-based learning. The subsequent recapitulation of critical systems thinking begins with a brief introduction to *systems thinking*, followed by *critical* systems thinking, and, finally, critical systems *heuristics*.

---

[1] For readers from outside South Africa: the South African 'honours' degree is an extension of the classical 'B.Sc.' degree which enables a student to commence with Master-studies thereafter. While already considered 'postgraduate' in South Africa, the 'honours' degree in South Africa is reasonably well comparable to the final study-year in the (longer) U.S.American 'B.Sc.' curriculum.

Systems thinking strives to understand problems in their wider contexts by viewing a situation as a set of related components [1]. Over time, different strands of systems thinking developed about how systems are perceived [6,7]. Critical system thinkers argue that hard systems thinkers are not able to model all complexities in the problem environment [19]. From an ontological perspective, critical systems thinkers view reality as socially constructed and subject to power imbalance [19]. From this perspective, [20] identified five commitments of critical systems thinking: critical awareness (critical self-reflection); social awareness (wider perspective); methodological pluralism (openness to methods from different paradigms); theoretical pluralism (fostering of understanding of assumptions guiding theories); and 'emancipation' which is achieved where participants can reach the maximum of their own potential [19].

Critical systems heuristics (CSH) [34] provides a method to reflect on assumptions from different perspectives [19]. CSH also provides structure to investigations into the assumptions of different stakeholders' perspectives in a given problem situation. The characteristics of a 'system' are described in [8]:

– Total system objectives: measurable, key features of the system that may not be compromised;
– Environment: phenomena that influence the operation of the system which cannot be controlled by the system itself;
– Resources: anything required by the system to reach the stated objective;
– Components: independently describable sub-systems that work together to reach the total system's objective.
– Management: Coordination of sub-systems' use of available resources within the constraints resulting from the environment in order to reach the total system's objective.

In developing CSH, [34] extended these characteristics of [8], whereby 12 questions must be asked to identify a stakeholder's assumptions—in other words: a specific conditioned view of the problem environment. Thereby the first 8 questions consider the 'involved', i.e.: those who are part of the process of investigation, and the remaining questions consider the 'affected', i.e.: those who are influenced by the decisions of experts who are not themselves part of the process. Students, for example, are affected by collaboration between faculty members and the industry without being part of the initial setup.

For this paper, the CSH questions in the normative mode ('ought to') were used in conjunction with principles of project-based learning and the stated module-outcomes to develop interview questions to understand the views of the stakeholders on the use of industry data for BI student projects.

3   Background: Project-Based Learning

The principles of project-based learning (PBL) align with [11] on 'learning by doing'. The application of PBL puts the learner into the centre of the learning process [12]. In this paper we support the definition of PBL by [2,16]: PBL is a method where the solution of a problem is involved; problem-solving is initiated by students and requires a variety of activities; the result is an accomplished project; it involves working over a longer period of time; the role of the teacher is that of a 'facilitator'. Key to PBL is 'reflection in action and on action' [3, 10, 31]. Reflective practice is a cycle that typical starts with an experience with unanticipated problematic consequences, followed by a reflection on or abstraction of the events, which leads to a reconceptualization (new strategy), and finally experimentation [29]. In the reflection stage, the reflective practitioner tries to understand the problematic events. If learners still cannot understand their experience they are motivated to extend their knowledge [29]. The learners' reflective skills are thus developed [3].

Literature on PBL (also in combination with 'organisational learning') is abundant—for a few examples see [3, 11, 21, 30, 31]—and the concept of PBL with all its advantages and disadvantaged can be considered as generally well-known. However, sustainable learning and new organisational knowledge are not automatic consequences of doing projects in an organisation, because short term pressure can prevent the required cycle of reflective practice [21]. This is not only a danger in a commercial organisational setting but also in an academic environment where we want to use PBL for our BI module. For PBL in higher education (especially in our 'digital age') see [4,5]. Moreover, if we agree with [9] that a facilitator in higher education should design the problem to be solved in PBL in such a way that all the desired academic outcomes are achieved, then it becomes difficult to involve the industry (outside academia) and industrial data in such projects.

4   Background: Business Intelligence

BI is an umbrella term for all activities in an organisation aimed at making better strategic decisions based on available information [23]. Business-relevant data from different internal and external sources are combined in *data warehouses* to support various end-user applications [23] in management, marketing, sales, etc. Traditionally there are two different approaches for data warehouse design: the data-driven approach [18] and the requirements-driven approach [23]. The data-driven approach simply aims to store all of an organisation's data in one central store. In the requirements-driven approach, organisational needs are formulated on the basis of key performance indicators (KPI) such as to make the additional value of a BI system (or data warehouse) for its organisation measurable [22]. Several techniques are known for the implementation of the various types of data warehouses in BI systems [22–24, 28].

The BI 4th-year module discussed in this paper has a strong focus on data warehousing as it forms part of an honours degree programme in which students can enrol for modules in artificial intelligence, decision support systems, project management, and computer security; for comparison see [26]. Advanced database management is a prerequisite. In this BI module, the students shall gain

- good understanding of the elements of the BI lifecycle [23];
- good data modelling skills in current modelling techniques;
- practical experience in developing a data warehouse [24].

A project is the module's main vehicle for learning whereby students cooperate in pairs guided by a lecturer as 'facilitator'. In [33] the need for additional industry partners was identified. To make such industry involvement academically feasible, a number of conditions must be met: see Table 1.

Table 1. Guidelines from literature for good industry data for BI students

#	Guideline	Motivation and BI/DW/PBL concepts
1	The requirements must be strategic in nature and not answerable from one OLTP system	Students struggle to distinguish between OLTP and BI decisions
2	The provider of the data together with the facilitator should be able to compile a list of specific requirements from the data	Specific requirements are easier to assess consistently amongst the different collaboration teams in the class group
3	More than one OLTP data set should be integrated	Data integration; data quality; overall understanding of BI/DW in relation to OLTP
4	Data quality problems must be present but limited	Students' time management; data quality correction; data stewardship
5	The given data set should be more than a million transactional records but less than 5 million	Students should experience the problems of larger data sets but do not have the infrastructure to handle too many records
6	The industry represented by the data should be known to the students	Prevent too much effort in the understanding of the data compared to the understanding of the BI learning concepts
7	Opportunities should be created for students to share their knowledge	The collaborative nature and reflection on new knowledge facilitates deep learning

5   Experiences from a Previous BI Student Project

For a previous BI student project a data set about electricity consumption of supermarkets was provided by a local company. Available weather data were used as a second data set to facilitate the retrieval of correlations between weather and electricity consumption. Another external data source, containing fixed values for emission statistics, was also provided. Thus equipped the students were told:

- Create a list of supermarkets where the consumption of a specified division in relation to the total consumption has a deviation of more than 10% from the average.

Table 2. Evaluation of data provided by the industry to BI students

#	Issue, with reference to Table 1
1	This requirement is debatable, but since requirements focused on trends and comparison with averages it can be argued the requirements are strategic rather than operational. Different source data were required to support query output, which gave it a BI rather than OLTP nature
2	Industry partner and facilitator were able to compile a set of requirements
3	Temperature and emissions data was combined with energy consumption data
4	The grain of the temperature data was different (hourly) compared to the half-hourly consumption data, which enabled integration programming
5	The given data set was more than a million records and supplied data per supermarket per line, thus the students had to create more data lines to facilitate atomic fact table modelling
6	The university has a moral responsibility to extend the frame of reference of students to include energy consumption. Students had difficulty at first with the units of the measurements but were able to use it eventually
7	The given data set enabled student reflection and collaboration

- Create a list of supermarkets and calendar dates where the relative consumption in a division, influenced by the temperature, is unexpectedly high; (propose parameter values).
- Create your own realistic requirement from the data and produce the result.

Soon after the beginning of this project the course's facilitator became aware of some shortcomings in the data provided for PBL. The identified issues are listed in Table 2 wherein the guideline numbers # refer back to Table 1.

In this educational scenario, the main shortcoming was the lack of *descriptive* data—the data set provided only contained numbers from consumption measuring devices. *Dimension design* in dimensional data modelling is guided by the answers to the 'W' questions: when, where, why, who, and what? In the student project described here, the provided data only supported 'where' and 'when' (and, perhaps, 'what', depending on the 'grain' of the facts table design). Thus the grain of the fact table also turned out to be problematic, i.e.: not well aligned with the table design guidelines of [22]. Consequently the students had to be advised to work with the the atomic fact table grain (in order to facilitate learning), although it is not an advisable solution. Yet another problem was that the additional data sets (temperature and emissions constants) only impacted the fact table and not the dimension tables. Students were not confronted with realistic changes in the dimensional data (slowly changing dimensions) which is a key concept in dimensional modelling. Last but not least: Data vault modelling entails the removal of descriptive data into satellite tables; however, since there were almost no descriptive data provided, this scenario could not serve to teach the students the basic concepts of data vault modelling. Consequently, another dataset had to be used. Alas, the students were frustrated by the additional work thus required. Although the additional work facilitated learning, the students would have preferred the simpler implementation of the 'per store' grain. From this experience it seems clear that the specific nature of the data has an impact on the students' learning experience.

Under the impression of those educational experiences, *three additional guidelines* (#8–#10) were stipulated: see Table 3. Because this amendment of the guidelines cannot be implemented without the cooperation of the data-providing industry partners, their opinions had to be taken into account: see below.

Table 3. Refined guidelines after the previous PBL experience

#	Additional guideline	Motivation and BI/DW/PBL concepts
8	The given data set should contain a good variety of descriptive textual fields and numeric fields associated with events	Rich dimensional- and data vault models
9	A second set of event (transaction) data that requires changes to initial dimensional data should be available	Dimensional design concepts of slowly changing dimensions. Implementation of differences between dimensional- and data vault modelling
10	The integration of different sources must be on dimension-level	Deduplication and attribute transformations such as the formatting of name fields from different sources that must be integrated

6 Stakeholder Perspectives from Interviews

Information gathering and analysis were guided by the principles for *interpretive case studies* in information systems [25]. Further methodological considerations [13,32] were taken into account as well. The interview participants represented three roles in this problem context: educators, alumni with several years of experience in the BI industry, BI and IT managers as possible providers of industry data. In total, eleven practitioners (P1–P11) participated. The main activity of P1–P3 is lecturing BI modules in higher education, whereby P2 and P3 also have BI industry experience. P1 is an expert in educational models in IT higher education. P4 and P8–P11 are experienced IT/BI practitioners with data-oriented professional roles. P5–P7 are alumni of the NWU with knowledge of our curriculum. All interviewees provided their answers to a questionnaire in writing. In accordance with the above-mentioned theories of CHS and PBL (see literature), questions were posed to the experts about the following thirteen points:

1. your role in the BI industry, and number of years therein;
2. your opinion about how the use of industry data will benefit the BI students;
3. your opinion about whether BI students are disadvantaged when they using industry data;

4. what BI aspects should be covered by the data and the requirements, and how does this influence the nature of the data;
5. your opinion about whether the use of industry data will benefit BI faculty members;
6. your opinion about whether BI faculty members are disadvantaged when their students use industry data;
7. your opinion about how the use of industry data may benefit the industry partner;
8. your opinion about whether industry partners are disadvantaged by providing BI students with industry data;
9. what should be the extent of industry involvement w.r.t.: student support (understanding of data and requirements), setting the scope of the envisaged project solution, industrial guest-lectures, project valuation;
10. measureable outcomes you hope your involvement will yield;
11. what limits your involvement;
12. what are your intellectual property protection or privacy concerns when providing BI students with industry data;
13. any other matters not already mentioned above.

The responses obtained from the interviewees about those topics and questions were analysed in relation to other responses given by the same participant, as well as other participants' responses for a specific question, with the aim of understanding different perspectives of different stakeholders.

The analysis of the answers revealed that the use of industry data in a well-planned BI student project will yield many benefits to the BI students (Questions 2), faculty members (Question 5), and to the industry partners (Question 7). The critical systems perspective made it clear that people in the same role can have different perspectives; this became especially evident when the responses of the first three participants were compared. P1 regarded the industry practitioner as a partner who can share the workload and who should take responsibility of own issues (not answering the industry focus section of the questionnaire even after a verbal conversation on the components of the questions). P2 with BI experience was sensitive about the privacy aspects of the industry and was willing to spend more time on data preparation. P3, also with industry experience, was very concerned about the privacy issues and suggested that the industry partner ought to remain anonymous to the students. This can indeed have an impact on the previously identified success factors (such as recruitment) of industrial-academic collaborations.

After the analysis of the participants' responses,[2] new guidelines for the use of industry data and requirements in students' BI projects were identified. These guidelines, listed in Table 4, were all informed by the received responses after their careful consideration in accordance with the above-mentioned methodological framework of CSH.

---

[2] For further details see the 'Preliminary Remarks' on the title page of this paper.

After further critical comparisons these new guidelines from Table 4 were then, finally, integrated with the already existing guidelines from Tables 1 and 3, in order to obtain a 'comprehensive' set of guidelines for future student projects in BI.

Table 4. New guidelines, informed by the interviewees' responses

#	New guideline	Motivation
1	The chosen data set must have a strong real-world character which is not destroyed by the removal of sensitive data	The most stated benefit to the students is real-world exposure and the nature of the data should reflect this
2	The chosen industry should provide students with transferrable knowledge	Students should be exposed to industry data that will assist them in other organisations as well
3	Students should be able to achieve new insights in business process activities by analysing the data without jeopardising the competitive advantage of the organisation which supplied the data	It is understandable that sensitive data should be removed from the data. However, doing so should not remove all trends from the data. This guideline supports the BI rather than DW component of the module
4	The chosen data set should require a reasonable amount of ETL work on a large data sets from different sources	The ETL process is known to take up to 80% of all DW efforts
5	The chosen data set should expose the students to the entire lifecycle process starting from requirements collection, ending with end-user applications	A requirements-driven project process is desirable, with requirements supplied be industry
6	Care should be taken to preserve the confidentially aspects in the data	All respondents refer to the possible misuse of data from the organisation and the risks involved. Legal issues must also be considered
7	Where possible the industry partner should present the requirements to the students in context of the business process and be part of the final evaluation of the students' work in partnership with the lecturer	Although the majority of the participants are in favour of the stated guideline, some restrictions may occur in which case the lecturer should represent the industry partner
8	An upfront agreement is required that describes how the intellectual properties of new insights provided by students or faculty members are handled	It is possible that students or faculty members make discoveries in the industry data set and present those during project evaluation to the industry partner

7 Integrated Guidelines from Different Perspectives

After the above-mentioned integration of the data usage guidelines from various sources, the final set of 12 guidelines is outlined as follows. These guidelines are divided into three groups, focussing on the *context* of the data, the *technical requirements*, and the *interests of the industry partner* respectively.

Guidelines w.r.t. the Context of the Data:

- The chosen data set must have a strong real-world character which is not destroyed by the removal of sensitive data.
- The chosen industry should provide students with transferrable knowledge within their frame of reference.
- The requirements must be strategic in nature and not answerable from one OLTP system.
- Students should be able to achieve new insights in business process activities by analysing the data without jeopardising the competitive advantage of the organisation which supplied the data.
- The chosen data set should require a reasonable amount of ETL work on a large data sets from different sources; sources should facilitate integration of dimension records.
- Opportunities should be created for students to share their knowledge.

Guidelines w.r.t. the Data Set's Technical Characteristics:

- The given data set should contain a good variety of descriptive textual fields and numeric fields associated with events.
- A second set of event (transaction) data that requires changes to initial dimensional data should be available.
- The chosen data set should expose the students to the entire lifecycle process starting from requirements collection, ending with end-user applications.

Guidelines w.r.t. the Interests of the Industry Partner:

- Care should be taken to preserve the confidentially aspects in the data.
- Where possible, the industry partner should present the requirements to the students in context of the business process and be part of the final evaluation of the students work in partnership with the lecturer.
- An upfront agreement is require that describes how the intellectual properties (IP) of new insights provided by students or faculty members are handled.

From a critical systems thinking perspective, care was taken to incorporate the interests of the students, faculty members, and industry partners. The first group of guidelines aims to give the student long-term benefits in terms of enhanced understanding of the industry, business processes, and performance measures. The second group of technical guidelines aims to enable the facilitator to guide the students to achieve all stated learning outcomes of the BI module. The limitations of data not adhering to these guidelines were mentioned above in Sect. 5. The third group of guidelines shall ensure that the relationship with the industry partner is well managed and that unintended consequences are limited. Guideline #9, in particular, assumes that it might be possible for an industry partner to remain anonymous.

8 Conclusions

In continuation of previous work [14,15,27,33,35] the aim of this paper was to develop guidelines for using industry data for BI student projects from a critical systems perspective. Guidelines were developed from different stakeholder viewpoints. Although the complete list of guidelines may appear to be idealistic, it may also serve to identify shortcomings in available data sets. Available datasets can be evaluated in terms of these guidelines in order for facilitators to anticipate difficulties and to augment the data to solve the anticipated problems. If it proves difficult to obtain a data set that adheres to the proposed guidelines, or if the data set cannot be extended in accordance to the guidelines, an alternative strategy is proposed. In such cases a facilitator can create a simulated set of data as major instructional tool into which industry data sets (not adhering to all the guidelines) may be then incorporated in order to achieve the benefits of using industry data without jeopardizing the module outcomes. The explicit use of critical systems thinking has several benefits:

- The importance of emancipation for all involved and affected was noted;
- The sensitivity towards assumptions, promoted by critical systems thinking, enables us to distinguish different perspectives of PBL and to highlight the role of reflection in deep learning;
- CSH gave structure to the interviews and identified aspects to investigate;
- CSH motivated the data analysis to better understand the holistic perspective of each of the participants, irrespective if their roles.

References

1. Ackoff, R.L.: Redesigning the Future: A Systems Approach to Societal Problems. Wiley, Hoboken (1974)
2. Adderley, K., Ashwin, C., Bradbury, P., Freeman, J., Goodlad, S., Greene, J., Jenkins, D., Rae, J., Uren, O.: Project Methods in Higher Education, vol. 24. Society for Research into Higher Education, London (1975)
3. Ayas, K., Zeniuk, N.: Project-based learning: building communities of reflective practitioners. Manag. Learn. 32(1), 61–76 (2001)
4. Blumenfeld, P.C., Soloway, E., Marx, R.W., Krajcik, J.S., Guzdial, M., Palincsar, A.: Motivating project-based learning: sustaining the doing, supporting the learning. Educ. Psychol. 26(3/4), 369–398 (1991)
5. Boss, S., Krauss, J.: Reinventing Project-based Learning: Your Field Guide to Real-World Projects in the Digital Age. International Society for Technology in Education, Washington (2007)

6. Checkland, P.: Systems Thinking, Systems Practice. Wiley, Hoboken (1981)
7. Checkland, P., Poulter, J.P.: Learning for Action: a Short Definitive Account of Soft Systems Methodology and its Use for Practitioners, Teachers, and Students. Wiley, Hoboken (2006)
8. Churchman, C.W.: The Systems Approach. Dell, Round Rock (1968)
9. de Graaff, E., Kolmos, A.: History of problem-based and project-based learning. In: Management of Change: Implementation of Problem-Based and Project-Based Learning in Engineering, pp. 1–8 (2007)
10. de Fillippi, R.J.: Introduction to Project-Based Learning, Reflective Practices and Learning. SAGE, Thousand Oaks (2001)
11. Dewey, J.: Experience and Education: the Kappa Delta Phi Lecture Series. Collier (1938)
12. Fernandes, S.R.G.: Preparing graduates for professional practice: findings from a case study of project-based learning (PBL). Procedia Soc. Behav. Sci. **139**(1), 219–226 (2014)
13. Glaser, B., Strauss, A.: The Discovery of Grounded Theory: Strategies for Qualitative Research. Aldine, Houston (1967)
14. Goede, R.: Listening to the affected: student views after starting a 4th-year module in data warehousing. In: Proceedings CSERC 2016, pp. 12–21. ACM (2016)
15. Goede, R., Taylor, E.: Using critical systems thinking in emancipatory postgraduate supervision. In: Proceedings 59th ISSS Annual Meeting, Berlin (2016)
16. Helle, L., Tynjälä, P., Olkinuora, E.: Project-based learning in post-secondary education-theory, practice and rubber sling shots. Higher Educ. **51**(2), 287–314 (2006)
17. Inmon, W.H.: Building the Data Warehouse. Wiley, Hoboken (1996)
18. Inmon, W.H.: Building the Data Warehouse. Wiley, Hoboken (2005)
19. Jackson, M.C.: Systems Thinking: Creative Holism for Managers. Wiley, Hoboken (2003)
20. Jackson, M.C.: The origins and nature of critical systems thinking. Syst. Pract. **4**(2), 131–149 (1991)
21. Keegan, A., Turner, J.R.: Quantity versus quality in project-based learning practices. Manag. Learn. **32**(1), 77–98 (2001)
22. Kimball, R., Ross, M.: The Data Warehouse Toolkit: the Complete Guide to Dimensional Modeling. Wiley, Hoboken (2011)
23. Kimball, R., Ross, M., Thornthwaite, W., Mundy, J., Becker, B.: The Data Warehouse Lifecycle Toolkit: Expert Methods for Designing, Developing, and Deploying Data Warehouses. Wiley, Hoboken (1998)
24. Kimball, R., Ross, M., Thornthwaite, W., Mundy, J., Becker, B.: The Data Warehouse Lifecycle Toolkit, 2nd edn. Wiley, Hoboken (2011)
25. Klein, H.K., Myers, M.D.: A set of principles for conducting and evaluating interpretive field studies in information systems. MIS Q., pp. 67–93 (1999)
26. Kotzé, E.: Augmenting a Data Warehousing Curriculum with Emerging Big Data Technologies. This volume: CCIS 730, (2017)
27. Kotzé, E., Goede, R.: Preparing postgraduate students for industry without neglecting scholarly development: using project-based learning and design science research to develop a software artefact. In: Proceedings International Conference on Computer Science Education Innovation and Technology, p. 57 (2016)
28. Lindstedt, D., Graziano, K.: Super-Charge Your Data Warehouse: Invaluable Data Modeling Rules to Implement Your Data Vault. CreateSpace (2011)
29. Osterman, K.F.: Using constructivism and reflective practice to bridge the theory/practice gap. ERIC (1998)

30. Schein, E.H.: Organizational Culture and Leadership, vol. 2. Wiley, Hoboken (2010)
31. Schön, D.A.: The Reflective Practitioner: How Professionals Think in Action. Basic Books, New York (1983)
32. Seaman, C.B.: Qualitative methods in empirical studies of software engineering. IEEE Trans. Softw. Eng. **25**(4), 557–572 (1999)
33. Taylor, E., Goede, R.: Using critical social heuristics and project-based learning to enhance data warehousing education. Syst. Pract. Action Res. **29**(2), 97–128 (2016)
34. Ulrich, W.: Critical Heuristics of Social Planning: A New Approach to Practical Philosophy. Wiley, Hoboken (1983)
35. Venter, C., Goede, R.: A critical systems approach to business intelligence system development. In: Proceedings 59th ISSS Annual Meeting, Berlin (2016)

A Survey of Data Scientists in South Africa

Eduan Kotzé[✉]

Department of Computer Science and Informatics,
University of the Free State, Bloemfontein, South Africa
kotzeje@ufs.ac.za

Abstract. Academic programmes at South African Higher Education Institutions have predominantly educated students in managing and storing data using relational database technology. However, this is no longer sufficient. South Africa as a country will need to educate more students to manage and process structured, semi-structured and unstructured data. The main purpose of this study was to examine the status of data scientists, a role typically associated with managing these new data sets, in South Africa. The study examined the skills, knowledge and qualifications these data scientists require to do their daily tasks, and offers suggestions that ought to be considered when designing a curriculum for an academic programme in data science.

Keywords: Big data · Data science · Curriculum · Professional skills

1 Introduction

Humans have generated electronic data ever since the first digital computer was invented. As technology improved, so did our capacity to generate data. Today, new technologies enable humans to generate data at an unprecedented volume and speed. It is estimated that we create 2.5 *quintillion* bytes of data *daily* [22], and that the digital world will grow to a total of 40 *zettabytes* by 2020 [13]. As technology becomes more affordable, more humans are using mobile computing, social networks and cloud computing, resulting in data arriving not only in larger volumes, but also in unprecedented ways, which include variety and velocity [36]. Volume, variety and velocity ('the three V'), are synonymous with the term 'big data' [3] and are compelling organizations to take a fresh look their ability to transform their data into competitive insights [31] and to use it for strategic planning [16].

Competitive insights cannot be created by machines, but rather by humans with deep analytical skills. Data scientists are typically known to possess these kinds of analytical skills to interrogate and analyze big data [43]. However, data scientists, a relatively new occupation [9], are still in short supply worldwide [33]. In South Africa, big data technologies have not been widely adopted [37], but some businesses, however, are beginning to pay attention to big data and will require the use of data scientists to extrapolate value from their data [23,42]. Higher Education institutions will play a critical role in providing the new set

J. Liebenberg and S. Gruner (Eds.): SACLA 2017, CCIS 730, pp. 175–191, 2017.
https://doi.org/10.1007/978-3-319-69670-6_12

of skills by delivering formal educational programmes in informatics, advanced analytics, information and data science [42]. However, undergraduate database system curricula, to a large extent, still focus on relational database technology [1,28]. In South Africa, only one qualification could be found on the South African Qualifications Authority (SAQA) database of an undergraduate degree in big data or data science [44,45]. Furthermore, to our knowledge, only two South African tertiary institution offers a fulltime Master degree in data science or big data [39,47]. With the rising demand for data scientists, this is inadequate for a developing country such as South Africa, and more tertiary institutions need to offer an undergraduate degree in data science. Although some work has been done investigating big data and data science education in South Africa [29,32,34], more investigations are still needed.

The purpose of this paper is therefore threefold: to provide an overview on the current status of data scientists in South Africa; to identify the technology, skills and knowledge they require to perform their daily tasks [28], and to make suggestions for an undergraduate degree that a tertiary institution can offer to students who seek the analytical skills required to interrogate and analyze big datasets.

2   Big Data

While 'big data' is still a somewhat loosely defined and fuzzy concept in both the scientific [16] and the commercial [48] realm, most experts agree that the three fundamental characteristics of big data are the three V's: *volume, variety* and *velocity* [3,5,8,30,31]. In order to be considered a big data environment, at least one of these three characteristics must be present to a very large extent. This environment will also exceed the capabilities of most relational database management systems in terms of storing and analyzing the data for decision support purposes [27,52]. In this paper we will adhere to the well-known 'big data' definition by [30]. Though volume, variety and velocity are most often regarded as the typical characteristics of big data, alternative characteristics include *veracity* [24,51], *verification* [3], *validation* [3] and *value* [51].

Data volume in big data systems are characterized by large amounts of data that is generated continuously by different sources, including sensor logs, machine logs and click-stream data [30]. Not only is the data volume very large (measured in exabytes), it also require unique and advanced storage and management technologies [5]. In big data, variety refers to the heterogeneity of data types where the data is represented in diverse structured, semi-structured and unstructured forms [3,5,7,24]. Data velocity refers to the high speed at which data is created, accumulated and processed [36], and the time frame in which it must be acted upon [24,26]. Data veracity is often associated with the biases, noise and abnormality in data [7,38,54], as well as the level of trust in the collection, extraction, cleaning and processing of data from various data sources [51]. Data verification refers to the processes followed to ensure that the data conforms to a set of specifications [3] while data validation refers to scrutinizing the data to ensure data is accurate and correct with regard to its intended use [38]. Finally, data value refers to useful information that is extracted from big data sets [25] and the ability to generate insights and benefits for an organization [7,51]. Types of value creation include improved performance, replacing human decision making with machine learning algorithms, and improving business performance [51].

Big data technologies have emerged in an effort to overcome the limitations that modern relational databases have. These limitations include that relational databases are based on pre-defined schemas, which means that storing the data needs to adhere to ACID (atomicity, consistency, isolation, and durability) compliancy in terms of transaction management [30,53]. Due to space limitation in this paper, a full discussion of big data technologies is omitted here; see [21] for a comprehensive overview. Here it is only important to note that big data technologies predominately focus on distributed file systems [14,20], NoSQL [4], MapReduce [10], YARN [50], Hadoop [53] and Spark [55]. Hive [46], an open-source data warehouse, is the predominant big data technology used for big data analytics.

3 Data Science

The term 'science' in data science implies experience gained through systematic study by building and organizing knowledge in the form of testable explanations and predictions [12]. Data science therefore implies a focus involving data, statistics and the analysis of data [12]. Data science 'leverages' big data technologies [36] to analyze rapidly growing data, which has become more *heterogeneous* and *unstructured* [12] with increased volume and variety. Analysis must be done on the combination of those two types of data, which requires integration, interpretation, and the ability to make sense of the data. However, data science is more than statistics and data analysis; it also involves implementing *algorithms* that automatically process data to provide predictions and actions [17].

There are numerous views of what a data scientist actually is, what the responsibilities of a data scientist are, and what skills the data scientist needs. In this paper the concepts of 'vertical' and 'horizontal' data scientists are adopted. A *vertical* data scientist has deep technical knowledge in a narrow field, while a *horizontal* data scientist is a blend of a business analyst, statistician, computer scientist and domain expert [17]. Data scientists are generally high-ranking professionals and key players in an organization [9]. They are also multidisciplinary thinkers who use many techniques to create analytic models to meet business objectives [36] and visualize their findings [8]. Data scientists require an integrated skill set of statistics, databases and other related areas [12] to create value in solving business problems such as the 'what if' scenarios [36]. What sets data scientists apart from other data workers, including data analysts, is their ability to create meaning behind the data that leads to business decisions [15].

Possession of traditional business intelligence (BI) and data warehousing (DW) skills does not imply that a BI/DW practitioner is already a data scientist, since traditional BI is declarative and does not necessarily require any comprehensive domain understanding [36]. A data scientist is also not a database administrator (DBA), nor a data engineer, nor a data analyst. Data science involves implementing algorithms that process data automatically, which requires computer programming skills, as well as mathematical and statistical knowledge over terabytes of data [17,36]. Finally, data scientists are not exclusively tied to big data projects, but will use big data technologies to explore the increased scope and depth of data being generated these days [3,22]. From the literature it appears the essential requirements regarding education, knowledge and skills for a data scientist are:

- advanced degree in computer science, computational science (scientific computing), or network-oriented social sciences [8];
- education across disciplines with domain knowledge [15,49];
- algorithms and software knowledge [2,11,18];
- problem solving skills [2,11];
- upgraded data management skills [2,8];
- computer programming, mathematics and statistics [2,8,9,11,35];
- machine learning [2,12,18,35,49];
- data or information retrieval [2,35];
- distributed computing [35,49];
- statistical model building and assessment knowledge [11,18];
- data warehousing, business intelligence and data mining [35];
- skill to visualize large data sets (computer graphics) [11,18,35,36,41,49];
- data munging of large data sets or streams [2,18,19,35,41];
- ability to communicate effectively with decision-makers [2,11,13,49];
- appreciation for social, ethical and legal issues [2,11].

4   Existing Big Data and/or Data Science Curricula

Since the focus of this paper is on *undergraduate* studies in data science, existing curricula for postgraduate programmes in data science will here be ignored. According to [2,6], big data and analytics training related to data science is viable at the undergraduate level. A foundation for data science programmes and recommends a skill set required for big data analytics is provided in [35]. In [11] it was pointed out that the key competencies for an undergraduate data science programme should be: analytical thinking (computational and statistical), mathematical foundations, model building and assessment, algorithms and software foundation, data curation, as well as knowledge transference including communication and responsibility. The need for a project-based capstone module, to provide the students with an opportunity to synthesize the concepts learned throughout the module, was also highlighted in [2,11]. Table 1 summarizes [2,6,11] and highlights recommended topics for preparing students to become productive data scientists.

Table 1. Literature summary of knowledge and skills required in data science

Theme	Topics according to [11]	Topics according to [2,6]
Statistics	Exploratory data analysis	Statistical methods
	Estimation and testing	Regression and correlation
	Simulation and resampling	Statistical learning
	Statistical models and selection	Machine learning
	Statistical learning	
	Machine learning	
Mathematics	Mathematical structures	Logic and counting
	Linear modeling	Discrete structures
	Optimization	Linear algebra
	Multivariate thinking	Modeling and simulation
	Probabilistic thinking, modeling	
Software and algorithms	Algorithm design	Computer programming
	Programming concepts	Data structures and algorithms
	Data structures	Advanced algorithms
	Tools and environments	Artificial intelligence
	Scaling of big data	Information retrieval
Data curation	Database and data management	Databases (design, storage, query,
	Data munging (data preparation)	modeling), Large data sets,
	NoSQL	streams (design, access, clean,
		analyze, aggregate, organize,
		visualize)
Knowledge transfer	Technical writing	Oral and written communication
	Public speaking, Ethics	Social, ethical and legal issues

5 Method

According to [9] the well-known 'LinkedIn' web site could be a useful resource to search for data scientists. It was indeed used to construct a convenient sampling frame of employed data scientists in South Africa since the population of data scientists is still relatively small. A total of 127 data scientists from South Africa were identified via LinkedIn where the individual job title contained the phrase 'data scientist' and the location matched 'South Africa'. The search included current and previous job titles of an individual. Only 81 of the identified potential participants' could actually be contacted for participation. A questionnaire was used for data collection which consisted of three main parts. Part 1 collected demographic information while Part 2 collected information about the participants' traditional and big data management practices. Part 3 collected information about their daily tasks and responsibilities, qualifications, knowledge and skills, and, finally, how they would describe a data scientist. Parts 2–3 of the questionnaire used free-text fields to allow for any possible response.

Then, QSR International's NVivo 11 software [40] was used for the qualitative data analysis. Word frequency queries were conducted on the open-ended questions related to analytical tools and big data tools. Since the number of keywords was relatively small, the query was set to search for *words of same (exact) meaning*. Similar word frequency queries were done on the 'type of business' question, 'type of tasks with analytical tools', 'type of tasks with big data tools', 'information usage', 'daily tasks and responsibilities', qualifications, knowledge, technical skills, soft skills, and personality. Those queries were set to search for *synonyms* as well as semantic similarities (thesaurus). Thereby, strings of words were also reduced to single word codes. NVivo weighted the frequency of the words relative to the total number of words counted. Common concepts were thus extracted from each question's answers with open coding.

6   Results

6.1   Demographic Information

Out of 81 invited participants 23 completed the survey. Table 2 shows the respondents' industry profiles. The largest number of respondents ($n = 11$, 48%) were between 31 and 40 years, 30% ($n = 7$) were between 23 and 30 years old, and 21% ($n = 5$) were older than 40 years. This suggests that on average, the respondents of the survey were still early to mid-career professionals. The majority of the respondents ($n = 13$, 56%) had less than 2 years of experience as data scientists at their current organizations, whereas 21.7% ($n = 5$) had more than 4 at their current organization. Sixty-one percent ($n = 14$) had less than 4 years total experience as a data scientist, irrespective of their current employment duration. Only 17% ($n = 4$) of the respondents had more than 8 years total experience as a data scientist. In terms of the number of employees in the organizations, the largest group of the respondents ($n = 10$, 43.7%) indicated that their organization employed more than 2000 employees. A total of 81% ($n = 9$) of the 10 results reported have more than 50 IT employees in the organization. These results suggest that the organizations, where the data scientists are employed, are large. The respondents were also questioned on the data size of their current organizations, and if they perceive their organization to be big-data-enabled. The majority of the respondents ($n = 12$, 52%) considered their organization as big-data-enabled, with 9 of those indicating that the total data size of their organization (which includes structured and unstructured data) was between 1 TB

Table 2. Survey respondents categorized by industry type

Industry type	Frequency	Percentage
Multiple industries	6	26.1%
Finance & banking	5	21.7%
Information technology	4	17.4%
Government	2	8.7%
Retail	1	4.3%
Other	5	21.7%

and 1024 TB. Of the remainder, 8 respondents considered their organization partially big-data-enabled. Only one respondent reported a total data size of more than 1 PB. This finding suggests that, on average, even though the data scientists viewed their organization as big-data-enabled, the volumes they reported did not correspond with the exabytes of data currently associated as 'big data' according to [5]. This suggests that there is still a disagreement on the volume that constitutes a big dataset.

6.2   Analytical Tools and Tasks

The respondents were also asked to elaborate on the type of analytical tools they use in the execution of their daily tasks. The results are shown in Table 3. In some instances, organizations used more than one analytical tool and these are all included in the presentation of information in all further tables. These results indicate that *SAS*, *R*, *Python* and *SQL* were most frequently used as analytical tools. All four of these analytical tools have scripting capabilities, which suggests that the tools are probably used to query and process data. SAS, R and SQL also need structure, which suggests that they are used on structured data, typically stored in a data warehouse environment. An interesting finding was that both Python ($n = 7$) and SQL ($n = 6$) were preferred in the 31–40 age group, while SAS ($n = 6$) and R ($n = 5$) were preferred in the 25–30 age group. Python and SQL are both considered programming languages, while R and SAS are software packages for statistical analysis of data. From the findings, the older groups were more involved in programming activities and less in data analysis. One might have expected this to be the other way around, with the 31–40 group more involved in data analysis, using their industry and domain experience; however this was not the case in this survey. The respondents were also asked to elaborate on the type of tasks they do using analytical tools. The majority of the respondents indicated that they used analytical tools for modelling ($n = 10$) and prediction ($n = 10$), whilst only five respondents indicated that they used analytical tools for data visualization.

Table 3. Analytical tools per age group

Tool	Age 25–30 ($n = 7$)	Age 31–40 ($n = 11$)	Age 41+ ($n = 5$)	(Total)
R	5	7	4	16
SAS	6	2	2	10
Python	1	7	-	8
SQL	1	6	1	8
Excel	3	3	-	6
Microsoft	-	2	4	6
Java	-	1	2	3
SPSS	2	1	-	3

Table 4. Big data tools per age group

Tool	Age 25–30 ($n = 7$)	Age 31–40 ($n = 11$)	Age 41+ ($n = 5$)	(Total)
Hadoop	1	6	1	8
Spark	-	7	1	8
SQL	1	5	-	6
Elasticsearch	1	2	1	4
Hive	1	2	1	4
Pig	-	3	1	4
R	-	2	2	4
SAS	3	1	-	4
DB	-	3	-	3
HBase	-	2	-	2

6.3   Big Data Tools and Tasks

The respondents were also asked to list and describe the big data tools they used. The results are summarized in Table 4. In some instances, organizations used more than one big data tool and these are all included in the tables below. Here, *Hadoop, Spark, Hive, Pig* and *R* were most frequently used as big data tools. Apache Spark is seen as an alternative to Apache Hadoop, and it was interesting to note that in four cases the respondents were using both. Nine respondents indicated they used SQL, R or SAS; eight respondents were not using any other big data technologies whilst only one respondent was using genuine big data technology. This suggests that they may be involved in analyzing structured data as opposed to the semi- or unstructured data typically associated with Apache Hadoop or Apache Spark big data technologies. The respondents were also asked to elaborate on the type of tasks they do using big data tools. The majority ($n = 9$) indicated that they used big data tools for Extraction, Transformation and Loading (ETL) tasks.[1] This finding was surprising and suggests that the 'data scientists' were actually involved with the data scrubbing and preparation activities usually associated with business intelligence and data warehousing professionals. Only one respondent used big data tools for data visualization and inferential statistics, and another for predictive modelling.

6.4   Type of Business Questions

The respondents were asked to elaborate on the type of business questions they would need to answer by using the big data and analytical tools at their disposal. The NVivo query returned 365 words (176 unique); the set was grouped into

---

[1] Example answers included: *"transforming and shifting data"*, *"processing of large volumes of data (ETL/data pipeline)"*, *"data preparation for statistical models"*, as well as *"data warehousing, reporting, ETL development"*.

Table 5. Business domains in which the experts are answerable

Keyword	Frequency	Weighted %
Customer	18	5.42%
Data	11	3.31%
Marketing	8	2.41%
Prediction	8	2.41%
Product	7	1.76%
Go	7	1.28%
Analysis	6	1.81%
Modelling	6	1.81%
Expect	6	1.31%
Churn	5	1.51%

synonyms. Weighted percentage values (frequency of the word relative to the total words counted) were used to filter the results: see Table 5. In order to gain a better understanding, each answer was further analyzed for key concepts. Marketing analytics, including targeted marketing, customer segmentation and customer churn were evident with the majority ($n = 11$) of respondents. Building models and predicting behavioral patterns such as *payment default* or *product purchases* where the other main theme. These findings suggest that the typical business question would be focused on customers' behavior, which customers are likely to churn, or the effect of a marketing offer on the market.

6.5   Information Usage

One of the survey questions asked the respondents to explain what the extracted information was used for and what their daily tasks and responsibilities entailed using this information. The NVivo query on *information usage* returned 224 words (114 unique) and for *tasks and responsibilities* 379 words (164 unique). Both sets were grouped into synonyms. The keyword 'data' was used in conjunction with cleaning, modelling, analysis and presentation: see Table 6. A deeper investigation revealed that information was predominately used for customer lifecycle management ($n = 6$) and to support decision-making ($n = 7$). One respondent indicated that the ultimate goal was to *"integrate models into the business process to automatically identify new cases"*. In terms of tasks and responsibilities, *modelling* ($n = 12$), *development* ($n = 12$), *reporting* ($n = 9$) and *business* (including engaging) ($n = 9$) were most evident in the results. Only one respondent indicated that he was responsible for managing a team of analysts, specialist and data engineers. All the other respondents used their technical abilities in either cleaning or loading data, building models, or integrating various technologies with each other. It was interesting to note that *modelling* played an important role in both questions.

Table 6. Information usage and tasks

Keyword	Information usage	Tasks and responsibilities
Data	11 (5.29%)	29 (8.29%)
Making	10 (3.77%)	-
Customer	9 (4.09%)	-
Operational	9 (3.12%)	-
Business (Engaging)	8 (3.37%)	9 (2.29%)
Modelling	8 (3.85%)	12 (3.43%)
Decisions	7 (3.37%)	-
Management	7 (3.12%)	-
Reporting	6 (2.16%)	9 (2.57%)
Development	-	12 (2.81%)
Analysis	-	7 (2.00%)
Responsibilities	-	6 (1.57%)
Building	-	6 (1.43%)
Cleaning	-	5 (1.43%)
Presentation	-	5 (1.29%)

6.6 Qualifications and Knowledge

The respondents were asked to elaborate on the type of qualifications and knowledge required to do their daily tasks. The NVivo query for *qualifications* returned 351 words (174 unique), and for *knowledge* 236 words (96 unique). Both sets were grouped into synonyms: see Tables 7 and 8. For Table 7, the word *Applied* included 'employment', 'implement', 'practical' and 'using' as similar words, whereby the term *Applied Mathematics* only occurred once. *Learning* referred to 'knowledge' as well as the ability to learn and to study, while *Good* included the similar words of 'just', 'practical', 'right' and 'skills'. Further analysis revealed that the majority of the respondents ($n = 16$) considered 'a degree' as the minimum qualification. Of these, seven respondents indicated that a *postgraduate* degree ought to be the minimum qualification. One respondent indicated no degree would be needed. In this context, *Knowledge* included 'learn' or 'learning' as similar words, and *Business* included 'engaging' and 'engagement'. Further analysis indicated a strong preference for machine learning ($n = 7$) and deep business or industry domain knowledge ($n = 12$). The majority ($n = 15$) indicated the importance of statistics, and also mathematics ($n = 9$). Programming ($n = 8$) was somewhat important, database management knowledge ($n = 5$) to a lesser degree. Only three respondents mentioned 'critical thinking' as crucial for their daily tasks. The results indicated that the respondents were competent in statistics, mathematics and programming, but at the same time needed deep domain knowledge about the business and how to apply their knowledge to integrate end-to-end solutions. This result is consistent with the above-mentioned

Table 7. Qualifications required

Keyword	Frequency	Weighted %
Mathematics	16	4.55%
Applied	14	3.79%
Science	13	3.41%
Learning	10	1.96%
Statistics	10	2.84%
Degree	8	2.13%
Data	7	1.99%
Good	7	1.23%

Table 8. Knowledge required

Keyword	Frequency	Weighted %
Statistics	15	6.30%
Business	10	4.20%
Mathematics	9	3.78%
Programming	8	3.36%
Databases	5	2.10%

curriculum guidelines [11] according to which the theoretical foundations of data science education should be primarily statistics, mathematics, and computer science.

6.7   Soft Skills

The respondents were asked to elaborate on the type of 'soft skills' required to do their daily tasks. The NVivo query for *soft skills* returned 259 words (122 unique)—again grouped into synonyms—as shown in Table 9. A deeper investigation into the respondents' answers revealed that the ability to present their findings ($n = 12$) in a 'story-telling' form ($n = 5$) was important. This included data visualization (*picture*) and communicating findings to a client in an 'interesting' manner. Furthermore, a data scientist must have the ability to listen and reason with a client as well as to present complicated concepts in a tractable way. Only three respondents indicated that the data scientist should be a 'team player'. Also this result is consistent with the above-mentioned curriculum guidelines [11], including includes oral, written, and visual modes of communication.

Table 9. Soft skills required

Keyword	Frequency	Weighted %
Presentation	15	6.09%
Skills	12	4.62%
Understanding	10	2.80%
Business	6	2.52%
Story	6	2.52%
Client	6	2.52%
Communication	5	2.10%
Picture	5	1.89%

6.8 Personality

Finally, the respondents were asked to elaborate on the type of 'personality' required to be a data scientist. The NVivo query for *personality* returned 194 words (110 unique)—again grouped into synonyms—as shown in Table 10. *Good* in this context included 'just' and 'respectful' as similar words and was used in conjunction with communication or interpersonal skills. *Work* included 'make' and 'solve' as similar words. *Learning* included 'knowing' and 'determined', whilst *Persistent* included 'continuous' and 'perseverance'. A deeper investigation into the results revealed that a data scientist should be inquisitive ($n = 6$), thrive on new challenges, and be persistent in solving a problem. This includes being curious and open to experimenting with new ideas, but at the same time staying focused and paying attention to details. Good communication and interpersonal skills ($n = 5$) were also mentioned, however to a lesser degree.

Table 10. Personality required

Keyword	Frequency	Weighted %
Good	8	4.14%
Work	8	4.14%
Inquisitive	6	3.31%
Learning	5	1.75%
Persistent	5	2.76%
Focused	4	1.93%

7 Discussion and Recommendations

The results of the survey suggest that a data scientist in South Africa is an early to mid-career professional working in a large organization (>2000 employees).

These data scientists have a wide variety of knowledge and skills and not all are necessarily experts in any one specific area. Concerning the minimum level of skills, knowledge and qualification, this study confirms that a data scientist should possess at least a bachelor degree to be employed, though a postgraduate qualification is strongly recommended. Some theoretical knowledge and skills in mathematics, statistics and computer science will be required if somebody wants to work as a data scientist. Consequently the following *curriculum suggestions* are made to fulfil these requirements:

- Knowledge of statistics is vital. Statistics could include: statistical tests, data distributions, maximum likelihood estimators, *p*-values, as well as insight and understanding to apply the appropriate statistical techniques on sets of data. The statistics must be presented in a way that will assist management and other stakeholders to make the best decisions possible, including predictions to create possible solutions to current problems.
- Knowledge of mathematics is strongly recommended, too. The curriculum could include linear algebra or matrix computations and multivariate calculus, since these theories are applied to create techniques for data processing. Many techniques are already built into some available packages and libraries of programming languages.
- Computer programming on a suitable level is a minimum requirement. Machine learning is strongly recommended to be included in the programming curriculum. Big data technologies like Apache Hadoop, Apache Spark and Hadoop Hive in conjunction with programming languages such as Python, Pig, R and Scala should be strongly considered.
- Computer programming should also include some lectures in data munging, which cannot be done without knowledge of regular expression and relational and/or distributed databases. Data munging (curation) should also include knowledge about data preparation and data management throughout an entire problem-solving process. This includes preparing data for use with statistical methods and models (commonly referred to as ETL in data warehousing [28]). Working knowledge of the business-relevant Structured Query Language (SQL) is also important.
- Algorithm optimization is important as well, since companies strive for the best possible solutions in all respects. In addition to algorithm design, software engineering is also recommended, whereby a theoretical background ought to be obtained together with practical implementations.
- Data analysis and data exploration are essential skills for a productive data scientist. Knowledge of an analytical tool such as SAS, SPSS or Excel is strongly recommended, because data scientists will need to apply their critical thinking abilities in solving business problems either by programming new machine learning solutions, or by using already existing business intelligence or data mining tools, or a combination of both.
- The ability to communicate effectively with decision-makers and to visualize large datasets are important skills as well. Furthermore, a data scientist must be competent in 'story-telling'; using data visualization tools such as *Tableau* or *Gephi* should be considered as part of their curriculum.

– Being able to work in a team will be required in any company. Interpersonal skills must therefore be learned, practiced and applied. This should include scenario training of how a data scientist should interact with other team members, such as managers, engineers or even data capturers. It is also important that they are able to identify the important aspects when starting a project on new data, how it must be handled and how to define a clear vision of what the outcome should be, (e.g.: the continuation of specific advertisements).

8   Conclusion

The purpose of this paper was threefold: to provide an overview of the current status of data scientists in South Africa; to identify the technology, skills and knowledge they require to do their daily tasks; and to offer suggestions for an academic programme that a tertiary institution can offer to students who seek to develop the necessary analytical skills required to interrogate and analyze big datasets. The results obtained indicate that a data scientist must a wide range of skills and knowledge *w.r.t.* different aspects of data processing and data management. According to the findings a data scientist should not be an expert in a single area, but rather have wider expertise in several areas. Soft skills are also important. A data scientist must have the ability to visualize analysis results, while at the same time having an inquisitive personality and continuously look out for new challenges and answers. The findings also suggest that a data scientist should be a dynamic and creative thinker. Because individuals with all of those skills are rare, interdisciplinary teams of database specialists, programmers, mathematicians and statisticians must work together in an organization's big data department. The study has some limitations that offer future research possibilities. A major limitation was the small sample size ($n = 23$) together with the fact that the respondents were conveniently sampled only via the 'LinkedIn' platform. Hence, the results from this study cannot be generalized, though they offer some 'qualitative' insight into the daily tasks and responsibilities of data scientists in South Africa. Their needs in terms of training and skills are highlighted, which could be used by Higher Education institutions inside and outside South Africa to come up with practically relevant data science curricula.

Acknowledgements. Thanks to all respondents who made this study possible. Thanks to *Anelize van Biljon* for her helpful comments on drafts of this paper. Last but not least thank the anonymous reviewers of SACLA'2017 for their valuable comments and suggestions.

References

1. ACM and IEEE 2013: Computer Science Curricula 2013: Curriculum Guidelines for Undergraduate Degree Programs in Computer Science. ACM (2013). https://doi.org/10.1145/2534860

2. Anderson, P., Bowring, J., McCauley, R., Pothering, G., Starr, C.: An undergraduate degree in data science: curriculum and a decade of implementation experience. In: Proceedings of the 45th ACM Technical Symposium on Computer Science Education, SIGCSE 2014, pp. 145–150 (2014)
3. Berman, J.J.: Principles of Big Data: Preparing, Sharing, and Analyzing Complex Information. Elsevier/Morgan Kaufman, Amsterdam/Burlington (2013)
4. Cattell, R.: Scalable SQL and NoSQL data stores. ACM SIGMOD Rec. **39**(4), 12 (2011)
5. Chen, H., Chiang, R., Storey, V.: Business intelligence and analytics: from big data to big impact. MIS Q. **36**(4), 1165–1188 (2012)
6. College of Charleston: Data Science Program Information (2017)
7. Daniel, B., Butson, R.: Foundations of big data and analytics in higher education. In: Proceedings of the International Conference on Analytics Driven Solutions, ICAS 2014 (2014)
8. Davenport, T.H., Barth, P., Bean, R.: How big data is different. MIT Sloan Manag. Rev. **54**, 22–24 (2012)
9. Davenport, T.H., Patil, D.J.: Data scientist: the sexiest job of the 21st century (2012). http://hbr.org/2012/10/data-scientist-the-sexiest-job-of-the-21st-century. Accessed 25 Nov 2013
10. Dean, J., Ghemawat, S.: MapReduce: simplified data processing on large clusters. Commun. ACM **51**(1), 107 (2008)
11. de Veaux, R.D., Agarwal, M., Averett, M., Baumer, B.S., Bray, A., Bressoud, T.C., Bryant, L., Cheng, L.Z., Francis, A., Gould, R., Kim, A.Y., Kretchmar, M., Lu, Q., Moskol, A., Nolan, D., Pelayo, R., Raleigh, S., Sethi, R.J., Sondjaja, M., Tiruviluamala, N., Uhlig, P.X., Washington, T.M., Wesley, C.L., White, D., Ye, P.: Curriculum guidelines for undergraduate programs in data science. Ann. Rev. Stat. Appl. **4**(1), 15–30 (2017)
12. Dhar, V.: Data science and prediction. Commun. ACM **56**(12), 64–73 (2013)
13. Gantz, J., Reinsel, D.: The digital universe in 2020: big data, bigger digital shadows, and biggest growth in the far east executive summary: a universe of opportunities and challenges. Technical report, EMC (2012)
14. Ghemawat, S., Gobioff, H., Leung, S.T.: The Google file system. In: ACM SIGOPS Operating Systems Review, vol. 37, no. 5, pp. 29–43 (2003)
15. Gittlen, S.: Could data scientist be your next job? Technical report, Computerworld (2012)
16. Gopalkrishnan, V., Steier, D.: Big data, big business: bridging the gap. In: Proceedings of the 1st International Workshop on Big Data, Streams and Heterogeneous Source Mining, BigMine 2012, pp. 7–11 (2012)
17. Granville, V.: Developing Analytic Talent: Becoming a Data Scientist. Wiley, Hoboken (2014)
18. Harris, J.G., Shetterley, N., Alter, A.E., Schnell, K.: The team solution to the data scientist shortage. Technical report, Accenture Institute for High Performance (2013)
19. Holtz, D.: 8 skills you need to be a data scientist. Technical report, Udacity (2014)
20. Howard, J.H., Kazar, M.L., Menees, S.G., Nichols, D.A., Satyanarayanan, M., Sidebotham, R.N., West, M.J.: Scale and performance in a distributed file system. ACM Trans. Comput. Syst. **6**(1), 51–81 (1988)
21. Hu, H., Wen, Y., Chua, T.S., Li, X.: Toward scalable systems for big data analytics: a technology tutorial. IEEE Access **2**, 652–687 (2014)
22. IBM: What is big data? Technical report (2015)

23. ITWeb: Business intelligence survey 2013 results. Technical report (2013)
24. Jagadish, H.V., Gehrke, J., Labrinidis, A., Papakonstantinou, Y., Patel, J.M., Ramakrishnan, R., Shahabi, C.: Big data and its technical challenges. Commun. ACM **57**(7), 86–94 (2014)
25. Jukić, N., Sharma, A., Nestorov, S., Jukić, B.: Augmenting data warehouses with big data. Inf. Syst. Manag. **32**, 200–209 (2015)
26. Kim, B.G., Trimi, S., Chung, J.H.: Big-data applications in the government sector. Commun. ACM **57**(3), 78–85 (2014)
27. Kim, W., Jeong, O.R., Kim, C.: A holistic view of big data. Int. J. Data Warehouse. Min. **10**(3), 59–69 (2014)
28. Kotzé, E.: Augmenting a data warehousing curriculum with emerging big data technologies. In: Liebenberg, J., Gruner, S. (eds.) SACLA 2017. CCIS, vol. 730, pp. 128–143. Springer, Cham (2017)
29. Kotzé, E.: An overview of big data and data science education at South African universities. Suid-Afrikaanse Tydskrif vir Natuurwetenskap en Tegnologie, **35**(1) (2016). https://doi.org/10.4102/satnt.v35i1.1387
30. Krishnan, K.: Data Warehousing in the Age of Big Data. Elsevier/Morgan Kaufman, Amsterdam/Burlington (2013)
31. Lopez, J.A.: Best practices for turning big data into big insights. Bus. Intell. J. **4**(17), 17–21 (2012)
32. Lutu, P.: Big data and NoSQL databases: new opportunities for database systems curricula. In: Proceedings of the 44th Annual Southern African Computer Lecturers' Association, SACLA'2015, pp. 204–209, Johannesburg (2015)
33. Manyika, J., Chui, M., Brown, B., Bughin, J., Dobbs, R., Roxburgh, C., Byers, A.H.: Big data: the next frontier for innovation, competition, and productivity. Technical report, McKinsey (2011)
34. Marshall, L., Eloff, J.H.P.: Towards an interdisciplinary master's degree programme in big data and data science: a South African perspective. In: CCIS, vol. 642, pp. 131–139 (2016)
35. Mills, R.J., Chudoba, K.M., Olsen, D.H.: IS programs responding to industry demands for data scientists: a comparison between 2011 and 2016. J. Inf. Syst. Educ. **27**(2), 131–141 (2016)
36. Minelli, M., Chambers, M., Dhiraj, A.: Big Data, Big Analytics: Emerging Business Intelligence and Analytic Trends for Today's Businesses. Wiley, Hoboken (2013)
37. Moyo, A.: South Africa snubs big data. Technical report, iWeek (2014)
38. Normandeau, K.: Beyond volume, variety and velocity is the issue of big data veracity. Technical report (2013)
39. North-West University: BMI (2016). http://natural-sciences.nwu.ac.za/bmi
40. NVivo: Qualitative Data Analysis Software (Version 11). QSR International (2016)
41. Patil, D.J.: Building Data Science Teams. O'Reilly, Sebastopol (2011)
42. Pieterse, I.: How big data is changing business. Technical report, iWeek (2014)
43. Rouse, M.: Data scientist. Technical report, Search Business Analytics (2011)
44. SAQA: http://www.saqa.org.za/
45. Sol Plaatjie University: Bachelor of Science in Data Science. Technical report (2016)
46. Thusoo, A., Shao, Z., Anthony, S., Borthakur, D., Jain, N., Sen-Sarma, J., Murthy, R., Liu, H.: Artefact data warehousing and analytics infrastructure at Facebook. In: Proceedings of the SIGMOD Conference, pp. 1013–1020. ACM (2010)
47. University of Pretoria: Master's Degree in Big Data Science at the University of Pretoria. Technical report (2016)
48. van Biljon, A., Kotzé, E.: How big is big data and where will you find it? Technical report, EngineerIT (2015)

49. van der Aalst, W.M.P.: Data scientist: the engineer of the future. In: Mertins, K., Bénaben, F., Poler, R., Bourrières, J.-P. (eds.) Enterprise Interoperability VI. PIC, vol. 7, pp. 13–26. Springer, Cham (2014). https://doi.org/10.1007/978-3-319-04948-9_2

50. Vavilapalli, V.K., Seth, S., Saha, B., Curino, C., O'Malley, O., Radia, S., Reed, B., Baldeschwieler, E., Murthy, A.C., Douglas, C., Agarwal, S., Konar, M., Evans, R., Graves, T., Lowe, J., Shah, H.: Apache Hadoop YARN. In: Proceedings of the 4th Annual Symposium on Cloud Computing, SOCC 2013, pp. 1–16. ACM (2013)

51. Wamba, S.F., Akter, S., Edwards, A., Chopin, G., Gnanzou, D.: How "big data" can make big impact: findings from a systematic review and a longitudinal case study. Int. J. Prod. Econ. 165, 234–246 (2015). https://doi.org/10.1016/j.ijpe.2014.12.031

52. Watson, H.J., Marjanovic, O.: Big data: the fourth data management generation. Bus. Intell. J. **18**(3), 4–9 (2014)

53. White, T.: Hadoop: The Definitive Guide, 4th edn. O'Reilly, Sebastopol (2015)

54. Yin, S., Kaynak, O.: Big data for modern industry: challenges and trends. IEEE **103**(2), 143–146 (2015)

55. Zaharia, M., Chowdhury, M., Franklin, M.J., Shenker, S., Stoica, I.: Spark: cluster computing with working sets. In: Proceedings of the 2nd USENIX Conference on Hot Topics in Cloud Computing (2010)

Industry versus Post-graduate Studies: CS and IS Alumni Perceptions

André P. Calitz[1(✉)], Jean H. Greyling[1], and Margaret Cullen[2]

[1] Department of Computing Sciences, Nelson Mandela University,
Port Elizabeth, South Africa
{andre.calitz,jean.greyling}@mandela.ac.za
[2] Business School, Nelson Mandela University, Port Elizabeth, South Africa
margaret.cullen@mandela.ac.za

Abstract. Students graduating from Computer Science (CS) and Information Systems (IS) departments face the question whether to go into the industry or whether to continue at university with postgraduate studies. The IT industry is currently experiencing a severe skill shortage, and recruitment agencies and recruitment websites are advertising posts for more than 10′000 IT job vacancies in South Africa. Academics, at the same time, encourage capable graduates to continue with their studies at university. This paper focuses on CS and IS graduates and postgraduates' (alumni) perceptions and their surveyed opinions regarding the decision to work in industry or complete post-graduate studies. Our advice, on the basis of our survey, could assist academic CS/IS departments in providing better guidance to students completing their B.Sc. studies from the perspective of graduates who already work in the IT industry.

Keywords: Industry career · Postgraduate studies · Alumni perceptions

1 Introduction

Academic departments at Higher Education Institutions (HEIs) are increasingly engaging with various external stakeholders. These include alumni, employers, professional accreditation bodies, and the like. Stakeholder engagement ensures closer university and alumni/industry collaboration and liaison. Stakeholder feedback obtained in various forms, specifically on academic matters, career choices and perceptions relating to working in industry or completing postgraduate studies is increasingly being analysed by academic departments. Academic departments have used surveys, mailing lists, web sites and social media, such as Facebook and LinkedIn, to maintain contact and acquire information specifically from graduates (alumni) working in industry [6,8,24]. Maintaining contact with alumni and obtaining feedback from alumni on academic programme quality have become an important activity at Computer Science (CS), Information Systems (IS) and Information Technology (IT) departments at HEIs [15].

© Springer International Publishing AG 2017
J. Liebenberg and S. Gruner (Eds.): SACLA 2017, CCIS 730, pp. 192–205, 2017.
https://doi.org/10.1007/978-3-319-69670-6_13

Alumni further provide an important perspective and valuable source of advice on academic matters, career choices and guidance for currently registered students. Presuming that they formally fulfill the entry requirements (in terms of marks or grades) for a possible continuation of their studies at the next higher academic level, students in CS, IS and other related IT fields face the question during their final undergraduate year: *shall I leave the university to work in the industry, shall I continue with postgraduate studies, or shall I attempt a combination of both in a part-time mode of postgraduate studies as a student-employee?* Presently, 'Indeed'[1] is advertising about 17'000 IT jobs, and 'Careers24'[2] more than 9'000 in South Africa. 'CareerJunction'[3] has consistently reported over the past years that the greatest demand for recruitment is in the IT sector. Companies recruiting students on campus generally indicate that students should enter the 'real world' as soon as possible and gain industry experience. A number of companies are further offering excellent graduate and postgraduate students starting salaries plus benefits. Recruitment agencies and industry representatives recruiting graduates generally indicate that students should consider working in industry as they would gain on-the-job training and get the relevant industry skills. Academics, on the other hand, encourage capable students to complete postgraduate studies. Completing postgraduate studies can have various advantages, such as higher starting salaries, greater exposure to different study fields in IT, improved conceptual thinking skills and assistance with promotions later in the students' IT careers. Postgraduate students are further exposed to the research process, which can assist them with academic writing, working with new technologies and obtaining various business skills. However, academia is competing with industry [9] as well as with the financial realities of students who need to start earning a living and who have limited funding to pay for postgraduate studies. Meanwhile, a number of ICT departments at HEIs have established closer collaboration with industry. Industry Advisory Boards have been established at the academic institutions in order to address the industry ICT graduate skills requirements and establish closer collaboration [5].

This paper provides new insights regarding graduate and postgraduate alumni perceptions on the value of postgraduate studies. Our results highlight the importance of alumni feedback and the advantages of completing postgraduate studies. In the subsequent sections we explain our method (Sect. 2), review the literature (Sect. 3), present our findings (Sect. 4), and conclude our discussion with some possibilities for future work (Sect. 5).

2 Research Problem and Survey Design

Industry is presently experiencing an IT skills shortage. In March 2017 alone, 'CareerJunction' advertised 2'872 IT jobs, including 1'901 Developer jobs, 1'292

---

[1] https://www.indeed.co.za/IT-jobs.

[2] http://www.careers24.com.

[3] http://www.careerjunction.co.za.

Information Specialist jobs, 611 Computer Science jobs, 169 Business Intelligence jobs, 864 jobs requiring a B.Sc. degree, 473 IT Analysts, and 242 Business Analysts. At the same time, academics are generally advising capable undergraduate students to complete postgraduate studies [6]. Thereby, departments of CS, IS and IT generally do *not* use the information provided by alumni to evaluate the alumni's perceptions of completing postgraduate studies.

The problem investigated in this paper is based on the realisation that *alumni have different perceptions*, which do not always match with academic opinions, of going to work in industry versus completing postgraduate studies first. Academic departments, thus, should 'learn' the views of alumni on completing a CS/IS Master degree or completing an IT-related MBA qualification. The Nelson Mandela University (NMU)'s Department of Computing Sciences offers CS and IS programmes. Students can complete an undergraduate programme, such as a B.Sc. in Computer Science, or a B.Com. in Information Systems. In our study, this group is referred to as 'graduates' with a completed 3-year qualification. 'Postgraduate' alumni are here understood as graduates who have completed either a B.Com. Honours,[4] B.Sc. Honours, M.Com., M.Sc., or a Ph.D. degree in Computer Science and Information Systems at NMU. This exploratory study focused on the perceptions of graduate and postgraduate alumni from the Department of Computing Sciences at NMU.

Research Question: *What are the CS/IS alumni perceptions of postgraduate studies?*

An alumni questionnaire was compiled using a number of already existing alumni questionnaires from similar studies [2,6,17,24]. In order to determine personal perceptions and honest information, it was decided to keep the survey anonymous. The NMU alumni questionnaire consists of the following sections.

- Degree details: highest CS or IS degree, date of graduation;
- Five open-ended questions relating to industry and postgraduate studies:
  1. Why should (or should not) a CS/IS graduate do an Honours degree in CS/IS?
  2. Are there reasons why an Honours student should stay on for another 1–2 years and complete a CS/IS Master degree?
  3. Would you advise a postgraduate student to do the M.Sc. in CS/IS or the MBA?
  4. When should a Masters degree student consider doing a Ph.D.?
  5. Industry versus postgraduate studies: What advice would you give to a 3rd-year student?

---

[4] For readers from outside South Africa: the South African 'honours' degree is an extension of the classical 'B.Sc.' degree which enables a student to commence with Master-studies thereafter. While already considered 'postgraduate' in South Africa, the 'honours' degree in South Africa is reasonably well comparable to the final study-year in the (longer) U.S. American 'B.Sc.' curriculum.

A number of faculty members in the Department of Computing Sciences and the NMU Business School evaluated the questionnaire and suggested changes and improvements. A pilot study was conducted among three alumni working at NMU to validate the questionnaire initially. The questionnaire was captured using the NMU on-line survey tool. The next step was to contact alumni who were working in industry. Social networks are increasingly being used and a large number of graduates are on social networks such as Facebook or LinkedIn. The first call for participation was distributed via Facebook ($n = 1800$), LinkedIn, and the Department's alumni e-mail address list ($n = 600$). The *snowball* sampling technique was applied, requesting participants to forward the same survey request to further possible respondents. A total of 114 alumni completed the survey over a three-week period after three requests for participation. The qualitative results were thematically analysed using AtlasTi.

3   CS and IS Alumni Perceptions on Postgraduate Studies

Alumni surveys conducted by the Department of Computing Sciences at NMU assessed whether graduates believe that the academic programme adequately prepared them for their IT careers [6]; for comparison see [17]. The surveys focused on undergraduate experience and employment, and indicated that surveying alumni is an effective method of gathering information regarding their perceptions of job preparation, employment, skills development and programme effectiveness. Academic CS/IS departments must continuously monitor the employability of their graduates in order to evaluate the effectiveness of their academic programme offerings. Alumni employment rates and job position listings are useful indicators of the quality of a department's programme offerings. The information could be used for attracting prospective students and collaboration with industry.

On-line surveys are nowadays the most popular method of obtaining relevant information from alumni [2]. The surveys focus on obtaining information regarding programme quality and relevance, employability of the graduates, as well as alumnis' departmental and university experiences. Alumni addresses are typically obtained from a university's alumni office, departmental Facebook pages, or LinkedIn. Bulk e-mails are sent to the alumni requesting them to complete the surveys [2,17]. Requests are also posted publicly onto social media sites like Facebook and LinkedIn. Various departments have established alumni 'groups' on Facebook or LinkedIn to maintain contact with ICT graduates, to track graduate destinations, and to engage with graduates in the industry [25].

Few studies focused on decisions which students face regarding accepting a position in industry or completing postgraduate studies. The majority of studies focused on general alumni surveys [2,8,17,19]. The advantages of completing postgraduate studies are important factors which influence the employability of students [6]. An important question graduating 3rd-year CS/IS students ask is: *Why should I aim for a postgraduate Computer Science or Information Systems qualification?* Later in their IT careers, alumni may be faced with the question:

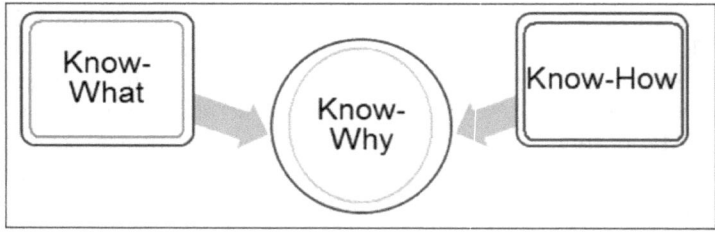

Fig. 1. Taxonomy of knowledge according to [12]

Should I do the M.Sc. or the MBA degree? Some studies have focused on the value of the MBA degree in comparison against the M.Sc. degree [1,3,26].

The competencies entrenched in postgraduate science and engineering education are a key factor contributing to technological competitiveness and economic growth [10,18]. A critical shortage of these skill was often noticed and reported in policy reports and 'White Papers' across various countries [5]. Recognisably, postgraduate qualifications improve career mobility and are noted as an important labour source for recruiters [13,15,19]. A postgraduate qualification offers individuals a source of highly portable skills, with a deep competency and developing an international network of professional contacts [18]. Additionally, employers demand postgraduate qualifications when hiring employees who own highly transferable skills for specialised and/or senior positions [11,13]. Postgraduate qualifications are becoming 'benchmarks of excellence' for employers because of the exposure to advanced knowledge and competencies. The competencies derived from postgraduate studies promote the value of students for employers looking to recruit and retain high-flying staff [18].

According to [12], the taxonomy of knowledge in the organisational context is described as the 'know-what' (understanding of processes and procedures), 'know-how' (deep understanding of best practices and methodologies), and 'know-why'. We expand on this taxonomy by identifying a postgraduate's ability to synthesise the former two descriptions. In [12] their synthesis into the 'know-why' is based on the ability to comprehend phenomena and to justify sustainable, applicable solutions; see Fig. 1.

Postgraduate qualifications are an advantage for the acquisition of more specialised jobs and streamline the recruitment process in the early selection stage [21]. Postgraduate studies create a distinct competitive advantage over other candidates, provide in-depth knowledge for specialised disciplines, and develop important transferable skills. Arguably, pursuing a postgraduate qualification displays an individual's personal cognitive abilities, characterised by: self-determination, achievement and a strong internal locus of control.

A postgraduate qualification also offers a platform for international career mobility through the portability of the skills set, which supports the assertion of the demand for skilled migrants in developed countries [27]. The opportunity existing from international exposure highlights benefits, namely: cultural capital, economic capital and social capital [27]. Social capital is the professional networks gained abroad; cultural capital refers to education, language and cultural knowledge, while economic capital refers to the monetary value that is a convertible commodity [27]. The diversity of attributes of a CS postgraduation qualification is apparently linked to the basic economics of demand and supply. The demand for this critical skill is high and asserted as a driver of technological competitiveness, notwithstanding the demand from employers for specialised and transferable skills [20,21].

3.1   Interplay of Industry and Postgraduate Studies

Industry leaders assert that the competencies derived from postgraduate studies are more valued in specific career categories. In the category of science, the Organisation of Economic Cooperation and Development (OECD) confirms that the demand for scientists and engineers with deep competencies and transferable skills is a matter of importance [20]. Arguably, the extent of the skills and competencies entrenched in the sciences and engineering emphasises the ideal of undertaking a postgraduate qualification [21].

Scientific and engineering occupations are key factors contributing to national, technological competitiveness and economic growth [10]. In light of this, the new economy requires competitiveness. Knowledge and intangible assets drive competitiveness [16,28]. The value of postgraduate studies in science and engineering develops a distinct competitive advantage and guarantees specialised and senior positions across employers due to the scarcity of these skills [20,21,28]. According to [4], knowledge resources reveal themselves in the form of technology, competencies and capabilities. Competence and capability are assets which promote organisational, technical and strategic skills to achieve a high level of performance and competitive advantage [4,16,28]. The human capital theory states that a correlation exists between education and productivity. Human Resources within organisations develop perceptions about employability based on the level of education, and this determines hiring decisions [22]. Thereby, an advanced qualification is a pre-condition for more specialised jobs and international career mobility [10,20,21,27].

The onset of globalisation has provoked organisations to work in complex environments with high degrees of risk. In light of this, specialised skills manifest competencies to reduce risk and ensure brand protection [28]. Therefore, the depth of knowledge and understanding embedded in postgraduate qualifications, through the interplay with government and private institutions, offers a specialisation sought-after across the world.

According to [20], especially the Ph.D. qualification serves as an exclusive education credential characterised by social prestige. The social prestige attached to the Ph.D. may prove a distinct competitive advantage that dictates a value proposition, which in turn announces an intangible resource that reinforces credibility. The literature indicates that students who leave academia to join private institutions fill senior management positions, where the value of knowledge is entrenched in the Ph.D. degree [20,21]. The suppositions of [20,21] explain that a

Ph.D. is at the centre of acquiring and developing new knowledge and correlates to a highly competent workforce with competencies necessary for knowledge-intensive environments; for the engineering disciplines see [10].

3.2 Advantages and Benefits of a Postgraduate Degree

Postgraduates possess academic knowledge and practical competencies in a domain of expertise [28]. The impact of advanced knowledge is regarded as economic capital as academic courses meet the requirements of professional bodies [21]. As economic pressures from globalisation increase, so does the nature of postgraduate education evolve to compete internationally and emerge as a competitive advantage for those undertaking it [21,27]. The benefits of undertaking postgraduate studies create opportunities to deliver knowledge to industries trying to expand within their sector [21]. New knowledge explicates practical implications to industries and this emphasises that a mutually inclusive relationship exists [12]. According to [21] the knowledge economy is a key driver of innovation and change which ignites economic power. The relationship between universities and the private sector has increased and shows that the value of postgraduate research is a remedy to identify gaps within private institutions. Thereby, postgraduate studies develop the mind through the transfer and exchange of knowledge [21]. Moreover, a postgraduate qualification is a specialisation capable of identifying and addressing problems not previously encountered [16,28]. This leads to competitive advantages for the individuals owning the degree and promotes a complex skillset [4]. In complex environments, organisations face challenges and require highly talented skills to ensure protection of their brand to reduce risk from the external business environment [28].

The value inherent in postgraduate studies is the 'know-why' of processes and procedures relevant for an organisation to thrive [12,21]. This culminates as a stock of knowledge that will flow, transfer and intensify knowledge transactions. Knowledge stock as a concept is maximised through a collection of expertise and creative problem solving capabilities [23]. Significantly, postgraduate education has the advantage of offering deeper understanding, precision of information and the ability to extract root-causes of real-world dilemmas [12]. This identifies a postgraduate qualification as a taxonomy of competitive knowledge [27].

The importance of industry interaction, and the fact that students are aware that their ultimate goal would be employment after their studies, was highlighted in [5]. An important aspect of postgraduate studies was captured in a question related to how postgraduate studies prepared alumni for their careers. In addition to positive feedback regarding the academic foundation provided by certain modules and the training in technical skills, emphasis was placed on the soft skills acquired during postgraduate studies. These include:

- the ability to work independently, being self-motivated and taking responsibility for oneself;
- managing one's own time in order to meet different deadlines under pressure;
- the ability to explore unfamiliar topics when needing to solve a problem;

- the ability to think and to use logic to solve problems;
- the ability to communicate effectively through written reports and oral presentations.

Although not specifically asked in [6], a general theme that came through was the value (or otherwise) of doing further postgraduate studies, specifically M.Sc. and Ph.D. The obvious tension here is between being employed after the Honours degree and working on a career as well as earning an income, compared to staying on as a student for a further 2–5 years. The general consensus in [6] was that not much initial benefit seems to be gained in industry from having such degrees. On the other hand, many graduates highlighted the value of these qualifications 10–15 years later in their careers. They stated that such graduates are better positioned in companies at that stage. Similar findings were reported in [14]. This trend seems to be more prevalent in developed countries [10,14]. One respondent indicated that a Ph.D. degree has benefits when its holder is involved with negotiations with international companies.

In an earlier alumni study in our Department of Computing Sciences on the topic of higher education and international mobility [7] it was found that respondents who graduated in the past 8 years are still mainly residing in South Africa, with only 9% residing in other countries. By contrast, 32% of the respondents who graduated before 2008 live now outside South Africa [7]. Moreover, the respondents were also asked to tell any additional degrees completed after their graduation from our Department of Computer Sciences. The results showed that only a small number of the respondents (15%) completed another qualification. Thereby the most popular second degree chosen was the Master of Business Administration (MBA); those respondents all moved into management positions at IT organisations [7].

4   Alumni Postgraduate Survey Results

Our *Alumni Postgraduate Perceptions* survey was completed by 114 NMU CS and IS alumni working in the ICT industry. A total of 390 viewed the survey, 213 started the survey and 114 completed the survey, i.e.: a 54% completion rate. The response distribution was from 23 countries: South Africa (62%), USA (7%, including California, Colorado, Florida, Georgia, Indiana, Minnesota, Nebraska, Washington), Great Britain (7%), Canada (2%), Continental Europe (3%, including Italy, Germany, Netherlands, Norway), Australia and New Zealand (3%); in 11% the locations were not disclosed. Standard biographical data (like gender and race) were not included in the survey. Out of the 114 respondents, 111 indicated their highest qualification. The distribution: B.Com. and B.Sc.: $n = 24$ (22%), Honours: $n = 58$ (52%), M.Sc.: $n = 22$ (20%), and Ph.D.: $n = 7$ (6%); see Fig. 2.

4.1   Reasons For or Against the Honours Degree in CS/IS

The results from the thematic analysis of our first question indicated a positive response to completing postgraduate Honours degree. The respondents indicated

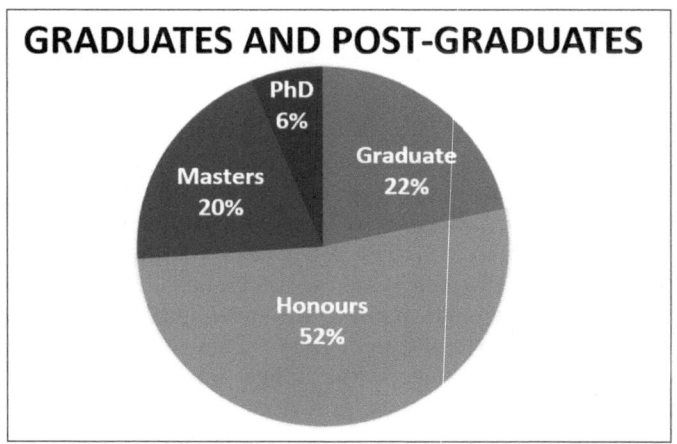

Fig. 2. Number of graduates and postgraduates in our survey

that completing an Honours degree provides better job opportunities (5%), provides in-depth knowledge of specific subject areas (14%), distinguishes the holder in the market place (13%), and is important for overseas job opportunities (10%). Negative comments relating to this question mainly focused on finance and highlighted the importance of relevant industry experience (9%).

Those graduates ($n = 24$) who decided to go and work in industry and did not complete any postgraduate qualification indicated that doing an Honours degree offers better job opportunities, faster growth in the market, obtaining more specialised skill sets and the broadening and deepening of IT knowledge. They did recommend that graduates working in industry should consider conducting postgraduate studies in part-time mode. The graduates who chose to work in industry indicated that they decided to go into industry due to financial circumstances and that industry provided experience which is more relevant to them. Five graduates who completed their studies after 2007 highlighted the importance of finances and that they went to work due to family expectations.

The Honours alumni ($n = 58$) highlighted the broader and more in-depth knowledge gained (17%), the importance of overseas job opportunities (14%) and that an Honours degree prepares the student for a more specific career path (12%). The Master degree and Ph.D. respondents ($n = 29$) highlighted the fact that an Honours degree provides the holder with better job opportunities (10%), distinguishes the holder in the market place (48%), broadens and provides in-depth knowledge (52%) and is important for overseas job opportunities (38%).

4.2   Reasons For or Against the M.Sc. Degree in CS/IS

The overall response to this question (with $n = 111$) was that the advanced degree provides better job opportunities, allows a student to specialise in a specific field, results in higher salaries, that it is important for overseas job opportunities and that it prepares a student for managerial positions. On the negative side, regarding the completion of the M.Sc. degree, the respondents indicated that one could become over-qualified, that one should rather gain industry experience (15%), and that one should better study it in part-time.

The Honours degree alumni ($n = 58$, 58%) mentioned better job opportunities, broadening their knowledge, higher salaries and overseas job opportunities. However, 10 Honours alumni indicated that students should rather gain industry experience and only choose to do the Master degree if their wish was to become an academic. The Master degree alumni ($n = 22$, 20%) stated the advantages of a Master degree as higher salaries, importance for managerial positions and employability. It was also highlighted that a Master student must choose a relevant research topic. The Ph.D. alumni ($n = 7$, 6%) mentioned the importance of completing the M.Sc. degree because it provided overseas job opportunities, made graduates more employable, improved analytical and writing skills and promoted greater independence.

4.3   M.Sc. in CS/IS versus MBA

A total of 109 respondents answered the question relating to the completion of a CS/IS Masters degree versus the MBA: see Table 1. The majority indicated not to do either ($n = 40$, 36%), whereas $n = 20$ (18%) indicated the M.Sc. degree, $n = 18$ (16%) indicated the MBA, and $n = 31$ (28%) stated that it depended on where one were in one's career and whether somebody wanted to move into management (MBA) or stay technology-focused (M.Sc. degree). The graduate alumni ($n = 24$) indicated it depended on the situation (42%, $n = 10$); 4 (17%) recommended the MBA, 1 (4%) the M.Sc. degree, and 8 (33%) indicated not to do either. The Honours alumni ($n = 58$, 52%) indicated it depended on the situation ($n = 12$, 21%); 9 (15%) recommended the MBA, 13 (22%) recommended the M.Sc. degree, and 22 (40%) indicated not to do either. The M.Sc. and Ph.D. alumni ($n = 29$, 26%) indicated it depended on the situation ($n = 9$, 31%); 5 (17%) recommended the MBA, 6 (20%) recommended the M.Sc. degree, and 10 (34%) indicated not to do either. The majority of the respondents highlighted the value of completing an appropriate master-degree, keeping in mind one's personal future career advancement.

Table 1. Alumni recommendations: M.Sc. versus MBA degree

Respondents/recommend →	"M.Sc."	"MBA"	"it depends"	"neither"
All: $n = 111$	18%	16%	28%	36%
Graduates: $n = 24$	4%	17%	42%	33%
Honours: $n = 58$	22%	15%	21%	40%
M.Sc. and Ph.D.: $n = 29$	20%	17%	31%	24%

4.4   When Should the Ph.D. Degree be Considered

The overall opinion of the alumni respondents indicated that completing the Ph.D. degree makes the holder an expert in a specific field of study. The group indicated that one must do a Ph.D. if one wants to pursue an academic career (31%) or pursue research (12%); 19% were unsure. The Honours alumni highlighted that a Ph.D. is done when a person wants to pursue an academic career (33%) or research (10%). The Master- and Ph.D.-holding alumni stated that a Ph.D. degree is suitable for an academic career, or a specialised industry career. Completing a Ph.D. degree further allows extensive research to be conducted, but also enables an individual to become an entrepreneur.

4.5   Advice to 3rd-Year Students About Postgraduate Studies versus Immediate Industry Employment

The entire group of respondents (n = 109) indicated that postgraduate studies separate individuals in the job market (16%), create better job opportunities (20%), and, most importantly, improve career opportunities (53%). Going into industry provides valuable technical experience (8%), is better for one's career (9%), and provides a better platform for gaining practical experience and knowledge (12%). The graduate respondents ($n = 24$) indicated that completing postgraduate studies or going into industry depends on somebody's financial situation. Of this sub-group, 29% emphasised the importance of finance, repaying study loans, or the un-affordability of postgraduate studies. They indicated nevertheless that postgraduate studies improve career opportunities and that if one can obtain a bursary it is the option one should choose. However, they also indicated that industry experience is valuable (33%) and that studying in part-time is a better option and more financially affordable (29%) if one does not have a grant or bursary. The advice provided by Honours alumni ($n = 58$) mainly focused on the importance of a 4-year qualification when considering overseas employment opportunities (24%), and that postgraduate studies improve one's career opportunities (34%). Three Honours alumni also indicated that completing postgraduate studies limits one's time in industry. The advice provided by Master- and Ph.D.-holding alumni focused on skills demanded by industry and improved career opportunities by completing postgraduate studies first. They highlighted the fact that studying part-time while working is very difficult and that the correct industry experience could be advantageous for a 3rd-year student's career.

5   Conclusions

CS/IS graduates at HEIs are faced with decisions to be made regarding postgraduate studies and different career opportunities. Industry is experiencing a vast skills shortage, and recruitment agencies and industry representatives are encouraging young graduates to enter the IT job market quickly. Academics try to encourage young graduates to complete postgraduate studies in order to enhance their skills, i.e.: to complete at least a 4-year qualification that will allow them to apply for positions internationally.

CS/IS graduates *perceive academics to be biased* and that academics encourage them to complete a postgraduate qualification only in order to conduct research and to increase their own postgraduate student numbers. 3rd-year students thus do not receive information about going to work in industry or completing postgraduate studies from their peers and graduates already working in the industry. Our study focused on this gap, providing valuable information from alumni, with different postgraduate qualifications, regarding questions faced by many 3rd-year students.

The results from the thematic analysis of our first question (relating to completing the Honours degree) indicated a positive response to completing postgraduate Honours studies. The respondents ($n = 111$) indicated that the qualification provides better job opportunities, provides in-depth knowledge of specific subject areas, distinguishes the degree-holder in the market place, and is important for international job opportunities. On completing a Master degree, the Honours alumni mentioned that a Master degree provides better job opportunities, broadens its holder's knowledge, enables the earning of higher salaries, and is also important for overseas job opportunities. However, 10 Honours alumni also indicated that a student should rather gain industry experience and only choose to pursue a Master degree if wishing to become an academic.

The question relating to completing a CS/IS M.Sc. degree or a MBA provided interesting feedback. The literature highlighted the importance of both [1, 7, 26]. The results of our study indicated that alumni must consider their career options when deciding to complete either a CS/IS M.Sc. degree or the MBA. The advice provided by alumni was that it depended on somebody's current career position and whether one wanted to move into management (MBA) or remain technology-focused (M.Sc. degree). Our findings support the claim [26] that the MBA is more advantageous for careers in management.

The advice the different alumni graduates would provide to a 3rd-year student raised some interesting points. The advice provided by graduate respondents indicated that completing postgraduate studies or going to industry depends on somebody's current financial situation. This respondents emphasised the importance of money [9], repaying study loans, or the un-affordability of postgraduate studies. The advice provided by Honours alumni mainly focused on the importance of a 4-year qualification when considering international employment opportunities. The advice provided by Master- and Ph.D.-holding alumni focused on industry skill demands and improved career opportunities by completing all postgraduate studies before employment. They emphasised the importance of completing a 4th-year Honours CS/IS qualification and further indicated that studying in part-time was very difficult.

Alumni can be a rich source of information about career advice and industry trends. Obtaining input from alumni on the quality and relevance of their education can provide a department with valuable information and assist with continuous improvement and self-evaluation. Alumni surveys can provide information relating to academic programme quality and relevance, alumni experiences and specifically employment success. Maintaining contact with alumni is essential for obtaining information on course relevance and industry requirements. Departments are utilising social media platforms such as Facebook and LinkedIn to establish alumni groups, maintaining contact with graduates and providing reasonably accurate information about their current place and position of employment.

All in all our exploratory study has provided the foundation for continuous alumni feedback and stakeholder engagement. Valuable opinions and information regarding career decisions and postgraduate studies were obtained from alumni working in the industry. The noteworthy findings from alumni participating in this survey indicated that postgraduate studies ought to be a very important consideration for graduating students and that capable 3rd-year CS/IS students should at least complete a 4-year qualification (if they academically meet the minimum progress requirements). The results of our study indicate that the financial sustainability of postgraduate studies relies heavily on the availability of bursaries. Future research may extend this exploratory study to include more quantitative data and to include more recently graduated alumni.

References

1. Baruch, Y., Bell, M.P., Gray, D.: Generalist and specialist graduate business degrees: tangible and intangible value. J. Vocat. Behav. **67**(1), 51–68 (2005)
2. Beidler, J.: Assessment: an alumni survey. In: 32nd Annual Conference on Proceedings of the Frontiers in Education. IEEE (2002)
3. Blass, E., Jasman, A., Shelley, S.: Postgraduate research students: you are the future of the academy. Futures **44**(2), 166–173 (2012)
4. Boisot, M.: Knowledge Assets: Securing Competitive Advantage in the Information Economy. Oxford University Press, Oxford (1998)
5. Calitz, A.P., Greyling, J.H., Cullen, M.D.: South African industry ICT graduate skills requirements. In: Proceedings of the 43rd Annual Conference on Southern African Computer Lecturers' Association, SACLA'2014, pp. 25–26 (2014)
6. Calitz, A.P., Greyling, J., Glaum, A.: CS and IS alumni post-graduate course and supervision perceptions. In: Gruner, S. (ed.) SACLA 2016. CCIS, vol. 642, pp. 115–122. Springer, Cham (2016). doi:10.1007/978-3-319-47680-3_11
7. Glaum, A.: Alumni perceptions of the NMMU department of computing sciences. Technical report, Nelson Mandela University (2016)
8. Gonçalves, G.R., Ferreira, A.A., de Assis, G.T., Tavares, A.I.: Gathering alumni information from a web social network. In: Proceedings of the 9th Latin American Web Congress, LA-WEB, pp. 100–108. IEEE (2014)
9. Gruner, S.: How big is your IT department? S. Afr. Comput. J. **38**, 62–63 (2007)
10. Hippler, H. (ed.): Ingenieurspromotion—Stärken und Qualitätssicherung: Beiträge eines gemeinsamen Symposiums von Acatech, TU9, ARGE TU/TH und 4ING. Springer, Heidelberg (2011). doi:10.1007/978-3-642-23662-4
11. Kiley, M.: Developments in research supervisor training: causes and responses. Stud. High. Educ. **36**(5), 585–599 (2011)

12. Kim, H.W., Kwak, S.M.: Linkage of knowledge management to decision support: a systems dynamics approach. Technical report, National University of Singapore (2002)
13. Le, Q.: E-Portfolio for enhancing graduate research supervision. Qual. Assur. Educ. **20**(1), 54–65 (2012)
14. Lee, H.F., Miozzo, M., Laredo, P.: Career patterns and competences of PhDs in science and engineering in the knowledge economy: the case of graduates from a UK research-based university. Res. Policy **39**(7), 869–881 (2010)
15. Lending, D., Mathieu, R.G.: Workforce preparation and ABET assessment. In: Proceedings of the 48th Annual Conference on Computer Personnel Research, pp. 136–141. ACM (2010)
16. Litalien, D., Guay, F., Morin, A.J.: Motivation for PhD studies: scale development and validation. Learn. Individ. Differ. **41**, 1–13 (2015)
17. McGourty, J., Besterfield-Sacre, M., Shuman, L.J., Wolfe, H.: Improving academic programs by capitalizing on alumni's perceptions and experiences. In: 29th Annual Conference on Proceedings of the Frontiers in Education. IEEE (1999)
18. Mijic, D.: Design, implementation, and evaluation of a web-based system for alumni data collection. E. Soc. J. Res. Appl. **3**(2), 25–32 (2012)
19. Mijic, D., Jankovic, D.: Towards improvement of the study programme quality: alumni tracking information system. In: Kocarev, L. (ed.) ICT Innovations 2011. AINSC, vol. 150, pp. 291–300. Springer, Heidelberg (2012). doi:10.1007/978-3-642-28664-3_27
20. Mueller, E.F., Flickinger, M., Dorner, V.: Knowledge junkies or careerbuilders? A mixed-methods approach to exploring the determinants of students' intention to earn a PhD. J. Vocat. Behav. **90**, 75–89 (2015)
21. Ng, A.K., Koo, A.C., Ho, W.J.: The motivations and added values of embarking on postgraduate professional education: evidences from the maritime industry. Transp. Policy **16**(5), 251–258 (2009)
22. Pinto, L.H., Ramalheira, D.C.: Perceived employability of business graduates: the effect of academic performance and extracurricular activities. J. Vocat. Behav. **99**, 165–178 (2017)
23. Rudež, H.N.: Intellectual capital—a fundamental change in economy: a case based on service industries. In: 5th International Conference on Proceedings of the Intellectual Capital and Knowledge Management. University of Primorska (2004)
24. Schneider, S.C., Niederjohn, R.: Assessing student learning outcomes using graduating senior exit surveys and alumni surveys. In: Proceedings of the Frontiers in Education Conference. IEEE (1995)
25. Steele, A., Cleland, S.: Staying LinkedIn with ICT graduates and industry. In: Proceedings of the ITX, pp. 8–10 (2014)
26. Vioreanu, D.: 7 very good reasons to do an MBA. Technical report, MastersPortal (2017)
27. Wiegerová, A.: A study of the motives of doctoral students. Proc.-Soc. Behav. Sci. **217**, 123–131 (2016)
28. Wybo, J.L., van Wassenhove, W.: Preparing graduate students to be HSE professionals. Saf. Sci. **81**, 25–34 (2016)

Computer Programming Education

Problem Solving as a Predictor of Programming Performance

Glenda Barlow-Jones[1](✉) and Duan van der Westhuizen[2]

[1] Department of Applied Information Systems, University of Johannesburg,
Auckland Park, South Africa
glendab@uj.ac.za
[2] Department of Science and Technology Education, University of Johannesburg,
Auckland Park, South Africa
duanvdw@uj.ac.za

Abstract. The purpose of this paper is to establish what correlation exists between students' problem solving ability and their academic performance in 1st-year programming courses. The students' achievement in the programming courses is specified as the dependent variable and four programming aptitude tests for logical reasoning, non-verbal reasoning, numerical reasoning and verbal logic are specified as the independent variables. The study group consists of 379 students. Our findings show a correlation between students' logical reasoning, numerical reasoning and verbal logic and performance in computer programming modules. The correlation between students' non-verbal reasoning and performance in computer programming modules was, however, not significant.

Keywords: Computer programming · Problem solving · Logical reasoning · Numerical reasoning · Verbal logic · Non-verbal reasoning

1 Theoretical Background

In this paper, we firstly explore some theoretical constructs around the notion of 'problem-solving'. We particularly invoke Bloom's revised taxonomy and literature around critical thinking. We then explore the difficulties that students typically experience in learning how to problem solve. Finally, we report on a project during which the correlation between the dimensions of problem-solving and student performance was calculated. It is difficult to determine what knowledge and skills first year programming students possess prior to their programming course. The main objective of computer programming is to implement programs that solve computational problems. In [3,10] we can find that problem solving ability is an indicator of programming performance. Critical thinking, also referred to as problem solving, reasoning or higher order thinking skills, can be defined as *"disciplined, self-directed thinking which exemplifies the perfections of thinking appropriate to a particular mode or domain of thought"* [18] and also as *"a process of gathering and evaluating data to make decisions and*

© Springer International Publishing AG 2017
J. Liebenberg and S. Gruner (Eds.): SACLA 2017, CCIS 730, pp. 209–216, 2017.
https://doi.org/10.1007/978-3-319-69670-6_14

solve problems" [15]. Taxonomies of learning have been implemented worldwide to describe learning outcomes and assessment standards reflecting what learning stage a student is at. The original learning taxonomy developed by Bloom and several of his colleagues in 1956 identifies six levels of thought [2]:

- *Knowledge:* rote memorization, recognition, or recall of facts;
- *Comprehension:* understanding what the facts mean;
- *Application:* correct use of the facts, rules, or ideas;
- *Analysis:* breaking down information into component parts;
- *Synthesis:* combination of facts, ideas, or information to make a new whole;
- *Evaluation:* judging or forming an opinion about the information or situation.

These levels of thought start from the lowest order process to the highest order process with higher levels building on lower levels [2]. Once a student reaches the highest level they can be said to have grasped a subject matter. Bloom's Taxonomy was revised in 2011 to address the differences between comprehension and application and to better define the term evaluation. The changes made to the revised taxonomy are as follows [1]:

- *Remember* (previously 'knowledge');
- *Understand* (previously 'comprehension');
- *Apply* (previously 'application');
- *Analyse* (previously 'analysis');
- *Evaluate* (previously 'evaluation');
- *Create* (previously 'synthesis').

According to [9,12], students learn to write complete computer programs in their first year of a programming module which falls within the top two levels of Bloom's Revised Taxonomy of teaching and learning [1]. These two levels however, depend on the first four levels before a student is said to be able to grasp computer programming. For example, in computer programming, 'learning syntax' is the lowest order process [11] and efficiently utilising syntax in order to 'produce effective computer programs' is the highest order process [4]. Accordingly, lecturers expect students to be able to write programs within the first few weeks of their programming module [6]. These programs may be basic and get more difficult as the module progresses, however, many students may be left behind whilst still struggling to find solutions to basic problems. This means that the difficulty level at which the programming module starts, is already at too high a level for a novice programmer, which can lead to a lack of motivation and ultimately a student failing the module. Although novice students may have little experience programming, they do have experience solving problems in everyday life [19]. Problem solving is a mental process of analysing a given problem, developing a solution to the problem and presenting the solution [13]. When students solve problems either independently or in collaboration with other students they are learning by doing. While learning by doing is synonymous with problem solving [13] computer programming as a discipline is also synonymous with problem solving. However, according to [7], students struggle to solve problems for the following reasons:

- Students do not fully understand the problem either because they have not interpreted the problem statement correctly or they just want to start writing code.
- Students fail to transfer the knowledge that they have already acquired from past problems over to new problems.
- Students who take too long to find a solution just give up trying and wait for the solution to be given to them.
- Many students do not have enough mathematical and logical knowledge.
- Students lack specific programming expertise and struggle to detect simple syntactical and logical programming errors.

According to [3], more attention should be paid to novice programming students' problem-solving abilities by encouraging them to practice problem solving, as learning to solve problems algorithmically contributes to learning to program. Students need to think about the processes they go through in solving everyday life problems and look at how to use the same processes to develop algorithms, for example: *"they need to identify things that are familiar to them, divide the problem into smaller problems and use existing solutions"* [5] — the very same things that [7] identifies as what students struggle with. The purpose of this paper is to establish what correlational relationship exists between students' problem solving ability and their academic performance in first-year level programming courses.

2   Method

The participants of the study were a group of 186 first year students enrolled for the National Diploma in Business Information Technology (NDBIT) at the University of Johannesburg (UJ), and 193 first year students enrolled for the National Diploma in Information Technology (NDIT) at the Tshwane University of Technology (TUT).[1] The research process was preceded by a thorough literature review.

2.1   Instrumentation

Four programming aptitude tests (which measured a student's ability to problem solve) were used with permission from the University of Kent Careers and Employability Service Department. All tests were completed in a test-like setting with hired venues and appointed invigilators.[2] Ten items from each test were used to reduce the load on students and due to time constraints.

---

[1] Explanation for readers from outside South Africa: the 'National Diploma' in South Africa consists of a vocational 2-year curriculum below the level of a Bachelor of Science degree.

[2] Students were required to complete consent forms stipulating what was expected from them during the research process. Permission to conduct the research was sought from the Ethics Committee at the Faculty of Education at the UJ and from the Research Ethics Committee at the TUT. In all cases, data were collected responsibly and recorded as accurately as possible.

A B C D E F G H I J K L M N O P Q R S T U V W X Y Z
Question:
What is the missing letter in this series: c c d ? e f g g h
Answer: e

Fig. 1. Logical reasoning test questions

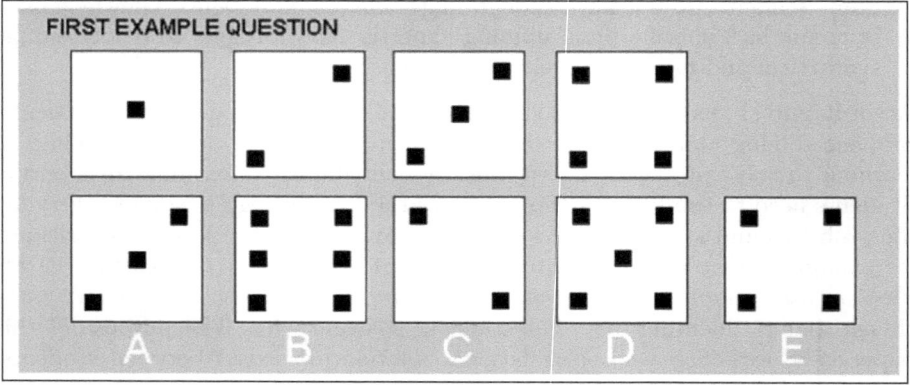

Fig. 2. Non-verbal reasoning test questions: In the first example question the top row of four boxes make up a series from left to right. You have to decide which of the 5 boxes underneath, marked A to E, will be the next in the sequence. For example in the first example, the top four boxes have 1, 2, 3, and 4 dots respectively. Obviously, the next box in the sequence will have 5 dots, which is box D. Answer 1 = D

Logical Reasoning Test. The first programming aptitude test was the logical reasoning test which involved letter sequences and tested the students' ability to think logically and analytically. The test involved looking at a specific sequence of letters and working out the next letter of the sequence: see the example in Fig. 1. The average score for the logical reasoning test was 6.7/10.

Non-verbal Reasoning Test. The second programming aptitude test was the non-verbal reasoning test which determined a student's ability to understand and analyse visual information and solve problems using visual reasoning — for example: identifying relationships, similarities and differences between shapes and patterns, recognizing visual sequences and relationships between objects, and remembering these. The non-verbal reasoning test enabled students to analyse and solve complex problems without relying upon or being limited by language skills. The test involved looking at a specific sequence and working out the next member of the sequence from the pictures given: see the example in Fig. 2. The average score for the non-verbal reasoning test was 4.65/10.

Question:
A taxi driver works 46 weeks of the year and gets an average of 70 customers per week averaging 4 kilometers each at 90 cents per kilometer. His expenditure is as follows:

Car service/repair/insurance:	R1,250,00 per year
Petrol costs:	R0.06 per kilometer
Mortgage costs:	R250,00 per month
Other expenditure — food, electricity, etc.: R125 per week

What is the total income in Rands of the taxi driver for the whole year?

Answer:
Average fare = 4 × 90c = R3.60
Income per week = 70 fares at R3.60 each = 70 × 3.60 = R252
Income for 46 weeks work = R252 × 46 = R11 592

Fig. 3. Numerical reasoning test questions

Question:
Simon, Cheryl and Dannii are going by train to Pretoria to watch a singing competition.
Cheryl gets the 2.15 pm train.
Simon's train journey takes 50% longer than Dannii's.
Simon catches the 3.00 train.
Dannii leaves 20 minutes after Cheryl and arrives at 3.25 pm.

When will Simon arrive?

Answer:
Dannii leaves at 2.35, arrives 3.25, therefore 50m journey.
Simon's journey takes 75m, therefore arrives at 4.15

Fig. 4. Verbal logic test questions

Numerical Reasoning Test. The third programming aptitude test was the numerical reasoning test which included mathematical questions: see the example in Fig. 3. The average score for the numerical reasoning test was 3.24/10.

Verbal Logic Test. The fourth programming aptitude test was the verbal logic test which included verbal logic puzzles, some of which had a numerical element. This test, tested the students' ability to think logically, analytically and numerically, and also to extract meaning from complex information: see the example in Fig. 4. The average score for the verbal logic test was 2.79/10.

Programming Examination Results. Data were also collected from the examination results of the students programming module, Development Software 1: UJ — Development Software 1A (DSW01A1) and Development Software 1B (DSW01B1) and; TUT — Development Software 1A (DS0171AT) and Development Software 1B (DS0171BT). Student numbers were used as the key field to link the data sets. The Development Software 1 (DS1) results were used as the dependent variable throughout the study.

3 Data Analysis

The four programming aptitude tests for logical reasoning, non-verbal reasoning, numerical reasoning and verbal logic were correlated with the DS1 final mark of the students. The results of the computation are presented in Table 1.

Table 1. Correlation of programming aptitude tests and DS1 mark

Test type	Correlations	DS1 mark
Logical reasoning test mark	Pearson correlation	.199
	Sig. (2-tailed)	.000
	N	341
Non-verbal reasoning test mark	Pearson correlation	.095
	Sig. (2-tailed)	.078
	N	347
Numerical reasoning test mark	Pearson correlation	.257
	Sig. (2-tailed)	.000
	N	348
Verbal logic test mark	Pearson correlation	.143
	Sig. (2-tailed)	.008
	N	341

3.1 Logical Reasoning

A Pearson product-moment correlation coefficient was computed to assess the relationship between the logical reasoning test mark variable and the students' performance in DS1 variable. Logical reasoning refers to a student's ability to think logically and analytically. There was a small, positive correlation between the two variables, $r = .199$, $n = 341$, $p = .000$. Overall, there was a small, positive correlation between the non-verbal reasoning test mark and performance in DS1.

3.2 Non-verbal Reasoning Test Mark

A Pearson product-moment correlation coefficient was computed to assess the relationship between the non-verbal reasoning test mark variable and the students' performance in DS1 variable. Non-verbal reasoning refers to a student's ability to understand and analyse visual information and solve problems using visual reasoning. There was no correlation between the two variables, $r = .095$, $n = 347$, $p = .078$. The results for this group show an insignificant correlation between students' non-verbal reasoning ability and performance in DS1.

3.3   Numerical Reasoning Test Mark

A Pearson product-moment correlation coefficient was computed to assess the relationship between the numerical reasoning test mark variable and the students' performance in DS1 variable. The numerical reasoning test included mathematical questions. There was a small, positive correlation between the two variables, $r = .257$, $n = 348$, $p = .000$. Overall, there was a small, positive correlation between the non-verbal reasoning test mark and performance in DS1.

3.4   Verbal Logic Test Mark

A Pearson product-moment correlation coefficient was computed to assess the relationship between the verbal logic test mark variable and the students' performance in DS1 variable. The verbal logic test included logical, analytical and numerical questions. There was a small, positive correlation between the two variables, $r = .143$, $n = 341$, $p = .008$. Overall, there was a small, positive correlation between the verbal logic test mark and performance in DS1.

4   Conclusion

The University of Kent's Careers and Employability Service Department assesses student's computer programming aptitude with tests measuring competencies such as numerical reasoning, logical reasoning, verbal reasoning and non-verbal reasoning which are required in computer programming jobs. These tests were adapted for this study. The findings show that there is a correlation between a student's logical reasoning ($r = .199$, $p = .000$), numerical reasoning ($r = .257$, $p = .000$) and verbal logic ($r = .143$, $p = .008$) and performance in computer programming modules. This supports findings of earlier studies [8, 14, 16] telling us that problem solving ability is a major predictor of performance in programming courses. The correlation between students' non-verbal reasoning and performance in computer programming modules was, however, not significant. This could be because the ability to use pictures in thinking is to a large degree a matter of practice, not aptitude [17]. A similar finding was reported in [14]; in this study Computer Science 1 (CS1) students were identified as having experienced difficulties with: decomposing problems, developing sufficient solutions, and re-using previously seen solutions (even for elementary problems). To this end they introduced the course Development of Algorithmic Problem-Solving Skills (DAPSS) to be taken in parallel to studying CS1. The main focus of DAPSS was to set aside the details of the programming language and concentrate on reflective processes, awareness to problem-solving behaviour and development of cognitive skills. Results showed that the DAPSS course had a positive effect on students' problem-solving skills which in turn improved their programming skills [14]. It is thus recommended that the teaching of problem-solving skills at the University of Johannesburg (UJ) and the Tshwane University of Technology (TUT) be introduced as part of the programming module, to provide opportunities to enhance students' programming performance and thinking processes.

References

1. Anderson, L.W., Krathwohl, D.R., Bloom, B.S.: A Taxonomy for Learning, Teaching, and Assessing: A Revision of Bloom's Taxonomy of Educational Objectives. Allyn & Bacon, Boston (2001)
2. Bloom, B.S., Engelhart, M.D., Furst, E.J., Hill, W.H., Krathwohl, D.R.: Taxonomy of Educational Objectives, Handbook. 1: Cognitive Domain. Longman, London (1956)
3. Chao, P.: Exploring students' computational practice, design and performance of problem-solving through a visual programming environment. Comput. Educ. **95**, 202–215 (2016)
4. Cooper, S., Dann, W., Pausch, R.: Alice: a 3-D tool for introductory programming concepts. Comput. Sci. Coll. **15**(5), 107–116 (2000)
5. Dale, N., McMillan, M., Weems, C., Headington, M.: Programming and Problem Solving with Visual Basic.NET. Jones & Bartlett Learning, Burlington (2003)
6. Gomes, A., Mendes, A.J.: Bloom's taxonomy based approach to learn basic programming. In: World Conference on Educational Multimedia, Hypermedia and Telecommunications, Association for the Advancement of Computing in Education (2009)
7. Gomes, A., Mendes, A.J.: Learning to program – difficulties and solutions. In: Proceedings International Conference on Engineering Education (2007)
8. Kimmel, S.J., Kimmel, H.S., Deek, F.P.: The common skills of problem solving: from program development to engineering design. Int. J. Eng. Ed. **19**(6), 810–817 (2003)
9. Lister, R.: On blooming first year programming, and its blooming assessment. In: Proceedings Australasian Conference on Computing Education (2000)
10. Liu, C., Cheng, Y., Huang, C.: The effect of simulation games on the learning of computational problem solving. Comput. Educ. **57**(3), 1907–1918 (2011)
11. Mayer, R.E.: Teaching and Learning Computer Programming: Multiple Research Perspectives. Routledge, Abingdon (2013)
12. Meerbaum-Salant, O., Armoni, M., Ben-Ari, M.: Learning computer science concepts with scratch. Comput. Sci. Educ. **23**(3), 239–264 (2013)
13. Muller, O.: Pattern oriented instruction and the enhancement of analogical reasoning. In: 1st Proceedings of the International Computing Education Research Workshop, pp. 57–67 (2005)
14. Muller, O., Haberman, B.: A Course Dedicated to Developing Algorithmic Problem-Solving Skills – Design and Experiment. PPIG, Limerick (2009)
15. Paul, R.W.: Critical thinking: what every person needs to survive in a rapidly changing world. Technical report. Center for Critical Thinking and Moral Critique, Sonama State University (1990)
16. Reed, D., Miller, C., Braught, G.: Empirical investigation throughout the CS curriculum. In: Proceedings 31st SIGCSE Technical Symposium on Computer Science Education, pp. 202–216 (2000)
17. Reynolds, C.: Intelligence Testing (2009)
18. Smith, C.J.: Processing thoughts: critical thinking. In: Ethical Behaviour in the E-Classroom: What the Online Student needs to know, pp. 31–43. Elsevier (2012)
19. University of Kent: Programming Aptitude Tests (2013)

Pre-entry Attributes Thought to Influence the Performance of Students in Computer Programming

Glenda Barlow-Jones[1]([✉]) and Duan van der Westhuizen[2]

[1] Department of Applied Information Systems, University of Johannesburg,
Auckland Park, South Africa
`glendab@uj.ac.za`
[2] Department of Science and Technology Education, University of Johannesburg,
Auckland Park, South Africa
`duanvdw@uj.ac.za`

Abstract. This study attempted to isolate seven pre-entry attributes that were thought to influence the performance of students in the module Development Software 1 (programming). The pre-entry attributes included students' problem solving ability, socio-economic status, educational background, performance in school Mathematics, English language proficiency, digital literacy and previous programming experience. We asked to what extent these pre-entry attributes influence our students' performance in computer programming. We found a correlation between the problem solving, digital literacy and previous programming experience with performance in programming. No correlation was found between socio-economic status, educational background, Grade 12 Mathematics and English marks with performance in programming.

Keywords: Computer programming · Pre-entry attributes · Student performance

1 Introduction

The failure rate of students studying to become computer programmers is high [1,9,12,13,25]. As the high failure rate is a cause for concern, we conducted a survey among 379 students from two higher education institutions across a variety of dimensions in order to establish how these dimensions correlate with their performance. In this paper, we focus on the set of pre-existing attributes with which students enter the programming qualification. Dimensions of the socio-economic and education background of students have been reported in the literature to be predictors of students' success in computer programming courses. If we could verify that these pre-entry attributes indeed correlate with student performance, we could then adapt the curriculum and pedagogical approach to account for the deficits that pre-entry attributes bring. This process is illustrated in Fig. 1. It needs to be noted that this paper does not address pedagogical interventions or

© Springer International Publishing AG 2017
J. Liebenberg and S. Gruner (Eds.): SACLA 2017, CCIS 730, pp. 217–226, 2017.
https://doi.org/10.1007/978-3-319-69670-6_15

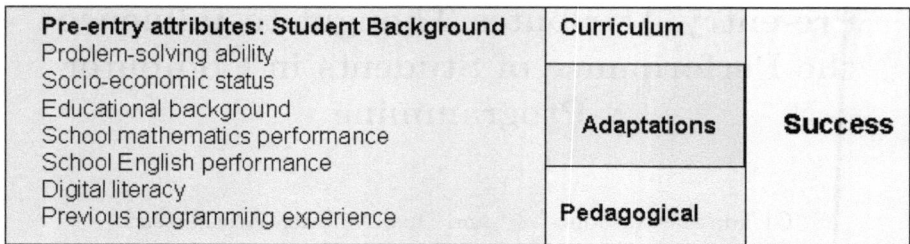

Fig. 1. Pre-entry attributes that may influence a student's success in programming

curriculum adaptations. Reference to these rationalizes the selection of the pre-entry attributes. Considering the identified pre-entry attributes of the students, we formulated seven *null hypotheses:*

H_{01}: There is no relationship between novice South African programming students' *problem solving abilities* and their performance in computer programming modules.

H_{02}: There is no relationship between novice South African programming students' *socio-economic status* and their performance in computer programming modules.

H_{03}: There is no relationship between novice South African programming students' *educational background* and their performance in computer programming modules.

H_{04}: There is no relationship between novice South African programming students' *performance in school mathematics* and their performance in computer programming modules.

H_{05}: There is no relationship between novice South African programming students' performance in *English at school level* and their performance in computer programming modules.

H_{06}: There is no relationship between novice South African programming students' *digital literacy* and their performance in computer programming modules.

H_{07}: There is no relationship between South African programming students' *previous programming experience* and their performance in computer programming modules.

2   Method

Annually, the University of Johannesburg (UJ) has an intake of 100 students for the National Diploma in Business Information Technology and the Tshwane University of Technology (TUT) has an intake of 500 students for the National Diploma in Information Technology.[1] 186 students from the UJ (2 cohorts

---

[1] Explanation for readers from outside South Africa: the 'National Diploma' in South Africa consists of a vocational 2-year curriculum below the level of a Bachelor of Science degree.

2013/2014) and 193 from the TUT were sampled. Repeating students were excluded from the data set. The total sample size for the study consisted of 379 students. 51% of the students were from the TUT and 49% from the UJ.

2.1 Demographics of Respondents

The data that were collected during this study yielded the following demographic description of the respondents. The majority of students were in the age range of 18 to 21 years of age (84%). Their home language was either Sepedi, isiZulu or Xitsonga, with only 9% of students being English, which is the language of instruction at both universities. The sample included 66% male students and 34% female students. It has been reported in many computer education studies [8,16, 20,30] that female enrolment in computer science is relatively low. The gender distribution in the sample is thus considered as typical of gender distributions relating to programming module enrolment elsewhere in the world.

2.2 Instrumentation

Several instruments were used to collect data from the students. Firstly, a set of four programming aptitude tests was identified and completed in February 2013 and 2014. The purpose of each of these tests was to determine students' problem solving abilities. Secondly, a 'Student Profile Questionnaire' (SPQ) was developed by us as an instrument to gather data from the respondents on the various pre-entry attributes (as independent variables) that were evaluated in this study, namely: (1) problem solving skills, (2) social economic status, (3) educational background, (4) performance in school Mathematics, (5) performance in school English, (6) digital literacy, and (7) previous programming experience; see Fig. 1. The SPQ was completed in November 2013 at the UJ and the TUT and in February 2014 for the next intake at the UJ. Thirdly, examination results of 2013 and 2014 students were tabulated.

Programming Aptitude Tests. The programming aptitude tests assessed competencies such as numerical reasoning, non-verbal reasoning, logical reasoning, and verbal logic which are required in technical computing jobs. These tests were used with permission from the University of Kent Careers and Employability Service Department [26]. All tests were completed in a test-like setting with hired venues and appointed invigilators. Ten items from each test were used to reduce the load on students and due to time constraints.

Student Profile Questionnaire. The questions contained in the SPQ were mainly based on a review of the international literature enhanced by studying the South African context. The review of the international literature yielded seven pre-entry attributes that have been shown to influence student academic performance in computer programming courses. These seven pre-entry attributes were converted into thirty four questions. These questions were enhanced by uniquely

Table 1. Variables addressed in the SPQ

Variable	Example
Problem solving skills	I try to understand problems before I attempt to solve them. AN = almost never true for me, S = sometimes true for me, HT = true for me about half of the time, O = often true for me, AA = almost always true for me and NA = I cannot respond to the statement/I do not understand the statement
Socio-economic status	How would you describe the immediate environment in which you grew up? (i) informal settlement, (ii) village/rural, (iii) township, (iv) town, (v) suburb of a city, (vi) inner city
Education background	I was encouraged to read a lot to improve my knowledge. The response options for the items ranged from SD = strongly disagree, D = disagree, A = agree, SA = strongly agree, NA = not applicable
Performance in school Mathematics	Students were asked to indicate their Grade 12 Mathematics mark
Performance in school English	My English ability prevents me from performing well academically. The response options for the items ranged from SD = strongly disagree, D = disagree, A = agree, SA = strongly agree, NA = not applicable
Digital literacy	An example of an item is "I used the Web to make phone calls, e.g. Skype". AN = almost never true for me, S = sometimes true for me, HT = true for me about half of the time, O = often true for me, AA = almost always true for me, and NA = I cannot respond to the statement / I do not understand the statement. This section of the questionnaire which consisted of 24 questions was adapted from a survey developed by Kennedy et al. [18] from the Lingnan University in Hong Kong. Students were asked to indicate their use of technology and access to technology in order to determine how digitally literate they were
Previous programming experience	How much programming experience did you have before enrolling at university? (i) I had no programming experience, (ii) I had some programming experience, (iii) I had quite a bit of programming experience, (iv) I had advanced programming experience

South African factors selected on the basis of local knowledge and personal experience gained working in a South African Higher Education Institution (HEI). During the development of the questionnaire, we attempted to create items to measure the independent variables using Likert-type items. The SPQ was piloted with five third-year tutors who were asked to complete the SPQ and comment on anything that they did not understand. Tutors were also asked whether the questions were pitched at the correct language level. Their feedback was used to make adjustments to the SPQ which was then shared with colleagues for comments. The questionnaires were completed during the second semester of 2013 at the UJ and the TUT and the first semester of 2014 at the UJ only. The data from the SPQ at the UJ were merged with data from the university's data base to include data about students' grade 12 English and Mathematics marks, as well as demographic data such as their age, gender, and ethnicity (Table 1).

Software Development 1 Exam Results. The examination results of the students programming module, Software Development 1: UJ—Software Development 1A (DSW01A1) and Software Development 1B (DSW01B1), as well as Tshwane University of Technology—Software Development 1A (DS0171AT) and Software Development 1B (DS0171BT) were included in the data set. Student numbers were used as the key field to link the data sets (SPQ, Programming Aptitude Tests and Software Development 1 Exam Results). The Software Development 1 results were used as the dependent variable throughout the study.

3   Discussion of Findings

3.1   Problem Solving

The results revealed some noteworthy findings. There is a correlation between a student's logical reasoning ($r = .199$, $p = .000$), numerical reasoning ($r = .257$, $p = .000$), verbal logic ($r = .143$, $p = .008$), and performance in computer programming modules. This supports findings of earlier studies [19,22,23] which show that problem solving ability is a major predictor of performance in programming courses. The correlation between students' non-verbal reasoning and performance in computer programming modules was, however, not significant. This could be because the ability to use pictures in thinking is to a large degree a matter of practice, not aptitude [24].

3.2   Socio-Economic Status

Very little research could be found on the relationship between a student's socio-economic status and computer programming ability. We considered a students' residential area, dwelling, level of education and employment status of the mother or female caregiver, level of education and employment status of the father or male caregiver. It could be concluded that, for this sample, there was *no* significant correlation between the socio-economic factors mentioned and students' performance in programming modules. The full results can be viewed in Barlow-Jones and van der Westhuizen [5].

3.3 Educational Background

With regard to students' educational background, the literature review high-lighted a lack of facilities, poor educational resources, overcrowded classrooms and a lack of qualified teachers as being a contributing factor of a learner's poor performance. Encouragement in the area of critical thinking at school and critical thinking at home, teaching received at school, students' learning, and hours spent studying were addressed—however, there was *no* significant correlation between a student's educational background and performance in programming modules.

3.4 Mathematical Ability

Another main pre-entry attribute is mathematical ability. The analysis of the results suggests that there is *no* correlation between a student's Grade 12 mathematics performance and performance in computer programming modules ($r = -.126$, $p = .079$) and South African National Senior Certificate mathematics performance and performance in computer programming modules ($r = .139$, $p = .192$). This is in contrast to [7,10,14,28] which claimed that performance in mathematics can predict programming performance. There is a global belief that the concepts which a student has to comprehend in order to master mathematics problems are similar to those for computer programming [10]. Mathematics aptitude is thus often a pre-requisite for acceptance into computer programming courses [11]. We believe that a student's mathematics ability may correlate with a student's performance in computer programming modules, however, in the context of this study, students' Grade 12 mathematics marks do not seem to correlate with their mathematics ability. Interestingly there is a correlation between a student's Grade 12 mathematics mark and numerical reasoning test mark ($r = .256$, $p = .000$), Grade 12 mathematics mark and logical reasoning test mark ($r = .109$, $p = 0.49$) as well as Grade 12 mathematical literacy mark and performance in computer programming modules. The full results can be seen in [27].

3.5 English Ability

58.7% of students disagreed that their English ability prevented them from performing well academically even though only 9% of students' first language was reported as being English. There is *no* significant evidence to show a relationship between students' Grade 12 English marks and their DS1 marks. We believe that students' Grade 12 English mark may not be a true reflection of their English ability. The study did find however that students' Grade 12 English marks correlate with their numerical reasoning test marks ($r = .159$, $p = .004$) as well as with their verbal logical test marks ($r = .125$, $p = .026$). The full results can be seen in [4].

3.6 Digital Literacy

42% of the students in our local domain of investigation either used a computer for the first time at University (26%), or had 1 to 2 years computer experience (16%) before embarking on their studies. 47% of the students reported that they did not have access to a computer while growing up. The relationship between students' computer experience and their DS1 marks was found to be significant ($r = .155$, $p = .003$) as was the relationship between students' computer access and their DS1 marks ($r = .149$, $p = .004$). The relationship between students' basic use of technology and their DS1 marks was also found to be significant ($r = .188$, $p = .000$). Paper [2] revealed that any kind of computer experience, no matter how basic, is helpful in learning to program.

3.7 Previous Programming Experience

80% of the students had no previous computer programming experience. The relationship between student's previous knowledge of programming and their DS1 marks was significant ($r = .186$, $p = .000$) which is in agreement with [15, 17, 21, 29]. The full results can be seen in [6].

3.8 Summary

Table 2 summarises all the findings described in the foregoing sub-sections.

Table 2. Acceptance ($\checkmark$) or rejection ($\times$) of the hypotheses

H_{01}	H_{02}	H_{03}	H_{04}	H_{05}	H_{06}	H_{07}
$\times$	$\checkmark$	$\checkmark$	$\checkmark$	$\checkmark$	$\times$	$\times$

4 Limitations

The following limitations of our research were identified:

- The questionnaire was lengthy, and the variables were informed by many items. It is not unreasonable to assume that the length of the questionnaire had an adverse effect on the quality of the responses due to respondent fatigue.
- The selection of the variables that represented pre-entry attributes may not be representative of all possible pre-entry attributes that influence student performance in programming courses. For example, personal attributes like self-efficacy, meta-cognitive awareness, or others were not measured.
- The mediating influence of variables post-enrolment on performance in the programming courses were not accounted for. The extent to which students may have participated in learning support programmes, their time-on-task during the courses, their engagement with tutors, and the pedagogical design of the courses are but a few of possible post-enrolment variables that influenced their performance.

- The study relied on quantitative data only. Engaging students in focus group interviews may have provided for a richer data set for an enhanced and more complete understanding of the influence of the variables on performance.
- Much of the data was self-reported. Therefore, there may be misalignment in the conceptual understanding of students of items related to critical thinking, teacher pedagogy, and so forth, and the actual theoretical constructs that underpin those items.
- The sample was comprised of students who enrolled for programming courses at the South African National Diploma level. Students who enrolled for degree level courses (B.Sc., above National Diploma) were not sampled.

5   Recommendations for Practice

Considering the findings of our study, and the limitations of the study identified above, the following recommendations for practice are given:

1. The HEI may reconsider, for South Africa, the use of Grade 12 Mathematics and English marks as admission requirements to programming courses. A better measure of a students' mathematical and English abilities could be the National Benchmark Tests (NBT) which were introduced into South African universities in 2009 because of the national concern of the quality of students entering universities.
2. NBTs should be made compulsory for first year students studying computer programming courses in South Africa.
3. Grade 12 Mathematics marks should be replaced by the NBT mathematics test scores as admission criterion into higher-education programming courses.
4. Grade 12 English marks should be replaced by the NBT academic literacy test scores as admission criterion into higher-education programming courses.
5. Introducing problem solving support modules in conjunction with computer programming may be beneficial [3].

6   Closing Remarks

This study examined the relationship between seven pre-entry attributes and performance in computer programming modules at two universities in South Africa. The dataset comprised of four programming aptitude tests, a student profile questionnaire and Software Development 1 examination results of 379 students studying the NDBIT and NDIT at the UJ and TUT. Correlations were made between the seven independent variables and the dependent variable (DS1 examination marks). The data analysed indicated that there is a correlation between the variables problem solving, digital literacy and previous programming experience and performance in programming modules. There was no correlation found between the variables socio-economic status, educational background, Grade 12 Mathematics mark and English mark, and performance in programming modules. In conclusion the mark achieved for school Mathematics and English cannot be considered as valid admission criteria for programming courses in the South African context, and an alternate requirement such as the NBTs should be implemented.

References

1. Ali, A., Shubra, C.: Efforts to reverse the trend of enrollment decline in computer science programs. Issues Informing Sci. Inf. Technol. **7**(16) (2010)
2. Allan, V., Kolesar, M.V.: Teaching computer science: a problem solving approach that works. SIGCUE Outlook **25**(1/2), 2–10 (1997)
3. Barlow-Jones, G., van der Westhuizen, D.: Problem solving as a predictor of programming performance. In: Liebenberg, J., Gruner, S. (eds.) SACLA 2017. CCIS, vol. 730, pp. 209–216. Springer, Cham (2017)
4. Barlow-Jones, G., van der Westhuizen, D.: The correlation between students' English proficiency and their grasp of computer programming. In: Proceedings of the Global Learn, pp. 569–573. Association for the Advancement of Computing in Education (2016)
5. Barlow-Jones, G., van der Westhuizen, D.: The role that a students' socio-economic status plays in their performance in computer programming modules. In: Proceedings of the Global Learn, pp. 143–148. Association for the Advancement of Computing in Education (2016)
6. Barlow-Jones, G., van der Westhuizen, D., Coetzee, C.: An investigation into the performance of first year programming students in relation to their grade 12 computer subject results. In: Proceedings World Conference on Educational Multimedia, Hypermedia and Telecommunications, pp. 77–83. Chesapeake (2014)
7. Bergin, S., Reilly, R.: Programming: factors that influence success. In: Proceedings SIGCSE 2005, St. Louis (2005)
8. Biggers, M., Brauer, A., Yilmaz, T.: Student perceptions of computer science: a retention study comparing graduating seniors versus CS leavers. In: Proceedings SIGCSE 2008, pp. 402–406 (2008)
9. Butler, M., Morgan, M.: Learning challenges faced by novice programming students studying high level and low feedback concepts. In: Proceedings ASCILITE, Singapore (2007)
10. Byrne, P., Lyons, G.: The effect of student attributes on success in programming. In: ITiCSE 2001 Proceedings 6th Annual Conference on Innovation and Technology in Computer Science Education, pp. 49–52. ACM (2001)
11. Chumra, G.A.: What abilities are necessary for success in computer science? SIGCSE Bull. Inroads **30**(4), 55a–58a (1998)
12. Corney, M., Teague, D., Thomas, R.: Engaging students in programming. In: Proceedings 12th Australasian Computing Education Conference (2010)
13. Garner, S.: A program design tool to help novices learn programming. In: Proceedings ASCILITE, Singapore (2007)
14. Gomes, A., Mendes, A.: A study on students' characteristics and programming learning. In: Proceedings World Conference on Educational Multimedia, Hypermedia and Telecommunications, pp. 2895–2904, Chesapeake (2008)
15. Hagan, D., Markham, S.: Does it help to have some programming experience before beginning a computing degree program?. In: Proceedings ITiCSE 2000, Helsinki (2000)

16. Hill, C., Corbett, C., St'Rose, A.: Why so few? Women in science, technology, engineering, and mathematics. Technical report, AAUW, Washington (2010)
17. Holden, E., Weeden, E.: The impact of prior experience in an information technology programming course sequence. In: Proceedings CITC4 2003, Lafayette (2003)
18. Kennedy, G., Judd, T.S., Churchward, A., Gray, K., Krause, K.: First year students experiences with technology: are they really digital natives? Aust. J. Educ. Technol. **24**(1), 108–122 (2008)
19. Kimmel, S.J., Kimmel, H.S., Deek, F.P.: The common skills of problem solving: from program development to engineering design. Int. J. Eng. Educ. **19**(6), 810–817 (2003)
20. Krause, J., Polycarpou, I. Hellman K.: Exploring formal learning groups and their impact on recruitment of women in undergraduate CS. In: Proceedings SIGCSE 2012, Raleigh (2012)
21. Kumwenda, B., Rauchas, S., Sanders, I.: The effect of prior programming experience in a scheme-based breadth-first curriculum. In: ITiCSE 2006 Proceedings 11th Annual SIGCSE Conference on Innovation and Technology in Computer Science Education, Bologna (2006)
22. Muller, O., Haberman, B.: A Course Dedicated to Developing Algorithmic Problem-Solving Skills – Design and Experiment. PPIG, Limerick (2009)
23. Reed, D., Miller, C., Braught, G.: Empirical investigation throughout the CS Curriculum. In: Proceedings 31st SIGCSE Technical Symposium on Computer Science Education, pp. 202–216 (2000)
24. Reynolds, C.: Intelligence Testing (2009). http://www.education.com/reference/article/intelligence-testing/
25. Robins, A., Rountree, J., Rountree, N.: Learning and teaching programming: a review and discussion. Comput. Sci. Educ. J. **13**(2), 137–172 (2003)
26. University of Kent: Programming Aptitude Tests (2013). http://www.kent.ac.uk/careers/tests/computer-test.htm
27. van der Westhuizen, D., Barlow-Jones, G.: High school mathematics marks as an admission criterion for entry into programming courses at a South African university. Indep. J. Teach. Learn. **10/2015** (2015). ISSN 1818-9687
28. Wilson, B.C., Shrock, S.: Contributing to success in an introductory computer science course: a study of twelve factors. SIGCSE Bull. **33**, 184–188 (2001)
29. Zhang, X., Zhang, C., Stafford, T.F., Zhang, P.: Teaching Introductory programming to IS students: the impact of teaching approaches on learning performance. J. Inf. Syst. Educ. **24**(2), 147–155 (2013)
30. Zweben, S.: CRA Taulbee Survey Report 2009–2010 (2011)

Programming: A Wicked Subject?

Apostolos P. Giannakopoulos(✉)

University of South Africa, Florida Campus, Johannesburg, South Africa
gianna@unisa.ac.za

Abstract. Studying programming in an Open Distance Learning setup can be more challenging than in a contact setup. It can be characterised as a 'wicked problem'. Wicked problems are problems that are so complex that current problem techniques fail to solve it. Wicked problems require a kind of unorthodox, innovative or creative way. Programming can be considered a subject that presupposes the existence of a number of cognitive functions such as problem solving. Problem solving in general requires critical thinking, and critical thinking is characterised by logic, decision making, paying attention to detail, the availability of all different types of knowledge. All these are prerequisites in the learning of programming. This paper shows that treating programming as a wicked problem can shed some light onto the question why not many students can be successful in programming.

Keywords: Programming · Wicked problems · Types of knowledge

1 Introduction

The term 'wicked' was introduced into the theory of problem-solving by *Horst Rittel* in the 1970s. It was used to describe problems that are either difficult to solve or do not have a solution (though we might think that they have a practical solution but mostly only an ideal solution which is impractical). In the daily life of our country, poverty, shortage of education, crime, etc., could be considered as 'wicked problems'. Characterising *computer programming* as a wicked problem (rather than a problem) calls for unorthodox, innovative ways rather than using standard problem solving techniques. If we do so, once we thought we solved the problem, suddenly it resurfaces.

Programming is a multifaceted and complex combination of skills [14,18,19, 25]. Problem solving and high level thinking skills are prerequisites to programming. If we accept that the learning of programming is a wicked problem then we must also accept that we might never solve the problem but what we can do is minimise the impact that it has on a learner. But what makes the learning of programming a wicked problem? Perhaps one should look at its nature. In simplistic terms it is about converting a solution to a problem into a code that a computer is designed to understand, and to use the computer to solve similar problems. If we think of solving quadratic equations in Mathematics, it is possible to write a program that can solve any quadratic equation by entering the values of a, b and c to

© Springer International Publishing AG 2017
J. Liebenberg and S. Gruner (Eds.): SACLA 2017, CCIS 730, pp. 227–240, 2017.
https://doi.org/10.1007/978-3-319-69670-6_16

$ax^2 + bx + c = 0$? Furthermore, wicked problems require what [5] called 'collective intelligence', a natural property of socially shared cognition, a natural enabler of collaboration. A barrier to collective intelligence is 'fragmentation': a phenomenon that pulls something apart which is potentially whole. According to [5], problem wickedness is a form of fragmentation. This normally happens in a situation (academia included) where there are a number of people that have been faced with a problem for years and according to them they tried everything in their power to solve it, but failed. Even professors could belong to that category. If an attempt is made to create a collaborative approach to solve a problem, there is a tendency by such people to 'highlight' the problem rather than to propose viable solutions. Perhaps one of the reasons could be because they had treated that problem like other problems where usual problem solving methods were used. Such methods are of 'linear' nature. The complexity of a wicked problem, however, requires a multi-dimensional, multi-disciplinary approach. Collaboration should stand in the center of such a problem's solution.

At the University of South Africa (UNISA), having recognized attrition as a wicked problem, I created a project—'Project 2020'—which is a longitudinal study over a five year period to tackle attrition in general, in as many fields as possible, by following a multi-disciplinary collaborative approach. Together with other lecturers, we discussed the programming problem; each one resorted to research the problem in order to then combine the findings. This paper is the first step towards the solution of our problem. Like with the learning of any subject, it is recognized that there are various factors that determine academic success. These factors are general factors that are applicable to learning in general, and particular factors to a specific subject. Therefore, although one might have the aptitude to do programming, general factors could still lead to a student's failure. Alternatively it might possible that one might able to solve complex problems but finds it difficult to construct even a simple program. For example, [14] mentions aptitude with respect to programming. An experiment conducted at the University of Leeds found that there was *no correlation between the final results in programming and measured aptitude*. Other aptitude tests were also inconclusive [17]. Another important finding was the relationship between Mathematics and programming. In [14] we can find studies which found a relationship [1] while other studies found none. At this point of the paper it shall be accepted that 'some' Mathematics is necessary; the question of 'how much' will be argued later. Therefore the learning of programming is treated like any another subject where various general factors play a role as well as factors that are particular to programming. The latter factors are discussed in this paper; the general factors are only summarily recapitulated.

2   Literature

The wickedness of programming can also be highlighted by the fact that *many studies are inconclusive* with respect to factors that affect the teaching and learning of programming [14]. The statement that predicting student performance in a particular course (or even assessment within a course) is a difficult but useful undertaking

Table 1. References on factors that affect the learning of programming

Factors	References
Learning style	[14,16,25]
Thinking style	[14]
Cognitive style	[1,14,22]
Motivation	[1,14,25]
Problem solving	[25]
Attitude to programming	[25]
Self-efficacy	[1,25]
Programming language	[25]
Mathematics	[1,22]
Prior programming experience	[1,22,24]
Computer anxiety	[1,22]
High School academic achievement (M-score or SAT)	[22]
Comfort level	[1]
Effort in (engagement with) programming	[1]
Student attributes	[1]
Perceptions about programming	[1]
Self-regulation	[1]
Study habits (time management)	[16,24]
Number of assignments submitted	[24]

[2] could be used as a starting point of encouragement to tackle a wicked problem. Table 1 shows some identified factors together with their references. What is interesting is that many studies were done during the 1980s, perhaps because that was the time when Computer Science was recognized as a new and necessary subject for secondary and tertiary education. The factors in Table 1 relate to learning in general as well as to programming in particular. Another important study can be found in [11] wherein causal attributions and their underlying properties were investigated. The study concluded that learning strategy on the part of the student, accompanied by effort and teaching strategy by the teacher, played the most important role in the teaching and learning of computer programming.

Several factors that contribute to the learning of programming were identified in [14]: *multiple skills*, as programming requires many skills applied simultaneously; *multiple processes*, where the specification must be converted to an algorithm and finally to code; *language*, whereby there is no specific language that is better than any other. This means that, although a certain language is involved, the essence is not to learn the language, and it is *not* the language that dictates the difficulty. It is like learning English by a Chinese person compared to learning Chinese by an English person. It can be argued that a certain language might be perceived to be more difficult than another, but in essence they are equivalent as they share many aspects. According to [7], programming languages enable expressions of problem solving, and, like in all natural languages, practice is mandatory in order to express and master programming.

Another factor, according to [14], is *educational novelty* since, programming is problem-solving-intensive and requires a significant amount of effort in several skill areas. At the same time it is also 'precision-intensive'. *Interest* is another factor. At worst, programming can be very boring since there are so many 'dos and do nots'. At its best programming can be an enjoyable, creative activity, and many students derive great enjoyment from their programming. They enjoy it even more (and learn more) when they are allowed to work on assignments that *inspire* them.[1] It is a shame that so few assignments do indeed inspire. *Reputation* and image also play a role. According to [14] the fact that programming has a reputation of being difficult contributes to having negative effects. Although many authors agree that programming is one of the hardest subjects to learn [11,25], which ought to increase the social status of capable programmers, programmers are often ridiculed as 'nerds' [14], and (understandably) not many people like to be stigmatised as such.[2] Finally there is *pace*: according to [14] a programming course is like a high-speed train without brakes. It is not that the students cannot do programming but they cannot cope with the speed of the lessons. Such students will quickly come to believe that they just cannot program—the educationalists call this 'learned helplessness'—and will attribute this to the perceived difficulty of the subject itself [14].

'Blended learning' (face-to-face and online) was thus suggested in [25] as a means to overcome many difficulties in the teaching and learning of programming. Accordingly, online learning can assist particularly the introvert students. This of course is related to various thinking and learning styles. In [19] it was argued that learning to program should be addressed from a psychological-educational perspective. For this reason it was suggested to use the idea of novices and experts. Accordingly there is a transition from 'novice' via 'advanced beginner', 'competent' and 'proficient' to 'expert'. This is in line with [15,22] w.r.t. to knowledge acquisition, whereby [15] maintains that there are three stages: the introductory, advanced, and expert. In the introductory stage, the learners have very little directly transferable prior knowledge about a skill or content area. The second phase of knowledge construction is the advanced knowledge acquisition where the learner acts as an apprentice and is able to solve more complex problems, domain- or context-dependent. Expertise is the last phase whereby the learner has acquired more rich interconnected knowledge structures. Constructivist learning environments are said to be most appropriate for advanced knowledge acquisition, as experts need very little instructional support [8,15].

---

[1] For comparison see the 'π' *Pédagogies Innovantes* initiative with its 'FabLab' at Bordeaux INP, http://pi.espe-aquitaine.fr/eirlab-high-tech-fablab/.

[2] This was different in the early days of digital computing when the electronic hardware was scarce and expensive: in those days computer programmers were admired [13].

Furthermore, [19] distinguished between program 'knowledge' and program 'strategy'. For example, being able to state how a FOR-loop works (declarative) is different from being able to apply such knowledge (strategic). The authors of [19] agree that programming ability rests on a foundation of knowledge about computers, programming language, programming tools and resources and theory and formal methods. However, most of the literature is focused only on the declarative part of knowledge in programming [19], although there is more to it.

It was stated earlier that problem solving forms the core business of programming. If that is accepted then there should be a close relationship to Mathematics, since Mathematics is by its nature about problem solving. It is not the content (which is subject-specific) but the knowledge and skills necessary to solve a problem that the two subjects could have in common that give rise to their relationship. To solve any problem, certain knowledge and skills are necessary. What type of knowledge and what skills (predominantly cognitive skills) are necessary are dictated by the problem to be solved. Treating programming as a wicked problem and considering it in a serious manner could bring us to making us see the light at the end of the tunnel. It is not really programming that is really difficult but the learning of it. Like in Mathematics there is not much room to maneuver. But let us look at the structure of a computer program.

As a rule, a programmer must a problem solver. In short, the programmer must be able to identify the problem, understand the problem, devise a method to solve the problem (an algorithm), convert that into the appropriate programming language, and test if it works. That is to evaluate the solution. There are also the basic constructs which, combined, form a working program. Such constructs are variables, data values and data types, statements, control structures, subprograms and parameter passing mechanisms. Then each programming language, like the spoken languages, has *syntax*. The syntax rules require absolute precision. The programmer must also understand the *operational semantics.*[3] Here is some room for flexibility, as often as the programmer has more than one implementation choice in order to meet a given specification. Understanding the meanings of expressions, statements and constructs is also very important [10].

One of the most important aspects in programming is the use of algorithms. Mathematics, for example, is 'riddled' with algorithms. These could be a set of plans combined to form a complex abstract or general plan to solve a problem. Algorithms are sometimes used interchangeably with procedures. However, a procedure is a well-defined process (step 1 do this, step 2 do that, etc.) while an algorithm is more than that [4]. The importance of the necessity for a prospective programmer to be aware of standard algorithms was highlighted in [10]; the programmer must be able to adapt and apply them appropriately to solve a particular problem. Furthermore a program uses data. In Computer Science we get *data structures, abstract data types* (ADTs) and *classes*: programmers need to understand the usage of these and define each one accurately to the point of practical applicability. It was also mentioned in [10] that there are 'obscured' requirements which some novice programmers might consider as 'waste of time':

---

[3] Which is in most textbooks explained only informally and by means of examples.

principles such as robustness, reusability, extensibility, etc.; design, (appropriate designs are necessary for the successful solution of a problem); style, aiming for a good style; documentation, within the code and as a separate document; cooperation among programmers and to be able to understand and explain their own code or someone else's program code. Finally the two most important functions are testing and debugging a finished program. Being able to test the particular and the general case, 'fixing bugs' which could make the program crash, and detecting complex errors that might be difficult to locate, are the final criteria for being a skillful programmer [10].

Analysing the above-mentioned structure and function of a program reveals that different types of knowledge are involved. But knowledge cannot be dissociated from content. It is always knowledge *about* something. Having knowledge of the content, unique to programming, is a necessary but not sufficient condition. According to [3], content knowledge refers to the knowing about a subject, and the disciplinary knowledge of a subject. With respect to Mathematics, mathematical content knowledge includes information such as mathematics concepts, rules and associated procedures for problem solving [8]. Understanding the content implies that the learner possesses all the different types of knowledge. If not, then all content could be considered as declarative, pure information that anyone could 'regurgitate'—then everybody could (utopically) learn any subject.

In [20] knowledge was classified into four basic types: *declarative* knowledge (that includes conceptual knowledge other than facts), *procedural* knowledge, *strategic* knowledge, and *schematic* knowledge. Declarative knowledge answers the question 'what is necessary' (facts, definitions, theorems); the procedural 'the know how' to use the declarative knowledge; the strategic knowledge is about 'where, when and how' to use the previous two types; and the schematic knowledge which normatively *justifies* the use of the other three types—for comparison (in engineering) see [23]. According to [20] these types of knowledge are affected by emotions and motivation (which concurs with the findings in programming). Table 2 lists a conceptual framework of those dimensions [20].

This classification is very important to the teaching and learning of programming because when one deals with programs all four types are applicable simultaneously, whereby [21] refers to intrinsic cognitive load which is characterized in terms of element interactivity. Accordingly, the elements of most schemas must be learned simultaneously because they interact with each other whereby it is the interaction that is critical [18,21]. *Conceptual* knowledge (not mentioned above) is the knowledge that is constructed by identifying the correct relations between different related concepts (incorrect relations lead to misconceptions). In [12] conceptual knowledge is seen as knowledge rich in relationships. It can be thought of as a connected web of knowledge, a network in which the linking relationships are prominent as the discrete pieces of information. In [20] the idea of 'mind maps' is described; they also form a web of related concepts. One can go as far as to say that conceptual knowledge is a combination of all above different types of knowledge and that it is greater than the sum of them.

Table 2. Types of knowledge according to [20]

Proficiency Low ⟶ high	Declarative knowledge	Procedural knowledge	Schematic knowledge	Strategic knowledge
Extent (how much?)	Knowing THAT	Knowing HOW	Knowing WHY	Knowing WHERE, WHEN, *and* HOW
Structures (how organised?)	Domain-specific content:	Production-rules Sequences	Principles Schemes Mental models	
Others (how efficient? how precise? how automatic?)	Facts Definitions Descriptions			Strategies Domain-specific heuristics

Programming consists also of the above-mentioned types including conceptual knowledge. But it can also be argued that it is more than that. For example, syntax could be considered to be declarative knowledge, where one must know certain rules and apply them. But the syntax involves also schematic knowledge because one must know why this is the case [18, 19]. Therefore we can accept that for programming, although the various types of knowledge could appear individually, their inter-connectedness must be borne in mind. Such inter-connectedness is normally revealed once one has become almost an expert programmer. In fact, using the idea of novices and experts is one of the methods researchers have used in order to understand why some students can program while others cannot [19], whereby it is estimated that it takes about 10 years to become an expert programmer [19]. For more information on novices and experts see [9].

3 Method

In order to try and understand the problems that students encounter when learning programming in the first year of study of a Diploma in Information Technology, their final examination papers were analysed and the perceptions of their lecturers in course module XCT1512 (which uses Javascript) were collected. XCT1512 is at our institution a prerequisite for learning Web Design.

In order to identify the lack of a certain type of knowledge, script analysis took place. The sample was a convenience sample comprised of students from our academic 'extended programme' who take the module over a year instead of a semester (as the mainstream does). The total number of students who wrote the exam was 21; for this paper only 13 exam scripts were analysed. The reason was that although all scripts were processed, it was decided that only students who obtained a mark of 49% or higher shall be analysed. This was because of an inconsistency between various questions where a student would perform well in

XCT1512-December 2016

Stud.script	mark (%)	questions	1	2	3	4.1	4.2	4.3		5.1	5.2		6	%	attempted	%
		mark/quest	15	15	15	3	3	4		5	5		38	100	38	100
1	67		13	14	13	0	2	4		3	1		17	45	21	81
2	66		14	12	10	0	2	0		3	2		23	61	33	70
3	63		13	8	6	0		4		5	3		24	63	29	83
4	72		14	15	9	0	2	4		2	2		23	61	28	82
5	71		13	15	14	0	0	0		0	2		28	74	30	93
6	54		11	11	7	0		4		5	0		16	42	30	53
7	53		13	12	7	0	1	4		3	2		11	29	23	48
8	51		12	11	4			4			1		19	50	26	73
9	53		12	10	4	3		4		2	2		16	42	30	53
10	53		12	13	6	2	3	1		2	2		12	32	24	50
11	51		13	12	7								19	50	35	54
12	56		12	13	7					1	2		21	55	31	68
13	49		14	8	9	1	1	4		5			7	18	19	37
14	37		9	8	4		0			0	3		13	34	23	57
15	29		11	7	1								10	26	35	29
16	34		11	8	4			4		3	2		2	5	30	7
17	34		11	10	5	1				1	2		4	11	9	44
18	27		9	13	4	0							1	3	6	17
19	28		11	6	1	0				4	2		4	11	34	12
20	26		7	8	4	0	1			3			3	8	22	14
21	27		7	4	5	1		4		3	2		2	5	17	12
						6	11	33								
						39	39	52								
						15.38	28.21	63.46								

Fig. 1. Students' performance for XCT1512 in 2016 final exam

some questions and very poorly in others, especially in the most important one which required a good knowledge of programming. The use of script analysis has been employed by [24] in Mathematics and proved to be a very reliable and effective way of identifying gaps in knowledge of the students. Details of the results are in given in Figs. 1, 2, 3. Figure 1 contains the performance of the students in each of the questions and sub-questions as well as some indication as to what type of knowledge each question required. The final marks are also shown. The last two columns indicate the number of sub-questions attended to by the students (out of 38 for question 6) as well as the percentage of attendance. The various types of knowledge were codified as follows: Declarative knowledge = D, Procedural = P, Strategic = St, Schematic = Sc, Conceptual = C. In the questions of the exam paper, the following types of knowledge were needed:

- Question 1: True/False.
- Questions 1, 2, 7: C; the rest: D.
- Question 2: Multiple choice.
- Questions 2, 3, 7, 10, 15: C; the rest: D.
- Question 3: *Fill in the gap* _____
- Questions 5, 6, 7, 12: C; the rest: D.
- Questions 4a, 4b, 4c: D.
- Question 5a: *Given a program and an input, predict what the output will be.* This requires:
  - P : understanding the correct flow of the program,
  - C : understanding the meaning of all code used,
  - D : knowing the function of each concept,

- Sc : knowing why which principles are involved.
- Question 5b: Sc (why).
- Question 6: presented a complete program containing many errors for a debugging task; *all* types of knowledge were involved.

As mentioned above, of the sample of 21 students only those that achieved a mark of 49% or more were considered in the analysis. The rest performed well only in less than 30% of all questions. Thus is it is possible that a student achieved in one question 70% and in most of the others questions less than 20%. Hence only the top 13 students were considered for script analysis.

4   Observations and Discussion

Question 1. For this question there was an 85% success rate. The fact that it was a True/False question implies that the student has a 50% chance of getting the correct answer even if not understanding the question. The majority of the questions (12/15) were of declarative nature. For the questions that required conceptual knowledge (3/15) there was almost a 50% success rate. It can be concluded that the students were overall successful in answering questions that required declarative knowledge: see Fig. 2.

Question 2. These were multiple-choice questions. Since there were four alternatives, the student stood a 25% chance of getting it right even without knowing anything about the course. The success rate was overall almost 80% and again the majority of the questions (10/15) had declarative knowledge. The majority of the students obtained more than 75% on this section, and only 2 obtained less than 70%. In the conceptual questions students did relatively well, however mostly those students that did well in the whole exam. A sub-questions of special interest was sub-question 10, where 8/13 students gave the same incorrect answer. This sub-question asked: *What can you add to your code to effectively serve the same role as a breakpoint?* The often-given wrong answer was: "break"; perhaps the students did not grasp the question carefully and were misled by the word 'breakpoint'. Also in sub-question 13, 4/5 of the students gave the same wrong answer, namely *"query"*, to the question: *The "(...)" property of the Location object contains a URL's query or search parameters.* This might have happened due to the fact that search could be used for query or search, however it is the search property. For the rest of the incorrect answers there was no fixed pattern as though the students chose answers randomly: see Fig. 2.

Question 3. This question comprised of several fill-in questions. Obviously here one cannot guess. The success rate was about 53% and only 4/13 obtained more that 70%, the rest between 27% and 60%. The students did better in statements of declarative nature (73%) compared to conceptual (42%). Some points of interest are: sub-question 1 where it is clear that students confuse the word 'object' with types of numbers and arrays. Also only a 15% understood the meaning of 'path', while only 23% understood the meaning of 'quantifier': see Fig. 3.

Question 1

	memo	1	2	3	4	5	6	7	8	9	10	11	12	13		incorrect	%
1	T														C	0	0
2	F		T			T									C	2	15
3	T								F	F				F	D	3	23
4	T				F				F			F			D	3	23
5	T					F		F	F						D	3	23
6	T	F	F							F					D	3	23
7	T			F								F			C	2	15
8	F	T										·			D	1	8
9	F					T	T		T						D	3	23
10	F							T		T		T			D	3	23
11	T				F										D	4	31
12	F				T						T				D	1	8
13	T														D	0	0
14	T					F						F			D	2	15
15	T														D	0	0

Question 2

	memo/students	1	2	3	4	5	6	7	8	9	10	11	12	13		inccrect	100
1	b		a			c						a		a	D	4	31
2	b			c		b	b	a	c		c		c		D	6	46
3	c			b								a		a	D	3	23
4	b														D	0	0
5	a			b	d				d	b		b	d	b	C	7	54
6	b							d	a	d		a	c	a	C	6	46
7	c	d													C	1	8
8	b														D	0	0
9	a				d						d				D	2	15
10	d		a	a		a		a	a		a	a	a		D	8	62

Fig. 2. Students' performance in questions 1 and 2 for XCT1512 2016 final exam

Question 4. This question required declarative knowledge only. For sub-questions 4a and 4b there was a dismal performance. For 4a the success rate was 15% whereby only 1 student obtained 100%, 1 student 67% and 3 students 33%; merely 31% of the students attempted the question at all. For 4b an average of mark of 28% was obtained. Again only 1 student obtained 100%, 2 obtained 67%, 3 obtained 33%, whereby only 50% of the students attempted the question. For 4c the success rate was 64%; however 70% attempted this sub-question: see Fig. 1.

Question 5. For sub-question 5a, where the students had to be able to understand the program code (and what a line or group of lines of code does), all types of knowledge (except Sc) were needed. The average performance was 48%. There were 3/13 students who obtained 100%; 3/13 obtained 60%, and 2/13 obtained 40%. For sub-question 5b, which required only schematic knowledge, the success rate was merely 29%. Only one student obtained 60%, while 7/13 obtained 40%, and the rest between 0–20%. Lack of schematic knowledge implies that those students cannot think at *higher levels of abstraction* which belongs to the prerequisites for successful programming: Fig. 1.

Question 6. This question comprises of 150 lines of code in which there are various semantic and syntax errors. It also involves all types of knowledge. It requires the student to analyse every word and line to ensure its final correctness. Here it is necessary for the student also to understand the essence of the problem: what problem is the given program aiming at solving. The average success rate

NB 0 implies not attempted, blank = right

Question 3

Q No	memo	1	2	3	4	5	6	7	8	9	10	11	12	13	incorr ect	%
1	OBJECT			array			stringify	Numbe r	0	number	integer	0	integer		D	8	62
2	for	constrain valid				do		d0/wh ile	do						D	7	54
3	try/catch//tr y catch	constrain valid		tree				0	0	browser	html		error	html		D	7	54
4	CSS														D	0	0
5	variable/s			progra m	nod e	0		0	watch			watch	watc h		C	7	54
6	===||strict equal									conditional binary	?	logical end			C	3	23
7	anonymous		named	local			unamed	0		local variable	local	0			C	7	54
8	path	domain	encrypt	share	domain		public	0	0	0		Https	name	serv er	D	11	85
9	methods			proper ties			parame ters		0	classes		0	attribut es		D	6	46
10	quantifiers		matches	operat ors	counter		0	anchors	0	0	0	0	meta-data		D	10	77
11	identifier			value			object				name			0	D	4	31
12	decodeURico mp			constru ctor			0	0	0	0					C	5	38
13	Unicode		metachara cters	0					0	unicod URI					D	5	38
14	tracing		debugging	iritatio n			breakp oint	step in		stepping	debugging	eleme nt	json	0	D	9	69
15	fallthrough			switchi ng	iteratio n		break	0	through	pass- overflow					D	6	46

Fig. 3. Students' performance in question 3 for XCT1512 2016 final exam

was 47% where 5/13 of the students obtained between 55–73%, 7/13 obtained between 28–50%, and 1 student less than 20%: see Fig. 1 (the last 4 columns). In the 4th-last column of Fig. 1 the marks out of 38 are shown, and the 3rd-last the percent-mark obtained for that question.

Two of the lecturers who teach programming made a very valuable input into the performance of the students in programming. Their comments are more of technical nature but they, too, contain different types of knowledge. Their comments appear to correlate with the findings of the research, especially [7]. The lecturers are in agreement especially with the following remarks:

Programming language, as a rule, should not be difficult—see [14]—but critical thinking is. According to the lecturers, the students fail to see the 'environment' and concentrate on the 'surface', not on the 'deep' meaning that constructs convey. There were also problem in understanding the control flow of the given program. This issue is directly related to lack of procedural and conceptual knowledge. Misconception of 'functions' is another big problem. There was also a lack of understanding of the IF-THEN-ELSE statements whereby some students did not grasp that *either* the one *or* the other one is processed. Conditions in a program are not framed properly and there is lack of understanding the precedence of operators. When it comes to data structures, it is puzzling that although the majority of students can handle one-dimensional arrays they cannot cope with two-dimensional arrays any more. Manipulation of strings was also problematic as the students failed to see a string as a specific instance of one-dimensional arrays. In the practice of programming, syntax and operational semantics are tightly connected, although they are not the same. In our case, some students even fail to understand what the syntax of a programming language is; they have no language-theoretical *notion* of 'syntax' at all: for comparison see [1,18,19]. Debugging could be considered 'unnecessary' if the programs were perfectly constructed. Nevertheless even the most experienced programmers 'debug'. Novice programmers debug extensively by instrumenting their program code with output commands, such as to be able to see what actually happens at run-time. Finally, there appears to be lack of understanding of basic mathematics among our students. Proper mathematics (not just mathematical 'concepts') teaches one to solve problems in a systematic manner. Surely it can assist the student in doing programming if computer programming is considered to be a subject about *problem-solving* rather than about the computers themselves.[4]

The above-mentioned perceptions are all related to some kind of knowledge or another. Using the types of knowledge involved in programming may assist the lecturer to identify gaps in the knowledge of the students. Because programming contains many constructs (declarative knowledge) not knowing all the constructs, why are they used (schematic knowledge), when, where and how to use them (strategic knowledge), how to use them and in what sequence (procedural) and finally not understanding the principles and the inter-connection between the constructs (conceptual knowledge), the student will not be able to become a programmer. The results of this study indicate that in many places there is lack of declarative knowledge, even in a purely declarative knowledge question like *Question 4*. Declarative knowledge can be seen as the first link into a long chain that forms the program. A program is like a web of inter-connected concepts and principles. In other questions studied in this survey, declarative knowledge also featured to a large extent. Students who could not answer conceptual questions also lacked sufficient declarative knowledge. In *Question 3*, where the students had to show that they really know the various constructs and their function, students gave many answers that simply did not 'make sense' at all; this could indicate that they have possibly not even grasped the meaning of the question (from a linguistic-terminological point of view).

Question 5 was about understanding a program in its entirety: what it does, what problem it solves, and what actual output it yields. Describing and explaining what a program does is as difficult as the reverse process of writing a program to solve a problem. The reader of program code must concentrate on the various aspects and functions of the program, and at the same time translate that into 'natural' language which a non-programmer can understand. The poor performance by the students in that section of their exam highlights one more time the lack of declarative and conceptual knowledge. This might also be linked to the mental 'memory overload' of [21], whereby a reader must analyse and synthesise what is being read at the same time. Lack of sufficient practice could cause that. The student must do sufficient exercises so that certain intellectual functions of the mind can be (so to say) 'automated'. Critical thinking plays an important role here, too, as paying attention to detail, making inferences, drawing conclusions (and so on) is an absolute necessity when one interprets program code while converting such code into natural language.[5]

---

[4] Recall *Dijkstra*'s striking analogon about calling surgery 'knife science' [6].

[5] Shortage of formal-grammatical linguistic training in nowadays Secondary Schools (prior to University) might perhaps contribute to the above-mentioned problems.

Finally, in *Question 6*, which was about error detection in a program, it became evident that there is a noteworthy imbalance in most of the students different types of knowledge. The fact that only one student managed to get 73% on that section by attempting 93% of the question, while the second best obtained 63% by attempting 83% of the entire exam paper, is an indication that the group as a whole finds programming to be a subject difficult to learn: a statement which is confirmed by many other authors, too.

5 Conclusion

Programming is a difficult subject to learn. It has been characterized as a 'wicked' subject. Although this was a case study with a small convenience sample, it shed some light onto the performance of 1st-year students in a programming course. It was found that lack of declarative knowledge has a negative effect on performance in programming due to its connection to all other types of knowledge. Perhaps lecturers assume that the students will learn all the facts and the lecturer can do nothing about it. If this is true, then it is a bad assumption because it could be part of the teaching of programming to motivate students. The perceptions of the other lecturers confirmed the findings and added also some technical aspect to the interpretation of the results. Since this study is part of a longitudinal study into attrition at our university, 'Action Research' and collaboration among the lecturers in programming can contribute to improving the pass rates in computer programming. Analysis of programming also brought into the foreground the various factors that make it a wicked subject. The most important one could be the existence of all different types of knowledge appearing simultaneously when one either designs a program or interprets a program. Finally, the idea of using types of knowledge in programming to improve pass rates could be beneficial to the lecturers who teach programming as it will be possible to classify the knowledge and skills necessary for the course and to design the course around them.

Acknowledgements. Thanks to two of my colleagues, *A. Mathew* and *B. Esan*, for having made their submissions and having added value to this paper.

References

1. Bergin, S., Reilly, R.: Programming: Factors that Influence Success. ACM SIGCSE Bull. **37**(1), 411–415 (2005)
2. Chamillard, A.T.: Using student performance predictions in a computer science curriculum. In: ITiSE 2006 Proceedings, pp. 25–30. ACM (2006)
3. Chinnapen, M.: Mathematics learning forum: role of ICT in the construction of pre-service teachers' content knowledge. Technical report (2003)
4. Cleland, C.E.: Recipes, algorithms, and programs. Mind. Mach. **11**(2), 219–237 (2001)
5. Conklin, J.: Wicked problems and social complexity. In: Dialogue Mapping: Building Shared Understanding of Wicked Problems. Wiley (2005)

6. Dijkstra, E.W.: On a Cultural Gap. Math. Intell. **8**(1), 48–52 (1986)
7. El-Zakhem, I., Melki, A.: Identifying difficulties in learning programming languages among freshman students. In: Proceedings of 7th International Technology Education and Development Conference, pp. 1202–1206, Valencia (2013)
8. Giannakopoulos, A.: How critical thinking, problem-solving and mathematics content knowledge contribute to vocational students' performance at tertiary level: identifying their journeys. University of Johannesburg, Doctoral Dissertation (2012)
9. Green, A.J.K., Gillhooly, K.: Problem solving. In: Cognitive Psychology, Oxford University Press (2005)
10. Halland, K.: Assessing programming by written examinations. In: Gruner, S. (ed.) SACLA 2016. CCIS, vol. 642, pp. 43–50. Springer, Cham (2016). https://doi.org/10.1007/978-3-319-47680-3_4
11. Hawi, N.: Causal attributions of success and failure made by undergraduate students in an introductory-level computer programming course. In: Computers and Education, pp. 1127–1136 (2010)
12. Hiebert, J., Lefevre, P.: Conceptual and Procedural Knowledge in Mathematics. Laurence Erlbaum Associates, Mahwah (1986)
13. Hoenicke, I.: Programmierer am Ende. Die Zeit 7/1995, 10 February 1995
14. Jenkins, T.: On the Difficulty of Learning to Program. In: Proceedings of 3rd Annual Conference of the LTSN Centre for Information and Computer Sciences, pp. 53–58 (2002)
15. Jonassen, D.H.: Evaluating Constructivistic Learning. In: Constructivism and the Technology of Instruction: a Conversation. Lawrence Erlbaum Associates (1992)
16. Kiblasan, J.A., Abufayed, B.F.A., Sehari, A.A., Madamba, F.U., Mhana, K.H.K.: Analyzing the learning style and study habit of students in the faculty of nursing of Al Jabal Al Gharbi University, Gharyan, Libya. Clin. Nurs. Stud. **4**(2), 48–56 (2016)
17. Mazlack, L.J.: Identifying potential to acquire programming skill. Comm. ACM **23**, 14–17 (1980)
18. Mhashi, M.M., Alakeel, A.M.: Difficulties Facing Students in Learning Computer Programming Skills at Tabuk University: Recent Advances. Technical report (2013)
19. Robins, A., Rountree, J., Rountree, N.: Learning and teaching programming: a review and discussion. Comput. Sci. Educ. **13**(1), 137–172 (2010)
20. Shavelson, R.J., Ruiz-Primo, M.A., Wiley, E.W.: Windows into the Wind. High. Educ. **49**, 413–430 (2005)
21. Sweller, J.: Cognitive load theory, learning difficulty, and instructional design. Learn. Instruct. **4**, 295–312 (1994)
22. Ventura, P.R.J.: Identifying predictors of success for an objects-first CS1. Comput. Sci. Educ. **15**(3), 223–243 (2007)
23. Vincenti, W.: What Engineers Know and How they Know it: Analytical Studies from Æronautical History. John Hopkins University Press, Baltimore (1990)
24. Willman, S., Lindén, R., Kaila, E., Rajala, T., Laakso, M.J., Salakoski, T.: On study habits on an introductory course on programming. Comput. Sci. Educ. **34**(8), 1–16 (2015)
25. Yağci, M.: Blended learning experience in a programming language course, and the effect of the thinking styles of the students on success and motivation. Turk. Online J. Educ. Technol. **4**, 32–45 (2016)

Testing Test-Driven Development

Hussein Suleman, Stephan Jamieson$^{(\boxtimes)}$, and Maria Keet

Department of Computer Science, University of Cape Town,
Rondebosch, South Africa
{hussein,sjamieson,mkeet}@cs.uct.ac.za

Abstract. Test-driven development is often taught as a software engineering technique in an advanced course rather than a core programming technique taught in an introductory course. As a result, student programmers resist changing their habits and seldom switch over to designing of tests before code. This paper reports on the early stages of an experimental intervention to teach test-driven development in an introductory programming course, with the expectation that earlier incorporation of this concept will improve acceptance. Incorporation into an introductory course, with large numbers of students, means that mechanisms are needed to be put into place to enable automation, essentially to test the test-driven development. Initial results from a pilot study have surfaced numerous lessons and challenges, especially related to mixed reactions from students and the limitations of existing automation approaches.

Keywords: Teaching programming · Test-driven development · Unit testing · Automatic marking

1 Background and Motivation

In a Computer Science (CS) degree, students are taught how to develop software systems, in addition to other skills and techniques relevant to the discipline. Due to the rapidly changing nature of CS as a young discipline, the recommended techniques and methodologies to develop software (and therefore what is taught) change over time, for example, from the 'Waterfall' method in the 1980s to 'Iterative' in the 1990s and 'Agile' in the 2000s. Traditionally, CS courses would mirror the general academic practice whereby students submit work for assessment by tutors and lecturers. Many CS programmes have now shifted the assessment of computer programs to computer-based assessment—termed 'Automatic Marking' at the authors' institution [8]. While this assists in teaching the development of computer programs, it does not help students to develop their own mechanisms to assess their solutions. However, when students graduate, assessment is a key skill required of them by their employers, as many software development companies expect assessment of software, as a quality control mechanism, to occur hand-in-hand with software development. Frequently, however, new graduates either do not have experience or exposure to such software development skills.

© Springer International Publishing AG 2017
J. Liebenberg and S. Gruner (Eds.): SACLA 2017, CCIS 730, pp. 241–248, 2017.
https://doi.org/10.1007/978-3-319-69670-6_17

While there are numerous approaches to software testing, Test-Driven Development (TDD) [2] has emerged as a popular software development approach, where the tests of software are produced first and drive the software creation process. TDD has been shown to result in programmers learning to be more productive [4,5], while generating code that is more modular and extensible [6] and is of better quality [7]. Some early experiments have been conducted to show that test-driven development can be incorporated into the kinds of automated testing that is used in many large teaching institutions. Edwards opted to measure the code coverage of tests and the adherence of student code to student tests as indicators of appropriate use of TDD [3]. He found that students appreciated the technique and there was a clear reduction in error rates in code. However, testing required the provision of a model solution for coverage analysis, which is not always possible or accurate.

This paper describes a first attempt at incorporating TDD approaches into an automated marking system, using the alternative method of testing the tests. Where students have, in the past, only encountered advanced software development techniques in later courses, TDD was designed into an early programming course to avoid resistance from students who had already learnt a traditional approach. It was expected that this would improve the skills of students while better meeting the needs of industry. We present the initial integration in a 1st year CS course, observations on that, and the results of a survey among students.

2 Implementation

As a pilot study, elements of TDD were incorporated into specific parts of two assignments of a second semester course on object-oriented programming. Students had already learnt imperative programming and were exposed to the general notions of program testing in the previous semester. Assignments 2 and 3 of the course were selected because they marked the start of OO content proper. It was felt that TDD could assist with mastery of the concepts; with thinking about objects and behaviour.[1]

2.1 Assignment 2

In Assignment 2, software testing was introduced as a means of guiding problem analysis, solution design and evaluation. Students were:

- given tests and asked to develop code,
  and
- given code and asked to develop tests.

Question 2.1 asked the students to develop an implementation of a specified `Student` class. Three test requirements were described and a corresponding test suite provided.

---

[1] Assignment 1 concerned knowledge transfer—constructing imperative programs in the language Java which was new for the students in that phase of their curriculum.

Question 2.2 asked the students to develop an implementation of a specified `Uber` class. Twenty-three test requirements were described and a corresponding test suite provided.

Question 2.3 asked the students to develop test requirements and a corresponding test suite for a given `Collator` class (specification and code provided).

In order to reduce cognitive load, it was decided that tests should be presented using plain old Java rather than JUnit. However, to support automated testing of tests, a specific structure was designed for their specification:

```
// Test <number>
// Test purpose
System.out.println("Test <number>");
// Set up test fixture
// Perform mutation and/or observation of objects
// Check result, printing 'Pass' or 'Fail' as appropriate.
```

Example:

```
// Test 1
// Check setNames sets name and and getFullName returns name.
System.out.println("Test 1");
Student student = new Student();
student.setNames("Patricia", "Nombuyiselo", "Noah");
if (student.getFullName().equals("Patricia N. Noah")) {
        System.out.println("Pass"); }
else {
        System.out.println("Fail"); }
```

2.2 Assignment 3

Assignment 3 builds upon Assignment 2, aimed to have the students work on both components of TTD, being the red-green-refactor approach: they were asked to develop tests and then develop code.

Question 3.1 asked the students to identify tests requirements and, subsequently, to develop a test suite for a specified `JumpRecord` class.

Question 3.2 asked the students to construct a `JumpRecord` class.

Students were given a single opportunity to submit their answers for Question 2 to the automatic marker. They were told that getting 100% for Question 1 should assure that an implementation that passed their tests would also pass the automatic marker's tests.

2.3 Testing the Tests

Question 3 of Assignment 2 (A2.3) and Question 1 of Assignment 3 (A3.1) required that the automatic marker be used to evaluate a set of tests to assess how well they discriminate between a correct implementation of the class under test and a faulty implementation. For each question, based on the given specification, a set of test requirements was drawn up and then, from these, a correct implementation (gold standard), and a set of faulty implementations (mutations) were developed. *Mutations* [1] were developed by hand by taking each requirement in turn and identifying ways in which the correct implementation could be modified such that it would fail.

All in all, 19 test requirements were identified for A2.3, and 22 test requirements were identified for A3.1. From these, 23 mutations were developed for A2.3, and 54 mutations were developed for A3.1. The automatic marker was set up for a question such that each mutation served as the basis of a trial. A trial was conducted by:

1. compiling and running the student test suite against the given mutation;
2. capturing the output;
3. searching for the presence of 'Fail'.

In accordance with the theory of testing [1], the test suite *passed* a trial if it generated a 'Fail'. As a final trial, the test suite was run against the 'gold standard' implementation of the class under test. Successful completion of the trial required, naturally, that the suite did not generate a 'Fail'. This final trial was implemented using a special penalty feature, whereby all awarded marks were deducted in the event that it was not successfully completed.

3   Observations

3.1   Automatic Marker

In the normal course of affairs, an assignment requires that a student submit programs to the automatic marker. Each program is compiled once, and run may times, i.e., trials are conducted in which inputs are applied to the program and the actual output compared to the expected output. The student is informed of the outcome of each trial. Where the program fails a trial, typically, they are shown the inputs, the expected output and the actual output. Automatic marking of a typical 1st-year students' program usually involves about 10–20 trials. In the case of testing the tests of question 1 in assignment 3, there were 56 trials, resulting in a very large amount of feedback that was hard to process. A different approach to providing feedback on tests of tests needs to be considered. The user experience was impacted by a decision to have the automatic marker compile mutations rather than precompile them. This made configuration easier, but at the expense of submission response times.

3.2   Mutation Development

Mutations appear to make for effective characterisation of programming problem requirements, and thus serve well as a means of assessing student tests that purport to do the same. However, the effort involved in devising mutations by hand is not trivial. While offset by the fact that materials are reusable, using this approach to roll out TDD across a whole course would be a sizable task. One way of reducing the workload may be through a more discriminating application. For example, rather than developing mutations for all requirements, identify those that possess genuine complexity, or those that are most important in the context of the learning outcomes for the assignment.

3.3 Testing Can only Reveal the Presence of Faults, Never Their Absence

In Assignment 3, the students were assured that, should their test suite be passed by the automatic marker, and their implementation be passed by their test suite, then their implementation would also be passed by the automatic marker. They were given one opportunity to submit their class implementation. The intention was to encourage students to focus on tests first and then code later. A large quantity of mutations was developed to ensure that the assurance held. However, there were still some problems. A number students made a mistake in the implementation of a particular method that resulted in a double being returned instead of an integer; unfortunately, this was not picked up when testing their tests because of automatic typecasting. One of the students created an off-by-one fault in their code that should have been accounted for in the mutation set but was not. These problems were fixed by adding two additional trials, one involving compiling the student test suite against an implementation of the class with the wrong return type, and the other by generating an additional mutation. In future applications, it would be wise simply to limit the number of submissions to a small number, as even test harnesses are seldom perfect.

4 Student Attainment

Table 1 shows the distribution of marks attained by the students on the TDD questions. A mark of zero generally indicates that the student did not make a submission. Thus, for A2.1, for example, 6% of the cohort fall into this category. Students were permitted to make multiple submissions for the questions of assignment 2 and for the first question of assignment 3. The mark spread for A2.1 indicates that the students largely found it unproblematic. Students were able to perfect their answers. The same generally applies to A2.2. The spreads for

Table 1. Student mark distribution.

Achievement	A2.1	A2.2	A2.3	A3.1	A3.2
$mark = 0$	6%	8%	19%	33%	25%
$0 < mark \leq 50$	0%	0%	3%	1%	4%
$50 < mark \leq 90$	0%	6%	33%	10%	5%
$90 < mark \leq 100$	0%	0%	38%	21%	2%
$mark = 100$	94%	86%	7%	35%	64%

A2.3 and A3.1 suggest that the students found designing test requirements and tests to be challenging. A significant number appear to have accepted a 'good enough' mark once past 50%. It is possible that that they were not equipped to perfect their solutions.

It is not uncommon in an assignment for later questions to score lower, possibly due to time pressure. However, the spreads for A3.1 and A3.2 indicate that assignment 3 was challenging from the start. A high percentage did not submit a test suite for A3.1, or could not gain a mark past zero, while a relatively small percentage achieved perfection. Due to the issues noted in Sect. 3.3, the restriction for A3.2 was actually relaxed, and the number of permitted code submissions was raised from 1 to 5. Given that the percentage attaining a mark of zero goes down from A3.1 to A3.2, it would suggest some students simply skipped the first question.

The average number of submissions made by the students was 1 for A2.1, 2 for A2.2, 5 for A2.3 and 12 for A3.1. These figures seem to confirm that designing tests and test requirements was challenging. Also, the slight increase in the average between A2.1 and A2.2 suggests that some students did not fully utilise the supplied test suite.

5   Student Feedback

Students were asked to comment on their experiences on TDD in assignments 2 and 3, as well as unit testing that they were exposed to in later assignments. 23 students responded to a short survey at the beginning of the following semester from a class that comprises about half the number of students enrolled in the 1st year course. In summary, the responses were as follows:

- When asked *"how do you feel about having to do test-driven development"*, students had mixed reactions. About half were intimidated or unexcited by the prospect, some of whom felt that it was inappropriate or unnecessary for the sizes of projects. Other students, in contrast, felt that these were industry-preparation skills that were necessary and useful for programmers.
- Most students stated that using TDD took more time, but some correctly reasoned that this was because of the shift in mindset and setting up of appropriate code frameworks.
- Most students thought that TDD did not improve their code quality, with many unsure of the impact because of small projects. Only 4 students thought TDD definitely improved their code quality.
- When asked *"was TDD too much unnecessary work"*, 14/23 indicated that it was.
- About 50% of respondents understood how TDD worked in the assignments.
- Almost all respondents understood why TDD is important.
- Students were asked how the assignments could be improved. Answers included: better instructions/teaching; less emphasis on pedantic tests; more examples; a checklist for what could go wrong; more practice exercises; and less ambiguity.

– When asked for general comments, many students agreed that the principle of TDD was important but the implementation could be improved in various ways.

6 Discussion, Conclusions, and Future Work

The results from the early pilot study are mixed. Students appear to appreciate the importance of TDD but they are not necessarily experiencing the immediate benefits and it is perceived to be more of a hindrance than a help to them. Staff observations also suggest that more effort needs to go into the design of assessments and the design of assessment tools to support teaching using a TDD approach. Traditional automated testing environments are based on feeding input values or parameters to code and observing output/parameters. TDD may need more sophisticated or different techniques to assess tests rather than code. Instruction was mentioned by many students as an inadequacy and may have to include a better explanation on the TDD methodology [2] and aforementioned benefits [4,5,7]. This also extends to the training of those who design assessment of TDD assignments. Automated assessment of programs is often considered to be a 'black art', where experienced teachers encode their value systems into programs to enforce those values on students. Automated assessment of tests takes this one step further and not much has been written on formalisms, guidelines or best practices for those designing the assessments. Given the many issues and mixed results, the next step is a refinement and possibly a second pilot study. Technology enhancements, mode of teaching, timing of TDD in the curriculum and many other factors need to be determined before the next group of students encounters TDD, with hopefully a much more positive experience than that of their predecessors. It may also be of use to assess the pattern of submissions to the automatic marker, in particular on whether the undesirable testing of one's code through the automarker also holds for submitting tests, rather than the intended red-green-refactor of TDD.

Acknowledgments. This research was partially funded by the National Research Foundation of South Africa (Grant numbers: 85470 and 88209) and the University of Cape Town.

References

1. Ammann, P., Offutt, J.: Introduction to Software Testing. Cambridge University Press, Cambridge (2008)
2. Beck, K.: Test-Driven Development by Example. Addison Wesley, Boston (2003)
3. Edwards, S.H.: Improving student performance by evaluating how well students test their own programs. J. Educ. Resour. Comput. **3**(3) (2003). doi:10.1145/1029994. 1029995
4. Erdogmus, H., Morisio, T.: On the effectiveness of test-first approach to programming. IEEE Trans. Softw. Eng. **31**(1) (2005). doi:10.1109/TSE.2005.37

5. Janzen, D.S.: Software architecture improvement through test-driven development. In: Companion to 20th ACM SIGPLAN Conference, pp. 240–245, ACM (2005). doi:10.1145/1094855.1094954
6. Madeyski, L.: Test-Driven Development - An Empirical Evaluation of Agile Practice. Springer, Heidelberg (2010). doi:10.1007/978-3-642-04288-1
7. Rafique, Y., Mišić, V.B.: The effects of test-driven development on external quality and productivity: a meta-analysis. IEEE Trans. Softw. Eng. **39**(6), 835–856 (2013)
8. Suleman, H.: Automatic marking with sakai. In: Proceedings SAICSIT 2008, ACM (2008). http://pubs.cs.uct.ac.za/archive/00000465/

ICT Courses and Curricula

Factors Influencing Poor Performance in Systems Analysis and Design: Student Reflections

Henk W. Pretorius[✉] and Marie J. Hattingh

Department of Informatics, University of Pretoria, Pretoria, South Africa
{henk.pretorius,marie.hattingh}@up.ac.za

Abstract. Educators in higher education institutions (HEIs) are under constant pressure to improve their educational practices. This study investigates possible causes for the first-year System Analysis and Design (SAD) course failures of Information System (IS) students. First-year failure is a significant contributor to non-progression statistics, which is a major concern for many universities. For this paper, data was collected from 29 IS students who had failed the first-year SAD course at a university in South Africa. The respondents identified factors related to educators, chosen pedagogy, confusion about the course content, and lack of self-responsibility as contributors to poor performance. These experiences are used to make recommendations to educators, especially first-year educators, about how to improve their educational practices.

Keywords: System Analysis and Design · Higher education · Academic performance · Summer School

1 Introduction

Educators in higher educational institutions (HEIs) are under constant pressure to improve their educational practices [13]. Educators in Information and Communication Technology (ICT) are not exempted from this pressure. Educational practices include a diverse set of performance areas, such as teaching (facilitated learning), mentoring, supervision, assessment, and the like [13]. This study focuses on one of these performance areas, namely teaching. The research objective of this study was to investigate possible causes for first-year System Analysis and Design (SAD) course failures of Information System (IS) students. First-year failure is a significant contributor to non-progression statistics, which is a major concern for universities [17]. Even worse, the consequences for a student not completing a university degree may be severe at a later stage of life. Our study was conducted at a prominent university in the capital of South Africa. The research objective of this study translates into the following:

Research Question: *Why do first-year IS students fail their System Analysis and Design course?*

© Springer International Publishing AG 2017
J. Liebenberg and S. Gruner (Eds.): SACLA 2017, CCIS 730, pp. 251–264, 2017.
https://doi.org/10.1007/978-3-319-69670-6_18

The study does not only propose recommendations to educators to improve their own educational practices (educators), but also proposes recommendations to first-year IS students to decrease the risk of failing SAD.

2   Literature

Students do not learn in isolation. Their ability to master the contents of the various modules offered as part of the curriculum is influenced by the students themselves and by the tools available to them, which includes educators. The educator can be seen as a mediator or tool [1,24] for the student to master the content of the subject—here: SAD. The student is influenced by the environment in which learning takes place, even though personal responsibility is required. Figure 1 illustrates the bidirectional relationship between the student, the educator and the content to be mastered in a dynamic learning environment. Each of these terms will be discussed in the subsequent sub-sections.

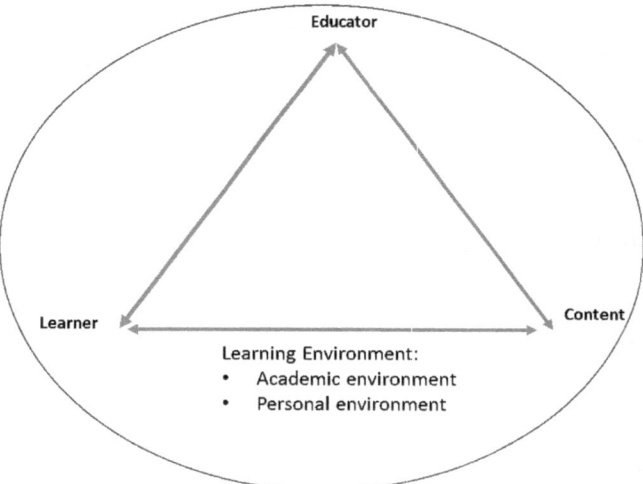

Fig. 1. The student-educator-content relationship.

2.1   The Role of the Educator

Teaching SAD can be quite challenging for educators. Rapid changes in the environment, changing industry demands, new market trends and changes in technology have a direct impact on how effectively students learn and how effectively they will be in the workplace [6]. Educators are therefore under constant pressure and are challenged to improve their educational practices [13].

Educational practices include a diverse set of performance areas, such as teaching (facilitated learning), mentoring, supervision, assessment, and the like

[13]. Many educational practices and teaching methodologies have been adopted and tested to respond to the educational challenges associated with teaching SAD [4,23]. Some of these practices and method(ologie)s are teacher-centered, while others are student-centered. All these share the same goal, which is to provide students with the best education and knowledge possible. However, in general (not just in SAD), the success of these methodologies is questioned. In 2013, the Organization for Economic Cooperation and Development (OECD), which is active in 34 member countries, published these statistics [12]:

- On average, 70% of the students who enter a university programme complete a first degree, although there are significant differences between countries. In countries like Hungary, New Zealand, Norway and Sweden this average falls below 60%. On the other hand, in countries like Australia, Denmark, France, Japan and Spain this average is above 75%.
- Full-time students have a better chance of graduating from degree programmes than part-time students. For example, in New Zealand, the completion rate of full-time students is 34% points higher than that of part-time students. Students may decide to leave the education system before completing their degrees. This can be attributed to attractive employment opportunities.
- Women are more likely than men to earn a tertiary degree, with a completion rate for women about 10% points higher than for men.

With regard to teaching SAD courses in particular, favourable arguments are made for teaching methodologies such a problem-based learning (PBL). PBL is a tried and tested student-centred teaching methodology with proven success. When applying a PBL methodology students are responsible, to a greater extent, for their own learning [20]. Learning becomes an act of discovery. The student examines the problem, does some research on the problem background, analyses possible solutions and proposes or produces a final result. In other words, the students develop a greater understanding of the course material and content because they are actively involved in the learning process [6]. This methodology automatically forces students to engage in academic content that leads to more self-discipline and mastery of academic content. PBL addresses some of the challenges raised with regard to SAD teaching. PBL raises the quality of education and further creates a contextual, collaborative and constructivist learning environment [3,5]. Therefore, PBL assists in developing the critical thinking skills needed by students in SAD. Finally, PBL presents an opportunity for the educator to become the facilitator or guide in group work [14], which may also become a critical skill for SAD students. However, even if PBL improves a student's understanding of SAD concepts and ideas, many failures in SAD courses remain.

This discussion therefore indicates that SAD educators continuously need to explore and investigate avenues to improve their teaching practices. Even more relevant to educators is to understand, from the perspective of students in the ICT disciplines of Information Systems, Information Technology and Computer Science, why they think they are performing poorly and even failing in SAD courses.

2.2   Systems Analysis and Design Contents

The field of SAD is dynamic and continually adjusting to the needs of organisational information systems [6]. Flexibility is needed in the field to ensure that students are learning new methodologies and the techniques necessary to deliver a product that meets the needs of industry, the client.

At the university in which our research was conducted, the SAD module is offered at undergraduate level in the first, second and third years of study. In the first year, the module covers a variety of concepts that are needed to understand the role of a business or systems analyst in the development of an information system. The curriculum is roughly divided into a theoretical element and a modelling element. The theoretical element includes systems theory, business processes, creative thinking, problem solving, feasibility analysis, systems development methodologies and requirements, and elicitation techniques. The modelling element includes use cases, process modelling and data modelling. In a similar context, Tepper reported on the challenges associated with teaching SAD to first-year students, including [21]:

– the need for students to develop analytical and inter-personal skills;
– students initially not really understanding the need for the module, which influences their motivation to learn.

An added challenge in the SAD module is that future employers have certain skills in mind when they employ SAD interns or graduates. Saulnier reported that employers still require students to have both 'soft skills' (such as the ability to work in a team and solve problems) and 'hard skills' that are organisation specific [16]. The challenges experienced by the students, in addition to the requirements of future employers, make the SAD course complex for both the students, who have a steep learning curve, and the educator, who needs to adopt the correct pedagogy to prepare the students for future employment.

2.3   The Role of the Student

As indicated above, students in SAD courses are challenged by course content that changes rapidly. Students need to stay relevant in a market where technology, methodology (techniques and approaches to develop new systems) and industry trends change quickly [6,16]. SAD courses typically discuss systematic methodologies for analysing a business problem and determining if and what role computer-based technologies can play in addressing the business need [6]. All this change in SAD courses directly impacts on a student's ability to effectively master SAD course content and skills. Students in software development need to become lifelong students.

Students must therefore rather focus on learning processes where learning becomes an act of discovery, rather than focusing on mastering the SAD course content that might already be irrelevant in the not-too-far future. Students should focus on understanding and examining the given problem, researching the problem background, analysing possible solutions, developing a proposal and producing a final result [6,25]. During this process, students develop a greater understanding of relevant and contextual SAD course content and skills, and the required critical thinking abilities to produce the final result [6,16]. In learning processes like these, students engage in active learning that leads to mastering changing academic content, such as SAD courses and content. Even though we recognise that mastering SAD content is challenging for students, there are still other factors that contribute to SAD first-year failures. Therefore, educators must investigate further and broader to gain insight into the possible causes of first-year SAD course failures from the perspective of the students.

2.4   The Learning Environment

Students are positioned in a dynamic learning environment. The environment in which learning needs to take place consists firstly of the academic environment provided by the university, and secondly the student's personal environment, which includes the socio-economic challenges that impact on student performance [10]. Each of these environments will be discussed next.

The Academic Environment. Students typically have two opportunities to pass a course (with applicable terms and conditions). In the first instance, students register for the course to complete it over the course of a semester or a year. Secondly, barring some entry requirements, if students are unsuccessful in their first attempt, they can register for a 'Summer School'. The subsequent discussion will explain the situation at the particular university on which this study reports.

Learning Throughout the Course. In the SAD course, students are presented with case studies of real-world business problems for which they have to develop ICT-related solutions. This approach allows educators to follow PBL [6] to SAD pedagogy. During a practical session, an educator may request that students' newly designed models be changed into database or business applications. Students therefore require a thorough understanding of solving a real-world problem by applying SAD concepts and techniques, and using ICTs to implement the solution.

Summer School, presented by the educator, mostly covers the theoretical and practical aspects of systems modelling. By taking the theoretical and practical dimensions that are involved to pass the course into account, the educator structures the Summer School learning opportunity as follows: First, the educator sets out the specific learning objectives and learning outcomes for the day. Then, the educator goes through the theoretical aspects involved in a given study unit. Typically, 60 to 90 min are used to go through the theoretical section. The students are then allowed a break of 15 min.

After the break, a shift is made from theory to practice. The educator typically starts with a guided exercise to demonstrate to the students how to apply the analysis and design concepts (theory) from the given study unit to a real-world scenario. Depending on the exercise, this activity may take anything from 60 to 90 min or longer to complete. Then the students are presented with a second real-world problem scenario. However, this time, the students have to try and solve the problem themselves. Due to the small number of students, they are encouraged to work in groups, which helps them build confidence and learn from each other.

In the last phase of the learning opportunity, the educator presents a solution to students for the given problem. Typically, in analysis and design, there may be more than one solution for a given problem. Students have the opportunity to reflect on their own designs and/or models to understand and enquire why some design solutions are better than others. The students then take a longer break for lunch, after which the process is repeated for the new study unit.

Every day, before the students go home, they are informed of a class test that will take place the following morning. Students use this opportunity to identify personal problems that can be resolved on a personal basis with the educator. The examination takes place on the last day of the Summer School week and determines if a student achieved the learning outcomes to pass the course. The examination is on the same standard as that of the year course.

The Student's Personal Environment. Any formal learning at an educational institution needs to be supplemented by learning at home, be it through homework, an assignment or preparation for an assessment. In South Africa, a number of factors influence a student's ability to extend their learning experience to an environment outside the formally structured classes.

In [10] it was explained how learning takes place on three levels: the cognitive level, the emotional level and the social level. Therefore, learning is firstly seen as a cognitive process whereby the abilities of the student (their knowledge) and their abilities to acquire and store new knowledge are essential. Secondly, learning is influenced by emotions, attitudes and motivations. Even emotions related to homesickness may impacts academic performance [26]. Finally, learning is seen as a social process. Students need to be considered in the environment from which they historically come and by which they are influenced on a daily basis.

In [10], universities were recommended to consider students within their environment holistically, which might influence their ability to learn in one way or another. For this reason the Council on Higher Education (CHE) states that *"all South African universities provide a range of student support services, through structures such as financial aid offices, counselling centres, health centres and writing centres. Generally, these support structures are managed through the offices of a Deputy Vice-Chancellor: Student Support, or similar"* [18]. However, the first round of CHE audits showed that the formal curriculum does not make provision for students who need the services mentioned above.

3   Method

In order to gain understanding and insight of first-year SAD failures, we decided to follow an interpretive research paradigm approach. This approach holds that social life is based on socially constructed meaning systems and social inter-actions. Therefore, people possess an internal experience of reality [22]. The methodological position of interpretive research is the qualitative approach, which does not exclude the positivist approach.

In this study, data were collected from 29 IS students (male and female) who failed the first-year SAD course. These students attended Summer School, the above-mentioned extra learning opportunity from 19 to 23 January 2015, to re-do the course that they had failed in 2014. To qualify for the extra learning opportunity, students had to prove that they had been officially registered for the course in the previous year (in this case: 2014), but had failed the course.

Most of the students questioned were aged between 19 and 21 years (typi-cal for first-year students) and represented different ethnic origins, social back-grounds and genders as indicated in Table 1.

Table 1. Numbers of respondents, with ethnic and gender attributes.

Ethnicity	Female	Male	(Total)
Black	5	7	12
Asian	1	3	4
White	6	7	13
(Total)	12	17	29

After the learning opportunity had been presented by the educator (see Sect. 4) and the students had completed their examination, they were presented with a questionnaire for data collection purposes. The questionnaire consisted of open-ended questions that resulted in richer responses from the respondents that suited the interpretive nature of this research study.

In the questionnaire, the students were asked how they had experienced the first-year lectures and lecturing staff. The students were then asked to share their positive and negative experiences of the Summer School learning opportunity. Finally, the students were asked to share their opinions of why they had failed the SAD course in their first year. Lastly, they were asked what they had learned through this experience. No time limit was set for the respondents, allowing them enough time to complete the questionnaire and think about the answers.

The data set was systematically coded into themes and categories as they emerged, whereby the constant comparative method [19] was applied. Figure 2 illustrates the major themes that emerged from the data. In our study, four major themes emerged as factors that contributed to poor student performance: the edu-cator, the pedagogy, the course content, and the student. The Summer School experience emerged as a supporting theme that assisted students in improving their overall performance. Each of these themes will be discussed in the following section.

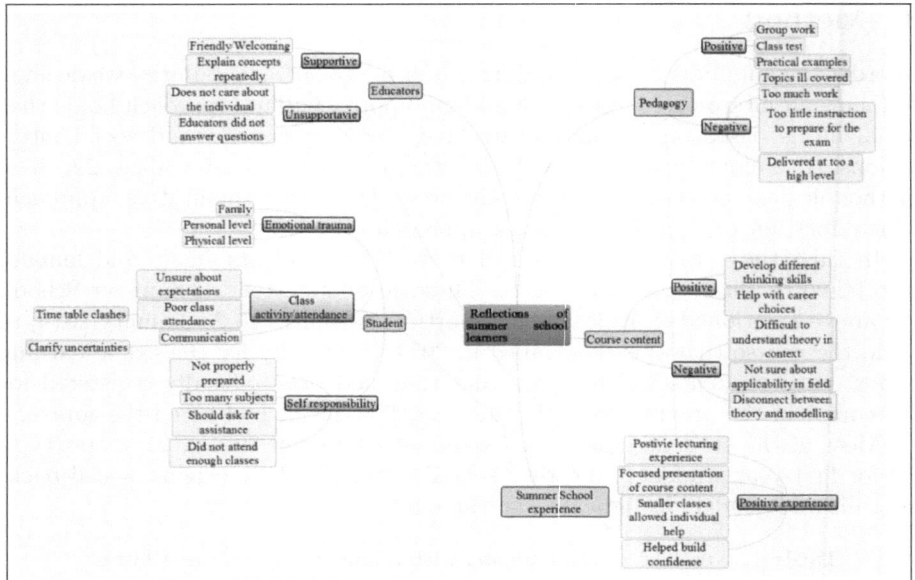

Fig. 2. Themes that emerged from the gathered data.

4 Causes of First-Year Failures in SAD

This section discusses the factors (as they emerged from the thematic analysis) that contributed to the poor performance of first-year SAD students. Firstly, the influence of the educator and their pedagogy will be discussed, followed by the influence of the course content and the student-related factors. Finally, aspects identified from the Summer School experience will be discussed in support of improving students' academic performance.

4.1 Influence of Educators and Their Pedagogy

The educator is a mediating tool [1] between the student and the content. According to [15], the most important factor in determining student experience in a course is the educator. The educator is therefore an important factor in the learning process. The data indicated that 45% of the respondents saw the first-year SAD lectures and lecturing staff in a positive light.[1] Conversely, 55% of the respondents indicated that they had a negative experience of the first-year SAD lectures and lecturing staff. The thematic analysis indicates that the most common causes of this negative response were that the students perceived the

---

[1] Examples of responses: *"The lecturing staff is extremely friendly and helpful"*. *"The staff was very welcoming and always willing to assist when needed"*.

lecturers as being 'unclear' in the way the content was presented, which led to the students experiencing some confusion.[2]

Regarding the educator's chosen pedagogy, individual students stated that the level at which the content was covered was too high and the topics were not well covered. This, in their opinion did not allow them to adequately prepare for the examinations.

As previously discussed, the content of an SAD course is dynamic and requires students to develop critical thinking and interpersonal skills [6,21], which is a challenge to students, especially on first-year level. Furthermore, according to [9], the challenges associated with large class sizes are exacerbated by the fact that first-year students have mostly been exposed to *"learning strategies constructed around the memorisation of facts and the simple reproduction of knowledge, or so-called surface learning"*. First-year students need to be shown how to implement the problem solving and critical thinking skills that are important to prepare them for their future careers. The higher order thinking skills required by students [21], coupled with the PBL approach that was followed (in the modelling part of our course) can compound the confusion experienced by students in mastering the content, especially students who experience challenges; (see Sect. 4.3 below).

Another common factor that emerged from the data was the large class sizes. Large class sizes have been attributed to low student performance [11]. Large class sizes should present educators with an opportunity to adapt pedagogy styles. However, in the context of this study, a respondent felt that the lecturer does not care about individuals when lecturing first-year SAD. Individuals tend to disappear, given the big classes and the large number of students that need to be reached. This can compound the communication problem experienced by a respondent who felt the educator did not respond to questions asked. Interestingly, only 11% of the respondents indicated that the educators and lectures were only partly to blame for the students' failure.

4.2   Influence of the Course Contents

In a survey completed by senior employees [16], 95% of the respondents preferred hiring graduates who are innovative in the workplace, 93% indicated that these graduates needed to demonstrate critical thinking abilities, communicate clearly, be able to solve complex problems, and apply their knowledge in real-world settings. Some of our respondents confirmed the results of [16] by reporting on the positive skills they had obtained from the SAD course:[3]

- 60% of the respondents indicated that the subject matter explains how to analyse and understand systems;

---

[2] Examples of responses: *"It was not as informative as I would have liked, I failed to understand some work". "I was often confused about tasks we had to do".*

[3] Examples of responses: *"The way we analyse a scenario or system... diagrams help us understand a system". "It is practical to a working environment". "The way of thinking is different".*

- 25% of the respondents indicated that the subject matter presents a different way of thinking about problems and work;
- 40% of the respondents indicated that the subject matter explains how to apply the theory in different real-world scenarios;
- 15% of the respondents indicated that the subject matter assisted them with logical thinking.

About 25% of the students were not sure how they would apply the theory in their future careers, while 15% of the students indicated that it was difficult to understand the theory.[4] The contradicting receipt of the content matter is not surprising if it is taken into account that, at first-year level, some modules are compulsory—like the SAD module at the university under consideration. The module might not have been chosen by the students if they had a choice. This can impact on the intrinsic motivation of the student to perform well in the module. For example, the *"intrinsic motivation is based on the fact that students are interested in what they are learning, and the driving force is not external rewards, but internal happiness and satisfaction"* [10].

The discussions of above underline the importance of the role of the educator in communicating the learning outcomes of the module. The importance of the module in terms of the value it will add to their future career choices needs to be communicated to the students. If the content is relevant in terms of industry needs [21], educators will get students' 'buy-in' in mastering the content.

4.3   Student-Informed Challenges

The students identified three subthemes that contributed to poor performances: lack of self-responsibility, lack of class activity, and emotional trauma. Most of the students (89%) indicated that they are also to blame for failing the course. Firstly, when considering the lack of self-responsibility, which sometimes also included lack of class activity, 22% of the students indicated that they had not prepared or studied hard enough for the module, while 26% of the students admitted that they had not attended classes frequently enough.[5] According to [8], however, class attendance (apart from competence in English and class participation) is indeed an important factor affecting student performance. Moreover, three students blamed administrative problems (e.g., time table clashes) for their failure.

When the students were asked what they learned from their experience, they indicated that they should not skip class and should further request and ask for assistance with subject matter when this is required.[6] Interestingly, only

---

[4] Examples of responses: *"It is difficult to understand". "I could not put the theory into context". "It will not help me in my particular field".*

[5] Examples of responses: *"I did not make enough time to study and revise". "I barely attended class". "I did not study hard enough".*

[6] Examples of responses: *"I need to address academic problems sooner so I can learn and comprehend the work". "Engage frequently with the educator and other students and explore more ways to implement SAD in the world".*

three students were 'battling' with mastering the content: two students indicated that they struggled to apply the theory to a practical solution, and one student indicated that it was unclear how the different diagrams fit together in one system. This is an indication that students need to understand holistically how the different parts of SAD fit together.

Another theme that emerged from the data was that some of the students experienced emotional trauma, which affected their performance. Taking into account the social processes of learning and considering the students' dynamic learning environment, 24% of the students indicated that they had experienced family or personal trauma that had affected their course outcome. The students were *not* asked to go into details of the emotional trauma they had experienced.

In two interesting cases, however, students indicated voluntarily that the amount of subject matter that they could not master was the cause of their emotional trauma. These students did not know how to emancipate themselves from this situation. In the second instance of emotional trauma, the educator needed to create a 'safe' environment' that would allow students to reach out if they did not understand the subject matter before it caused emotional trauma. Even though there have been reports that students fear their educators [7,15], students need to feel generally comfortable to approach them. Thus, students need to be provided with information about support structures available at the institution. As indicated above, the CHE confirmed that all South African universities have structures on site to support students in the various challenges they experience [18]. Educators are not always equipped to render specialised support, but they must have the knowledge to refer students to the specialised units and be sensitive to the implications of the traumatic experience.

In summary, even though students are influenced by the cognitive, emotional and social processes [10] in which learning takes place, they need to take self-responsibility by taking part in the learning opportunities provided by the educator. Furthermore, students need to have the courage to reach out if they 'fall behind' in mastering the content.

4.4   Lessons Learned from the Summer School Experience

Summer School is different from the SAD first-year learning opportunity in terms of time and class sizes. Timewise, Summer School is presented over five days, instead of a whole year. Furthermore, learning opportunities in the year course involve large class sizes, while typical Summer School classes do not exceed 30 students. In what follows we describe how students experienced our Summer School learning opportunity. All respondents indicated that they had a positive experience of the Summer School and educator, while 52% of the students indicated that the Summer School opportunity assisted them in clarifying and understanding subject matter better. About 61% of the students indicated that the educator played a major role in their positive and successful experience of the Summer School learning opportunity. Lessons learned from the Summer School experience included the following:

- a number of discussions and detailed explanations facilitated understanding;
- it was an advantage that the format was compact and clear (as opposed to a module taught over two semesters);
- small classes made for easier learning.

Especially the smaller class sizes allowed for more detailed discussions, a common critique against large class sizes [9]. In a semester system, according to [2], the burden of studies is less compared to an annual system, and students have more opportunities to score better. This might explain the success of our Summer School experience, as it makes allowance for smaller groups and focused discussions, which allows for better opportunities to obtain higher scores.

In summary, all students experienced the Summer School as positive. The Summer School experience assisted the students to better understand the subject matter and helped them to clarify the work. The students also complimented the educator who gave detailed explanations of the subject matter and encouraged class discussions. Students indicted that they preferred smaller class sizes that allowed for personal assistance and easier learning.

5   Conclusion

This paper investigated the factors that contributed to the poor performance of first-year SAD students at a prominent university in South Africa. In answering the research question, it was found that:

- Lecturers did not always simplify the manner in which the content was presented. This caused subject matter confusion and misunderstanding for students. One student indicated that it is not possible for educators to provide individual assistance to students because of the large class sizes.
- Students indicated that they preferred smaller classes that allowed for personal assistance and easier learning.
- Some students experienced emotional trauma (mostly on a family and personal level) during their first-year course that may have contributed to their failure. In two cases, the students indicated that the emotional trauma was caused by subject matter that they could not master. These students did not know how to ask for assistance to emancipate themselves from this situation.
- Students found it difficult to master (understand and apply) the subject matter. Some of the students experienced difficulty in envisioning how the subject matter could be applied to their fields.
- Students admitted that they needed to learn harder and prepare better for assessment opportunities. They further indicated that, when they are in trouble during the course, they should ask educators or fellow students for assistance and not fall behind.

The following *recommendations* are made towards the identified causes for first-year SAD course failures:

- Educators need to be sensitive to the complexity of the SAD course content. Learning outcomes and student expectations need to be set in the beginning to guide students through the content. This will decrease subject matter confusion and misunderstanding.
- To address the problem that students are generally not privileged with adequate individual attention, which is an unfortunate consequence of large classes, it is recommended that educators have consultation times and encourage students to use them to request personal assistance. If students experienced emotional trauma that may impact negatively on their courses and results, educators should make them aware of where to find professional help or counselling. Students should be encouraged to ask for assistance to emancipate themselves from their situation.
- In accordance with the PBL methodology students need to take responsibility for their own learning. Students need to learn harder and prepare better for SAD assessment opportunities. Educators should further encourage the students to ask for assistance (from educators and fellow students) when they need course or subject matter assistance. Students should not at any stage fall behind.

All students in our study experienced the Summer School learning opportunity as positive. Summer School assisted the students to understand subject matter better and helped them to clarify work, which they did not previously understand. They also enjoyed the individual help and assistance from educators during the Summer School. The contributions of this paper lie in the insights and experiences that were shared by students who failed their first-year SAD course. The experiences and insights of these students were used to make recommendations to educators, especially first-year educators, to improve their educational practices. The experiences and insights of the students can also be used by fellow students to identify possible pitfalls that may cause them to fail, when enrolling for an IS course.

We close this paper by encouraging other educators in the field of Information Technology to find creative ways to improve educational practices and to share this with other educators in the field.

References

1. Abdullah, M.R.T.L., Hussin, A.Z., Zakaria, A.R.: mLearning scaffolding model for undergraduate English language learning: bridging formal and informal learning. Turk. Online J. Educ. Technol. **12**(2), 217–233 (2013)
2. Aslam, H.D., Younis, A., Sheik, A.A., Maher, M.Z.A., Abbasi, Z.A.: Analyzing factors affecting students' satisfaction regarding semester system in Universities of Pakistan. J. Am. Sci. **8**(10), 163–170 (2012)
3. Berkel, J.H.C., Moust, H.J.M.V., Schmidt, H.G.: Signs of erosion: reflections on three decades of problem-based learning at Maastricht University. High. Educ. **50**(4), 665–683 (2005)
4. Cascone, L.: Co-operative learning, structured controversy and active learning. Technical report, George Mason University (2017)
5. Delisle, R.: How to Use Problem-Based Learning in the Classroom. ASCD, Alexandria (1997)

6. Fatima, S., Abdullah, S.: Improving teaching methodology in system analysis and design using problem based learning for ABET. Int. J. Mod. Educ. Comput. Sci. **7**, 60–68 (2013)

7. Haque, H.S., Hamid, H.S.A.: Detrimental external factors that influence performance of students who repeat psychological statistics. In: Proceedings of the 11th International Postgraduate Research Colloquium (2014)

8. Harb, N., El-Shaarawi, A.: Factors affecting students' performance. J. Bus. Educ. **82**(5), 282–290 (2007)

9. Hornsby, D.J., Osman, R.: Massification in higher education: large classes and student learning. High. Educ. **67**(6), 711–719 (2014)

10. McGhie, V.F.: Factors impacting on first-year students' academic progress at a South African University. Doctoral Dissertation, Stellenbosch University (2012)

11. Mulryan-Kyne, C.: Teaching large classes at college and university level: challenges and opportunities. Teach. High. Educ. **15**(2), 175–185 (2010)

12. OECD: Education at a Glance 2013: OECD Indicators. Technical report, Paris (2013)

13. Pretorius, H.W., Pretorius, J.J.: A higher education maturity framework for outcomes based education. In: Proceedings of the SAARDHE International Conference (2007)

14. Richardson, I., Delaney, Y.: Problem-based learning in the software engineering classroom. In: CSEET Proceedings of the 22nd Conference on Software Engineering Education and Training (2009)

15. Ruggeri, K., Dempster, M., Hanna, D., Cleary, C.: Experiences and expectations: the real reason nobody likes stats. Psychol. Teach. Rev. **14**(2), 75–83 (2008)

16. Saulnier, B.M.: Towards a 21st century information systems education: high impact practices and essential learning outcomes. Issues Inf. Syst. **17**(1), 168–177 (2016)

17. Scott, J., Graal, M.: Student failure in first year modules in the biosciences: an interview-based investigation. Biosci. Educ. **10**(1), 1–8 (2007)

18. South Africa Council on Higher Education: South African higher education reviewed—two decades of democracy. CHE **27**(6), 795–808 (2016)

19. Strauss, A., Corbin, J.: Basics of Qualitative Research: Grounded Theory Procedures and Techniques, 2nd edn. SAGE, Thousand Oaks (1998)

20. Tamblyn, R.M., Barrows, H.S.: Problem-Based Learning: an Approach to Medical Education. Springer Series on Medical Education, vol. 1. Springer, Heidelberg (1980)

21. Tepper, J.: Assessment for learning systems analysis and design using constructivist techniques. Technical report, The Higher Education Academy (2014)

22. Walsham, G.: The emergence of interpretivism in IS research. Inf. Syst. Res. **6**(4), 376–394 (1995)

23. Williamson, J., Pretorius, E., Jacobs, M.: An investigation into student performance in first year biology at the University of Johannesburg. In: Proceedings of the International Conference on Mathematics Science and Technology Education, p. 181 (2014)

24. Yamagata-Lynch, L.C.: Activity Systems Analysis Methods Understanding Complex Learning Environments. Springer, Heidelberg (2010). https://doi.org/10.1007/978-1-4419-6321-5

25. Zellars, K.L., Tepper, J., Duffy, M.K.: Abusive supervision and subordinates' organizational citizenship behavior. J. Appl. Psychol. **87**(6), 1068 (2002)

26. Zhang, Y.L., Sun, J., Hagedorn, L.S.: Homesickness at college: its impact on academic performance and retention. J. Coll. Stud. Dev. **57**(8), 943–957 (2016)

Cheap Latex, High-End Thrills: A Fantasy Exercise in Search and Seizure

Wai Sze Leung[✉][iD]

Academy of Computer Science and Software Engineering,
University of Johannesburg, Johannesburg, South Africa
wsleung@uj.ac.za

Abstract. Quite often Computer Forensics students are afforded little or no opportunity to gain experience when it comes to retrieving digital evidence from a crime scene. For our offering, the exclusion was attributed to a lack of resources in terms of time, physical space, and equipment. To circumvent the aforementioned short-comings, we tapped into the imaginations of our students to introduce a search and seizure exercise by means of role-play simulation, which is seen to be a form of active learning. This paper describes how we adapted the problem-based learning activities approach employed by the Universities of Sunderland and Glamorgan, to offer our students a similar student-centered learning opportunity, supplementing the lack of a professional crime scene environment with a storyteller to fill in the details of what the students experience during their role-play simulation. We review our version of the search and seizure role-play simulation against the lessons learnt at Sunderland and Glamorgan, examining how future implementations can be revised to reflect best practices in offering our students improved learning opportunities.

Keywords: Computer Forensics · Role-play simulation · Active learning

1 Introduction

When it comes to teaching what may be viewed as a rather voluminous amount of theory in Computer Forensic modules, lecturers are likely to require additional strategies beyond that of the traditional lecture-based format if they are to expect their students to grasp the work instead of resorting to rote memorization of the facts and processes.

The focus on teaching this subject field is particularly a concern in light of findings on the state of digital forensic practitioners in South Africa, where qualifications, training, and competences of practitioners generally *"do not match up to objective standards"* [7]; for comparison see also [10]. Given the prevalence of cybercrime [12] and the South African government's attempt to address shortcomings in legislation by means of the recently revised Cybercrime and

© Springer International Publishing AG 2017
J. Liebenberg and S. Gruner (Eds.): SACLA 2017, CCIS 730, pp. 265–277, 2017.
https://doi.org/10.1007/978-3-319-69670-6_19

Cybersecurity Bill [8], the aforementioned confirmation is unsettling. Computer Forensics lecturers must therefore revisit how the subject matter can be better taught to produce capable, potential practitioners.

An often overlooked learning unit in Computer Forensics modules relates to the crime scene [6], where one must be sufficiently knowledgable in order to identify, secure, and preserve digital evidence in a forensically sound manner for further examination at a later stage. As a result, students focus on the analysis and examination of digital evidence and do not always understand the intricacies involved in how that evidence got to the laboratory in the first place [6]. To fill this gap in students' impressions of the overall investigation process, Irons and Thomas adopted a problem-based learning approach in which through arrangements with the local police training college and university forensic science department respectively, students at the Universities of Sunderland and Glamorgan were given the opportunity to work through scenes set up at specialized crime scene facilities [6].

While the approach has been demonstrated to produce authentic learning gains [6], securing such a specialized physical space was not possible in our case. This paper describes our experiences in offering a similar learning opportunity despite a lack of the specialized resources by conducting a search and seizure of digital evidence at a crime scene by means of role-play simulation, with lecturers assuming the role of a storyteller to describe and manage what is happening in the simulated crime scene. By doing so, we ensured the inclusion of search and seizure at the crime scene as an active learning activity in an otherwise unforgiving lecture schedule. Our approach also encouraged students to carry out self-directed studies and conduct further research on the chain of custody guidelines in order to prepare for their attempts at digital evidence acquisition.

2   Literature

2.1   Problem-Based Learning and Role-Play in Education

A method of instruction that originated in medical school, Problem-Based Learning is an approach that makes use of real-world problems to provide students with the context and motivation for engaging in active learning activities—it is therefore seen to significantly promote student ownership, critical thinking, teamwork, problem-solving skills, and self-directed and life-long learning [1,3,5,11]. While numerous PBL variations have materialized since it was first implemented as a means of addressing student disenchantment with information overload and a failure for them to connect the dots between theory and envisaged practice, the original PBL model exhibited a number of characteristics [3]:

- problems are presented to form the focus and stimulus for learning;
- these problems are presented in a manner that develops students' problem-solving skills;
- students assume responsibility of their own learning;
- students were organised into small groups to facilitate learning;

– teachers guide learning within these groups;
– students acquire new information through their self-directed learning.

Similarly, role-play has also been accepted across multiple disciplines as an effective learning strategy in encouraging participation among passive learners while promoting autonomy and material retention in student learning [5, 14]. In role-play, students assume roles and immerse themselves in a given case scenario which enables students to utilize all four senses identified in the VARK (visual, aural, read/write, and kinesthetic) learning styles inventory [5]. In first-aid training, for example, students are usually presented with a scenario in which they arrive at a scene where several patients require emergency medical attention and the students must draw on their first-aid knowledge in order to assess and apply the appropriate aid to the various patients. Although role-play is seen to enjoy much success, it should not be seen as a 'silver bullet' in education—while research has shown that the use of role-play in PBL can help students achieve both active and experiential learning with several value adds values (such as creativity and student-driven learning) [5], it can also be counterproductive. To cite a few examples [4, 5, 14]:

– it can end up being time-consuming;
– it can distract students from the intended lessons;
– one student may hijack the scenario, overshadowing the contributions of their fellow students;
– not all students will be interested in participating due to their personalities and preparedness.

Greater planning and preparation must therefore go into determining the format of role-play exercises to ensure that the efforts are not undermined. One such aspect of the planning that must be considered is the type of role-play ('role switch', 'acting', or 'almost real-life') to use which is discipline-dependent [13]. For that reason, the next subsection examines the characteristics of Search and Seizure exercises to determine the appropriate role-play model.

2.2 Search and Seizure at Crime Scenes

While Computer Forensics lecturers do not neglect to cover the topic of search and seizure, Irons and Thomas observe that the learning unit is often done in a theoretical manner, with students rarely having the chance to obtain practical experience in the matter. Reasons for foregoing a practical learning opportunity in this subject vary, although common logistic motives entail, according to [6], a lack of:

– a physical crime scene area to perform the exercise in;
– technical knowledge on the lecturers' parts to develop or acquire realistic digital evidence;
– items necessary to perform the exercise with;
– opportunity to include the practical exercise.

Moreover, because most digital forensic practitioners are unlikely to visit a crime scene, the exclusion of the opportunity to practice search and seizure of digital evidence is not considered to be a serious concern. As a matter of fact, Irons and Thomas are of the view that practical assessments of students in search and seizure should not represent a significant weight in contributing towards a degree in Computer Forensics [6]. On the other hand, there is value in providing students with the opportunity to exercise their search and seizure skills through some practical activity as the opportunity would enable students to understand evidential continuity and integrity; as well as the need to separate the relevant data from all the 'noise' in digital evidence [6]; for comparison see [15].

As described in Sect. 1, practitioners at a crime scene tasked with the search and seizure tasks are required to draw on their knowledge (which is required to be rather expansive as many numerous technologies may be involved) in order to:

- survey the scene;
- identify the various hardware, software, and miscellaneous items that serve as sources of potential sources of evidence;
- devise a strategy for acquiring and preserving the identified evidence items;
- acquire the evidence (using the appropriate tools and techniques) in a manner that is forensically sound;
- bag-and-tag the evidence items collected (ensuring that they are preserved), documenting the necessary information.

This is similar to that of emergency medical technicians or paramedics who stabilise patients at the scene for transport to the hospital. Such parallels would thus indicate that the use of role-play in first-aid and emergency services training would be similarly, if not equally as suitable as an exercise for Computer Forensic students needing to study background knowledge in order to correctly carry out search and seizure tasks at a crime scene. Based on this comparison, we can establish that the 'acting' role-play model would be the most appropriate choice of the three types of role-play approaches to use.

Acting role-play can be described as 'playing out and/or exploring a scenario' which may need the use of props [13]. Our planning and preparation of a role-play search and seizure exercise must therefore consider the inclusion of props to support the acting; the inclusion of 'noise' from unrelated items at the crime scene; and some form of task(s) that will ensure that students do the necessary studying to prepare for the role-play.

3   Adapting the Search and Seizure Exercise

When it comes to learning about search and seizure, a number of learning outcomes are assessed. These include:

- the appropriate use of warrants;
- the relevant application of computer knowledge basics (e.g.: how computers work, how operating systems boot up, etc.);

- following processes correctly when conducting a search and seizure of evidence from a crime scene;
- being able to distinguish between techniques used in live and dead forensics;
- using appropriate tools and techniques to make forensic copies of data;
- maintaining a chain of custody.

Most of this information is published in several reference field guides published by law enforcement and judicial organisations such as the U.K. Association of Chief Police Officers (ACPO) [2] and the U.S. Department of Justice [9]. Traditionally, our students were advised to revise their computer knowledge basics from previous years of study, and to study the aforementioned field guides. Combined, the material is considerably lengthy and overwhelming, leading to students often skimming through the content, and memorising processes that were likely not retained after having answered the theory assessments.

Seeking to improve student engagement and knowledge retention in this learning unit, we adopt the problem-based learning approach employed by [6], adapting their exercise to suit our particular environment. To do this, we begin by examining the profile of our students.

3.1 Student Profile

Our Computer Forensics module is an elective offered to students studying towards an Information Technology or equivalent 'honours' (postgraduate) degree with the number of students registered for the module varying from year to year.[1] In terms of student profile, our students are in their twenties, mostly male, with a majority registered for full-time studies. Even so, there are considerably enough part-time students in the class to warrant scheduling our search and seizure role-play simulation during the official class times that are organized after office hours in order to accommodate the students who work. This does however impose a potential limitation in terms of the time we can allocate each group for conducting their search and seizure role-play simulation session.

3.2 Freeing Up Time

As is, our Computer Forensics curriculum is quite full due to the large volume of work that must be covered within the course of the single semester. This is in contrast to programs where students focus on acquiring a Computer Forensics-related degree across multiple years. Spanning 14 lectures with a learning unit being covered in each week, introducing a search and seizure practical would therefore come at the expense of our removing a single teaching unit from the module's programme. Guidelines for implementing successful role-play exercises

---

[1] For readers from outside South Africa: the South African 'honours' degree is an extension of the classical 'B.Sc.' degree which enables a student to commence with Master-studies thereafter. While already considered 'postgraduate' in South Africa, the 'honours' degree in South Africa is reasonably well comparable to the final study-year in the (longer) U.S. American 'B.Sc.' curriculum.

suggest that certain content or topics be earmarked for self-study to free up the time (in our case: one week) [4]. Because students will need to have knowledge of what to do with the digital evidence they come across in the crime scene, it makes sense to prescribe the ACPO and DoJ field guides (mentioned at the beginning of Sect. 3) as self-study. Unless they familiarise themselves with the content of the guides, students will not have the knowledge needed to carry out the search and seizure exercises correctly. Similarly, we can also consider freeing up additional time by setting the learning unit on basic computer knowledge aside as self-study content that students must cover in their own time. We believe that this is reasonable since our students are honours students in IT or similar. It would therefore be reasonable to expect all students to possess the necessary Computer Science knowledge that is regarded as a foundation in the field of digital forensics [7]. Our approach can be seen to be an implementation of the flipped classroom principle where our students learn at home in order to demonstrate how to do the 'homework' during the role-play exercise [16]. By casting students as members of a digital forensic investigation team that will be required to report to a crime scene in a few weeks to carry out a search and seizure of evidence items, we are motivating students to do that self-study with the promise of being able to problem-solve their way through a mystery as a 'reward'.

3.3 Role-Play Format

A second recommended guideline to ensuring a smoother role-play experience recommends that we supply students with sufficient information before the role-play itself takes place [4]. As such, we provide our students with an information sheet at the beginning of the semester, detailing the format of the role-play exercise.

Students are informed that in approximately two months, they will be given the opportunity to conduct a search and seizure exercise as a team of first responders who have arrived at a scene at the request of a client (either law enforcement or a private party). During this exercise, they will be required to apply the appropriate processes to ensure that they have acquired and preserved the digital evidence correctly. As with all such investigations, they will be expected to submit a report that documents everything that they have done, serving as their chain of custody records. At present, they do not know what the case will involve. However, they can be assured that they will need to be familiar with the different procedures that should be followed when it comes to searching and seizing the different types of digital evidence.

In preparation, students must first form and register their teams, (a maximum and minimum number of members per team is enforced). Students are strongly urged to form their own teams as we believe that familiarity between the team members will likely improve the team's ability to play out their roles more cohesively. We only step in and place students who have yet to register into teams should the registration deadline pass. Table 1 shows how student teams were formed for the search and seizure role-play simulations offered in the past two years. As the table shows, we were required to intervene during both years when students were not familiar enough with their classmates to form teams. Although these lecturer-formed teams were able to band together to complete the role-play exercise, we could not help but notice that they did not behave in as coordinated a manner as their peers that had formed teams on their own.

Table 1. Student search and seizure team compositions

Year	2016	2017
Class size	33	17
Teams	7	5
Student-formed teams	5	4
Lecturer-formed teams	2	1
3-member teams	-	1
4-member teams	2	2
5-member teams	5	-

Then, in addition to the weekly content taught during lectures, e.g.: legal requirements and warrants, students must do self-study and revision in order to build up their knowledge on what to do given various scenarios and requirements (e.g.: being creative when an investigation must be done secretly without alerting the suspect that their machine was examined). Students are further advised that the inclusion of props will help them create a more authentic experience during the role-play when they have to use various tools to carry out their tasks. Each team is also required to submit an inventory list that identifies the tools and equipment that they will be bringing with them to the crime scene. The requirement encourages students to do the necessary reading before hand.

Teams are graded based on their consideration of legal requirements (will they require a warrant?); how well they identify the various sources of evidence in their scenario (did they manage to find all the sources of evidence?); did they work together collaboratively as a team (were team members tasked to carry out specific roles? Did everyone participate?); did they use the appropriate acquisition strategy (given the circumstances, can they take that evidence item with them?); how the evidence is bagged and tagged (did they use a Faraday bag or cage for the mobile devices?); their chain of custody report (how detailed and meticulous is their record-keeping? Do they motivate their decisions?).

Finally, each team member will be required to complete a peer evaluation review in which the contributions of each team member towards the group's efforts are assessed. These results are used to calculate the individual marks of each team member, based on the overall group mark obtained.

The following subsections describe the various considerations that influenced how we adapted and developed the problem to suit our needs.

3.4   Resource Limitations and Overcoming Them

Whereas the Universities of Sunderland and Glamorgan were able to secure physical spaces designed specifically as crime scenes, we did not have such a luxury nor did we have the forensic equipment (including clothing such as gloves, overalls, masks, etc.) to use as props or tools. Against the backdrop of the notorious 'Fees Must Fall' protests in South Africa, it is unlikely that this resource issue would be addressed soon. As such, we decided to overcome this shortcoming by employing a role-playing simulation strategy in which:

1. Lecturers serve as the storytellers responsible for guiding the flow of the scenario. Very much like the 'Dungeon Master' role in a game of the 'Dungeons and Dragons' board game, the storyteller is equipped with the entire plot of the exercise, enabling them to narrate and interact with the students as the story unfolds based on what the students do at the crime scene.
2. Students act out how they would carry out their tasks at the crime scene, describing to the storyteller and fellow team members what they are doing and why they are doing so at each step.
3. Students are encouraged to make use of props to serve as substitutes of the equipment needed to carry out search and seizure tasks. The props serve as visual aids that assist all involved in the role-play to better visualize what is happening at the crime scene.

3.5   Setting Up Crime Scenes

For each group, we devised a different scenario, each with its set of evidence and items (such as books, laptops, USB storage devices, printouts, sticky notes, photographs, posters, and other hardware) for transforming a computer laboratory into the crime scene. In some cases, we had even made our own props to represent particular devices. A former paperclip container, for example, was converted into an 'Amazon Echo Dot' when we pasted the 'Amazon' logo on its side and attached a drawing of its familiar four-button interface to the container's top.

Scenarios were built around a number of basic activities that every team is expected to demonstrate competence of—these include knowing whether to use a warrant to conduct the search; imaging of a device; making use of appropriate tools to aid in the documentation of the scene; dealing with devices that are live. For example, in one scenario, the device that must be imaged is a hard disk located inside a laptop while it is a USB storage device in another scenario. In doing so, the difficulty of the different scenarios were standardized, with the principles that must be assessed assuming different guises. This approach is similar to assessments in first-aid training where students must be able to demonstrate competency in a number of first-aid basics such as the recovery position and CPR.

As a bonus, additional sources of evidence (such as a USB storage device disguised as a lanyard and a Yubikey) were also integrated into the scenario. These are in addition to the irrelevant items that are added to each scenario to add 'noise' to the crime scene, giving students a better understanding of how real crime scenes can be messy and difficult to navigate through. A stack of printouts, for example, can serve as a red herring that distracts them from completing their original goals.

To aid storytellers, a document containing the main plot, as well as a list of all the items in the scenario and what they represent is supplied, allowing storytellers to describe how events unfold as the students come across and interact with the different items at the crime scene. It should be noted that since students may pursue an action that is not relevant to the case, storytellers will need to be creative enough to accomodate deviations from the intended script and respond to student choices in an open manner. Ideally, storytellers should coach students back onto the right path should they stray.

4   Reflections

4.1   What Students Did

While the intention of the inventory list was to ensure that students prepared for the role-play and avoided the prospect of students suddenly 'magicking' items out of thin air during the exercise, we found that these lists were still hastily assembled. In some cases, these inventory lists were merely copied and pasted from some Internet source and then submitted late.

Every single team that participated brought props with them during their role-play session with the amount of props varying from team to team. The level of effort invested also varied, ranging from teams with just gloves (dishwashing gloves to meatpacking ones), to others that dressed up in reflector vests, and others donning surgical masks and scrub caps for the occasion. Likely heavily influenced by television, some teams were very eager (and adamant) about being able to 'dust for prints' or 'swab for DNA' despite repeated comments for teams to distinguish between digital and 'wet' forensics.

Throughout the session, team members had to talk us through with what they were doing with any piece of evidence they came across, voicing the reasoning behind their choices as they carried out the documentation, bagging, and tagging of evidence items for transport to their laboratories. While not every evidence item was identified in each scenario (despite prompting from the storyteller), student teams demonstrated that they had studied the theoretical content through their successful completion of the various tasks; (we regard a task as being completed successfully if the appropriate procedure as detailed in the field guides are followed). At the end of their session, the storyteller informs the team of the items they have missed and gives a quick rundown of any processes that may not have been done correctly. It should be noted that errors made arose from not finding the evidence and not paying attention to what was said—this suggests that the students need to work on their listening skills, rather than the students not having done the required self-studying. After the session the team gathers the recorded data (written notes, recordings, and photographs taken) to compile a report. The report allows students to reflect on their learning, identifying where they went wrong.

Of the 50 students who participated in the exercise, we noticed that only one student did not seem to be as enthused. This is in line with the feedback obtained from other researchers showing that role-play exercises are generally met with an overwhelmingly positive response [5,14]. We are therefore encouraged by the positive and eager attitude to our exercise. As such, we would like to engage with our students in identifying the positive and negative aspects of the exercise in order to refine future offerings of this learning opportunity.

4.2   Changes to the Role-Play (2016 versus 2017)

Over the span of two years, several aspects of our role-play simulation underwent revision. We detail these changes, indicating their impact on the effectiveness of the role-play as a result.

Noise at Crime Scene: In 2016, each scenario was prepared with their own set of items, including unrelated items serving as distractions at the crime scene. This required that we had many more devices and props, taking more time to prepare the unique scenes.

In 2017, we opted to reuse some of the props in other scenes. We did this by setting up the various scenarios close to each other. For example, books on a desk for one scenario then served as a bookshelf at the edge of another. This had the effect of creating additional 'noise' in the crime scene and more surface area for the teams to explore in their search for sources of evidence.

Team Size: Our experiences with five-member teams in 2016 demonstrated that larger numbers tended to be rather chaotic when the team members are not as well prepared. In such cases, members did not always work cohesively, talking over one another, repeating questions to the storyteller, and undoing work at times. In one case, a team member stood and watched while their four team members did everything.

We addressed this problem in 2017 by limiting the maximum number of team numbers to 4. This approach allowed us to better follow what the various team members were doing, making it easier for us to tell them how the story unfolds in response to their actions.

Staff per Scenario: Due to time constraints and many groups, one lecturer served as both the storyteller and observer (responsible for taking down notes and grading the team's activities) in 2016 for each group. This however proved difficult to follow (given the larger team sizes).

We rectified the problem by assigning a storyteller and a separate observer to each scenario (with smaller teams) in 2017.

Use of a Standardized Set of Tasks: A non-standard list of tasks that had to be completed in each scenario was used in 2016, leading to varying levels of difficulty for the different teams.

To standardize the difficulty in 2017, we started with a base set of requirements (tasks that had to be done) and built stories and evidence items around these requirements. In this way, every scenario required that the team carry out the same tasks.

Time: The time allocated to each scenario was reduced by five minutes in 2017 due to time constraints and venue availability. Although the smaller teams and shorter timeframes meant that teams focused on their tasks better, the overall sentiment was that students enjoying the role-play found the experience to be too short.

5   Conclusions

We set out to introduce a learning opportunity in our Computer Forensics module that would engage our students, prompting them to take ownership of their own learning and play a more active role in initiating research on the different (sometimes esoteric) hardware and software that they may come across during an investigation. Based on the quality of the reports, the extensive use of props, and their enthusiasm in tackling the problem, we believe that we have achieved some part of this. Informal feedback from students, for example, have been positive with certain students from the 2016 cohort indicating that the role-play represented one of the highlight of their honours studies. While we have observed visible improvements in our second offering of the search and seizure role-play simulation, we acknowledge that there is still much room for improvement in how we present the learning activity. Based on our observations, feedback from the students, and from colleagues at the SACLA'2017 conference, we list the following as possible considerations in future implementations.

Manageable Team Sizes: Future teams should comprise three to four members with one member responsible for documenting the process. This will allow us to follow what the team members are doing more effectively and give appropriate feedback; everyone participates meaningfully and in a coordinated fashion).

Increase Simulation: Make use of simulation on working desktops to imitate programs or processes that are shown to be running on the 'suspect machine' when the investigators interact with it. This will assist in helping students better visualize the context of the problem.

Red Team versus Blue Team Competition: By pairing teams against each other, the task of setting up scenarios can be delegated to student teams. This approach will allow students on the opposing team to consider the mindset of the organisations or individuals that are the subject of their investigations when it comes to hiding the evidence away.

Emphasis on Team Work: By stressing the importance of working as a coordinated team at the beginning of the module (and having smaller teams), we anticipate better division of duties amongst team members.

Detailed Inventory Lists: Instead of simply listing all the items that the team will be bringing with to the crime scene, teams will now need to detail what each item is meant to be used for. In addition, to avoid last-minute efforts, teams may need to formulate the list as an 'acquisition request' in which they motivate the 'purchase' of equipment, subject to a limited budget. These additional levels of preparation will force students to start their research earlier, potentially minimising the lack of participation from students who do not acquire the necessary background knowledge as required.

Finally, we intend to conduct a formal questionnaire that will enable us to qualitatively establish what worked and what had not worked from the students' perspective. Such a questionnaire will also allow us to compare our results with those experienced by students at the Universities of Sunderland and Glamorgan.

References

1. Akcay, H.: Learning from dealing with real-world problems. Education **137**(4), 413–417 (2017)
2. Association of Chief Police Officers: ACPO Good Practice Guide for Digital Evidence (Version 5) (2012). http://library.college.police.uk/docs/acpo/digital-evidence-2012.pdf
3. Barrows, H.S.: A taxonomy of problem-based learning methods. Med. Educ. **20**(6), 481–486 (1986)
4. Center for Learning Enhancement, Assessment, Redesign: Pedagogical Principles of Role-play in Learning (2017). http://clear.unt.edu/pedagogical-principles-role-play-learning
5. Chan, Z.C.Y.: Role-playing in the problem-based learning class. Nurse Educ. Pract. **12**(1), 21–27 (2012)
6. Irons, A., Thomas, P.: Problem-based learning in digital Forensics. High. Educ. Pedagog. **1**(1), 95–105 (2016)
7. Jordaan, J., Bradshaw, K.: The current state of digital forensic practitioners in South Africa. In: Proceedings of the Information Security for South Africa ISSA'2015, pp. 1–9 (2015)
8. Minister of Justice, Correctional Services: Cybercrimes and Cybersecurity Bill (2016). http://www.justice.gov.za/legislation/bills/CyberCrimesBill2017.pdf
9. Office of Legal Education: Searching and Seizing Computers and Obtaining Electronic Evidence in Criminal Investigations (2009). https://www.justice.gov/criminal/cybercrime/docs/ssmanual2009.pdf
10. Olivier, M., Gruner, S.: On the scientific maturity of digital Forensics research. In: Peterson, G., Shenoi, S. (eds.) DigitalForensics 2013. IAICT, vol. 410, pp. 33–49. Springer, Heidelberg (2013). https://doi.org/10.1007/978-3-642-41148-9_3
11. Prince, M.: Does active learning work? A review of the research. J. Eng. Educ. **93**(3), 223–231 (2004)
12. PWC: Global Economic Crime Survey (2016). https://www.pwc.co.za/en/assets/pdf/south-african-crime-survey-2016.pdf

13. Rao, D., Stupans, I.: Exploring the potential of role play in higher education: development of a typology and teacher guidelines. Innov. Educ. Teach. Int. **4**(49), 427–436 (2012)
14. Stevens, R.: Role-play and student engagement: reflections from the classroom. Teach. High. Educ. **5**(20), 481–492 (2015)
15. Tewelde, S., Gruner, S., Olivier, M.: Notions of hypothesis in digital Forensics. In: Peterson, G., Shenoi, S. (eds.) DigitalForensics 2015. IAICT, vol. 462, pp. 29–43. Springer, Cham (2015). https://doi.org/10.1007/978-3-319-24123-4_2
16. Tiahrt, T., Porter, J.C.: What do I do with this flipping classroom: ideas for effectively using class time in a flipped course. Bus. Educ. Innov. J. **8**(2), 85–91 (2016)

The Infinity Approach: A Case Study

Romeo Botes and Imelda Smit[(⊠)]

Institute TELIT-SA, North-West University, Vanderbijlpark, South Africa
{romeo.botes,imelda.smit}@nwu.ac.za

Abstract. As fourth-year Information Technology students make the transition from under-graduate studies towards a post-graduate degree, an honours research project may be a daunting task (For readers from outside South Africa: the South African 'honours' degree is an extension of the classical 'B.Sc.' degree which enables a student to commence with Master-studies thereafter. While already considered 'postgraduate' in South Africa, the 'honours' degree in South Africa is reasonably well comparable to the final study-year in the (longer) U.S. American 'B.Sc.' curriculum). At the Vaal Campus of the North-West University a series of lectures are offered in an attempt to guide students to make this transition. Students are firstly introduced to research paradigms and then each of the four prevalent information systems paradigms is focused on, to explain the methodology and methods applicable, and how it may be implemented. Each lecture in the lecture series is accompanied by an assignment. Although many obstacles present themselves in the completion of the honours research projects, a prominent one was identified. It concerns the interpretivist paradigm and specifically the collection of qualitative data with its accompanying data analysis. To assist students, the analogy of the infinity symbol (∞) with an example were used to explain the concept. Important focuses of the explanation include the crossroad faced by researchers when they need to decide on whether they have reached a point of saturation in a study or not. This facilitation supports the gaining of insight into a phenomenon and the implementation of the methodology in collaboration with the selected method to ensure the validity of the research. This is called the 'infinity approach'. This paper builds on earlier research by implementing a pilot case study as an evaluation of that approach.

Keywords: Interpretivism · Research method · Grounded theory · Saturation · Infinity method · Training of young researchers

1 Introduction

This study is part of a larger study on the evaluation of the so-called 'infinity approach', with a focus on interpretive research, amongst Information Technology (IT) students doing a post-graduate degree. The intervention with the intention to guide young IT researchers consists of the following steps: the first phase entails the lecture series intervention introduced to assist young and inexperienced people taking the first step towards doing research. The second phase

J. Liebenberg and S. Gruner (Eds.): SACLA 2017, CCIS 730, pp. 278–292, 2017.
https://doi.org/10.1007/978-3-319-69670-6_20

is an on-going process with the purpose to develop material that may clarify research methods that young IT researchers struggle with. The third phase is made up of the evaluation and refinement of the suggested material. The intention with the last phase is to do two studies, namely a pilot exploratory case study followed by an in-depth explanatory case study. The first phase is repeated every year to accommodate each new honours enrollment. The second phase, an on-going process of identifying challenging aspects of research and producing material explaining each concept, introduces the first challenging concept identified—how to determine the point of saturation in an interpretivist study. The material explaining this concept can be found in [5].

The focus of this paper is on the third phase where the intention is to refine the suggested material to ensure clarity and understanding regarding the concept addressed. Therefore, the paper on the infinity approach was made available to students and they were requested to read it before completing an assignment. In addition, upon handing in their assignments, they were requested to complete a questionnaire about their experience.

In subsequent sections the following topics are addressed: the context of the study, which is followed by a literature review on interpretive research. The suggested 'infinity approach' is then briefly introduced. The research design is explained before the pilot case study is discussed to provide clarity with regards to the usability of the approach. Finally, the paper is concluded with some recommendations for future work.

2 Context

After the successful completion of their undergraduate studies, many IT students continue with a post-graduate degree. An honours research project makes up one fifth of this course offering. Since a mini-dissertation needs to be completed as part of the honours research project, it is experienced by the IT students as a daunting task. To guide students in making the transition from being a learner of subject material to becoming a researcher who generates new material, a series of lectures by staff members involved in post-graduate supervision are offered on the topic of research methodology at the Vaal Campus of the North-West University.

Although students on this level are not required to address research philosophy extensively, they are introduced to the four research paradigms prevalent to the field of Information Systems, namely 'positivism', 'interpretivism', 'critical social theory', and 'design science'. Each of these paradigms is then scrutinized with the purpose to explain its associated methodology and applicable methods. Examples from previous years' research projects are used to support the explanations. In an attempt to facilitate understanding and to assure students understand all paradigms, methodologies and methods, and not only the ones they utilize in their own research projects, an assignment accompanies each lecture in the lecture series. The schedule covered to prepare these students for their research projects is listed in Table 1.

Table 1. Topics covered to guide honours students in their research projects

#	Topic	Motivation
1	Academic writing	Mind-mapping a topic, how to structure writing, paraphrasing, how to do in-text referencing and compiling the reference list
2	Research paradigms	Introducing the four research paradigms relevant in the field of Information Systems, namely positivism, interpretivism, critical social theory, and design science
3	Action research versus Design science research	Scrutinizing a method(ology) used in critical social theory and juxtaposing it with one used in design science research
4	The research proposal	Introduction of the research template to students; discussion of the sections to be addressed
5	Positivist (quantitative) research	Scrutinizing a method(ology) used in positivism
6	Interpretive (qualitative) research	Scrutinizing a method(ology) used in interpretivism
7	Ethics of research	The focus is placed on the importance of ethics in research; students are also guided to prepare an application for ethical clearance

During this lecture series, each student (or pairs of students) may approach an academic staff member to guide them throughout the completion of their research project. This process is in accordance with the research conducted by [7] where post-graduate alumni identified supportive academic staff as crucially important for success in post-graduate studies. The proposed chapter breakdown for the honours research project's mini-dissertation is listed in Table 2.

Table 2. Chapters supposed to appear in an honours mini-dissertation

#	Topic	Elaboration
1	Introduction	A proposal of the research project, addressing key concepts, applicable research methodology, the motivation of the study, its objectives, ethical considerations, and a chapter outline
2	Literature review	Related work addressing the key concepts used in the study
3	Research design (Meth.)	Literature review of the chosen paradigm and research methodology, as well as its application to the study
4	Study and results	The conducted research and its results described
5	Conclusion	The conclusions, also with regards to planned future research

Even with the guidance supplied, many obstacles present themselves in the completion of the honours research projects. A prominent one, concerning the interpretivist paradigm and specifically the collection of qualitative data with its accompanying data analysis, was already identified in earlier work. The refinement of that approach is the focus of this paper.

2.1   Interpretivism

Interpretivist research, with its focus on hermeneutic 'understanding', was promoted by the philosopher Dilthey in the late 19th century [20] on the basis of earlier work by the historian Droysen [8]; it arose in contrast to natural science in order to address the limitations associated with positivism [10,23]. The interpretivist researcher posits that it is important to understand the differences in humans' roles as social actors [19]. The essence of interpretivism requires the researcher to probe beyond facts towards meaning and eventually, understanding [15]. Emphasis is placed on research conducted among humans, rather than objects such as computers [19]. Qualitative data, such as narratives, are used in interpretive research. The metaphysical assumptions of interpretivist research include the following [13]:

Ontological assumptions originate from the nature of our being, whereby the researcher's view on reality is the focus. Social constructions such as language and the shared meanings conveyed are the interpretivist's access to reality.

Epistemological assumptions originate from the nature of knowledge. The relation between knowledge and truth is important; represented by our beliefs in relation to how we justify knowledge. The meaning people assign to a particular situation provide insight and understanding to the interpretivist.

Axiological assumptions appreciate what a researcher believes to be of value regarding a situation, therefore the interpretivist's context regarding a phenomenon drives the research.

The insurance of validity and trustworthiness of interpretive research was addressed in [11] by means of seven principles which are grouped in the following four categories:

- one *fundamental* principle, namely that of the *hermeneutic circle* [6,11];
- two *critical reflection* principles where the research needs to be *contextualized* and the participant-researcher's *interaction* recognized as the origination of data that is socially constructed;
- one principle based on the *philosophical framework* where the hermeneutic circle and the context are applied to the data to enable the researcher to *generalize and abstract* it;
- three *sensitivity* principles, including interpreting contradictions between literature and data, to utilize *dialogical reasoning* in order to tell a coherent story. Also there is always a possibility that participants have different interpretations of the same events: *multiple interpretations* need to be accommodated, and the researcher should be *suspicious* regarding distortions in participant narratives.

Interpretive data is classified as qualitative in nature and it is analyzed using qualitative data analysis techniques. The process of qualitative data analysis is defined in [3] as:

> "... *working with data, organizing it, breaking it into manageable units, synthesizing it, searching for patterns, discovering what is important and what is to be learned, and deciding what you will tell others*".

Much insight may be gained from the steps of grounded theory, where finding the point of saturation is important [22]:

- An introductory step states the problem and asks questions.
- The research procedure of grounded theory involves two actions; namely:
  data collection where interviews, questionnaires, observations and documents [16,19] may supply the researcher with data, and
  data analysis where sense is made of the data through three actions which include forming categories (each with its properties) of information called *open coding*, inter-connecting properties that supports a central theme called *axial coding*, and the use of *selective coding* that enables the researcher to tell a story. Qualitative data analysis processes such as tabulation [4] and software specific analysis tools including Hyper Research [17] and ATLAS.ti [1] may aid the researcher.
  Constant comparisons between data collection and analysis are suggested [21], until a state of saturation is reached.
- Finally, the intention is the discovery of a theory where this theory may be generated regarding a phenomenon or situation.

When doing interpretive research, especially with grounded theory, reaching saturation is a contentious issue since results found in one study may not necessarily be applicable to all situations [9]. In support, three factors were identified [18] that may influence saturation, namely the number and complexity of data obtained, the experience of the researchers, and the number of analysts involved in the research. These factors emphasizes the fact that a simple number (such as four, six or twelve interviews) cannot be applied to all situations. Also, the experience of young researchers is limited. In the research under discussion it is also possible for more than one student to work on one research project, which may influence the point of saturation.

2.2   The Infinity Approach

The infinity approach was formulated to assist young researchers doing interpretive research with grounded theory. This section provides a brief overview of this approach. For a more detailed explanation, and example, see [5]. This paper was inspired by [14] wherein the analogy of a dramaturgical model was used to explain the qualitative interview.

The infinity approach uses the infinity symbol (∞) to guide grounded theory, a qualitative research method used to obtain understanding of an environment

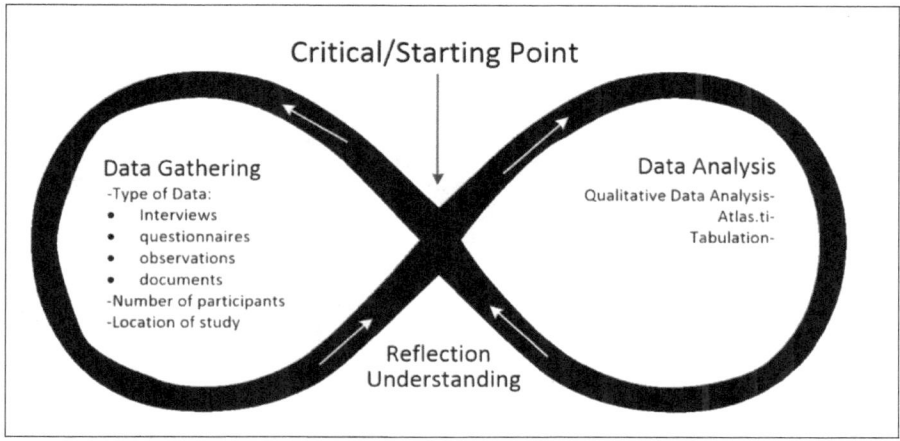

Fig. 1. The infinity approach

in which research is done. The two loops of the infinity symbol represent the two stages of the process: the loop to the left represents data gathering (DG), while the loop to the right represents data analysis (DA). These loops are grounded in hermeneutics, where it is important to understand the detail in the context of the whole. The middle of the infinity symbol, where the two loops meet, serves as the starting point, it represents a point of reflection after the gathering of data and its interpretation, and lastly it is also the point of eventual understanding of the phenomenon being researched. This crossing is referred to as the *critical crossing* (CX). The two cycles DG and DA, as well as intersection CX, are shown in Fig. 1.

The CX in the approach is the key to the successful implementation of the infinity approach since it is a point of departure regarding the research, a point of reflection during the research process, a point of evaluation on the progress of the research, and eventually the point where a decision on to how to proceed with another DG-DA cycle is made—therefore the CX is a point of conclusion w.r.t. the achievement of saturation.

The point of saturation draws closer as a study progresses with subsequent cycles of data gathering, analysis and reflection. Although the use of the infinity symbol (∞) implies that the DG-DA-CX may continue forever, in the suggested infinity approach each cycle is considered to be smaller than the previous one, with less interventions necessary to uncover new codes suggested during a previous cycle. Therefore, the study eventually converges into itself:

- At CX, interpreted data from the previous DA cycle are compared to that of the current cycle.
- Should the current cycle provide new insight, the researcher should initiate a new cycle to investigate.
- When the comparison between the last two cycles provides no new insight, the study may have reached saturation and this situation may guide the researcher to conclude the study.

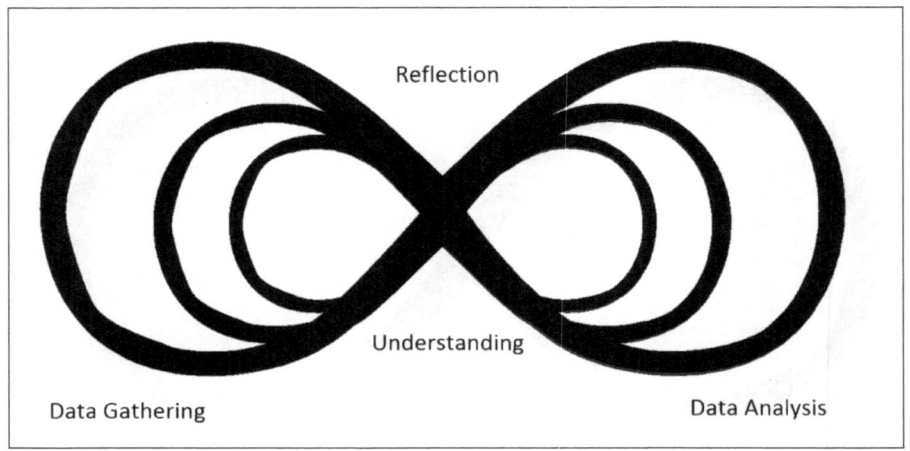

Reflection

Understanding

Data Gathering   Data Analysis

Fig. 2. Acquisition of knew knowledge with every cycle

The suggested convergence is reflected in Fig. 2, wherein the biggest outer cycle implies the first group of interventions and the smaller inner cycles implies subsequent groups of interventions where data may be gathered among fewer participants where suspected factors that are not yet confirmed may be either confirmed or rejected; (for comparison see [2] in Software Engineering). As a measure of caution one last cycle should be followed to ensure saturation, even though saturation may seem evident during a current cycle.

In the review of above the infinity approach no example was included. By explaining the infinity approach with an example [5] it becomes clear how the seven principles for conducting interpretive research—to ensure validity and trustworthiness of such research [11]—may be applied by researchers in 'qualitative' projects.

3   Research Design

Our pilot case study, designed to evaluate the suggested infinity approach, is exploratory [15]; for comparison see [24]. The intention is to explore features, factors and issues [12] that may guide future use of the infinity approach material. Also, in future (as a follow-up initiative), we intend a more in-depth study involving honours- and master-students, applying the infinity approach to their research projects and dissertations. With time the case studies may facilitate the discovery of an improved approach. The data obtained from the case studies may also be analyzed across cases with the intention to provide explanatory results [12]. Regarding the evaluation of the infinity approach material, we started by presenting a lecture on the interpretive methodology and methods. This lecture was followed by an assignment on the topic. Students were to investigate a topic unfamiliar to them and through data collection and analysis gain an understanding of the topic. As a first step, the whole group of students was divided into four more or less equal sub-groups. Each sub-group was allocated a general topic, namely:

1. electronics,
2. sports,
3. baking,
4. hacker spaces.

Each general topic was linked to a particular data collection method; live or video observation was required for the electronics topic, interviews were required for sports as topic, questionnaires were required for the baking topic, and the study of documents for hacker spaces. Each student in a group could select a more focused topic within that group to do the required data collection and analysis on. Thereby, our pilot study was driven by the following

Research Question: *Does the infinity approach assist students in gaining an understanding about interpretive research, as well as the process of doing interpretive research?*

Using a questionnaire with the questions listed in Table 3 we attempt to evaluate the approach and determine its feasibility.

Table 3. Survey questionnaire

#	Question to the students	Motivation
1	Did you read the provided article?	A filter to provide an answer as to whether a student actually read the infinity approach material
2	Did you apply the approach suggested in the article to the study you did for this assignment?	Determine to what extent the infinity approach material was applied to a student's study
3	Did the article help you in your study?	Determine if the infinity approach material guided the student regardless whether it was applied to the study or not
4	Reflect on your experience with regards to the article in the context of your assignment. Include (i) aspects that helped you in your study, (ii) aspects addressed by the article that became stumbling blocks in the completion of your study, and (iii) anything you did not understand—any improvements and/or corrections to the article that you may suggest would be much appreciated	Open-ended question for students to comment on their experience during the completion of the assignment and use of the infinity approach material

4   Pilot Case Study: Results

During 2017 the honours research project module had an enrollment of 38 students. A total of 30 students completed the assignment. The questionnaire, to be completed by the students upon finalization of the assignment, was completed by 28 students. The analysis of the questionnaire revealed that 20 students read the article uploaded to help them with the assignment, 19 students found it helpful in general, and 15 students applied the infinity approach in order to complete their assignment. The graph of Fig. 3 highlights those findings.

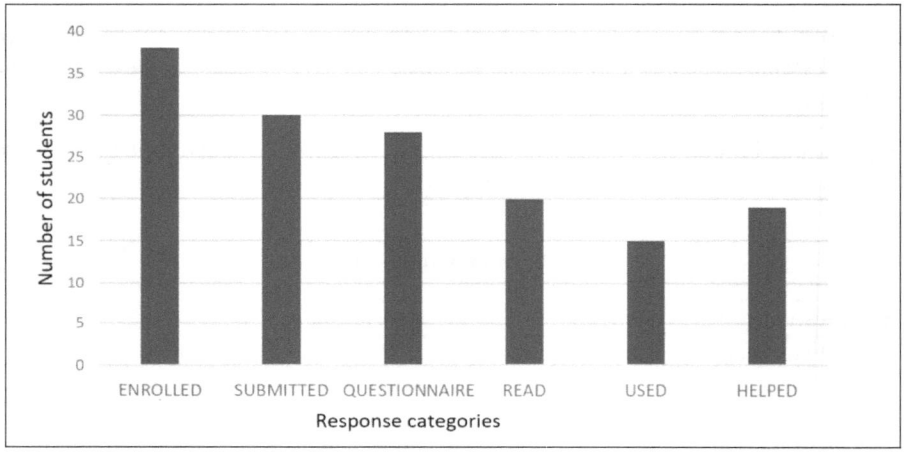

Fig. 3. Response categories of the pilot case study

Considering that 95% of our students found the approach helpful after reviewing it, it may be reasoned that the infinity approach material actually helped students. Also the fact that 75% of the students referred to our material in the process of completing their assignment is encouraging. Furthermore, from the assignment contents, we may conjecture that most students gained some understanding of their selected topics. In addition, with this being a formative assessment only counting a small percentage of the final mark, the process of doing the assignment and receiving feedback together with a mark may be a valuable learning experience for students performing well, as well as for those not getting a good mark for their assignments.

After comparing the assignment contents and results to the feedback from the questionnaire given to the students, it was clear that students reflected an improved understanding regarding the assignment contents; the average assignment mark was 56%, and the frequency graph shown in Fig. 4 indicates that most students achieved marks above 50% (i.e., 'pass').

Since the focus of our case study is on the evaluation of the suggested infinity approach, only the responses of those students who read the article are included

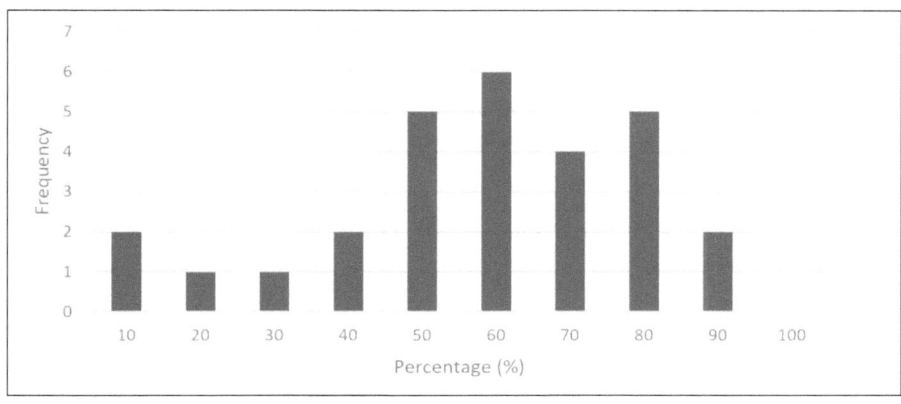

Fig. 4. Marks which students obtained after the assignment, with $m < 50\% =$ 'fail', and $m \geq 50\% =$ 'pass'

here. Also, with the assessment of the assignments, answers supplied by students on the subsequent questionnaire that requested the students to reflect on their experience, were compared to the assignment reports where relevant. The main focus of this paper is on the answers students supplied to the open-ended questions posed by our questionnaire. The first cycle of analysis produced a coding table. This table shows the suggested codes related to the identified categories and its corresponding descriptions: Table 4.

Table 4. Code explanations

Code	Description
ART:STR	How to structure reporting of interpretive research
STU:EXP	Students' experience with regard to the support the infinity approach material provided
THE:7PS	Guidance w.r.t. the seven principles (interpretive meth.)
STU:UND	In which ways did the infinity approach material facilitate students' understanding of interpretivist research?
STU:SUG	Any suggestions made by students that may improve the clarity of the infinity approach material

In Subsect. 4.1, responses from our students are categorized according to the given codes of Table 4. These responses are subsequently discussed in Subsect. 4.2.

4.1 Responses

In the following sub-sub-sections, the various respondents are represented as 'R_i'.

ART:STR

R$_1$: *"It assisted me on the layout/template I needed to use to do my assignment"*
R$_2$: *"The article was very clear and straight to the point"*
R$_3$: *"The article helped in regards to knowing how to approach the study and creating a structure"*
R$_4$: *"The tabulated representation of the grounded theory helped to clarify what the structure of the research should follow as a lot of the terminology is confusing at times"*.

STU:EXP

R$_1$: *"My experience with regards to the article was fine"*
R$_2$: *"The article guided me with analysing the data consumed from the respondents. I then could make a final judgment based on the analysed data"*
R$_3$: *"This was a great experience for me because it actually gave me a guideline of how to do this research with regards to interpretivism correctly"*
R$_4$: *"If you follow research guidelines and read literature on the subject it opens up your horizons and gives your insight to a whole new world of possibilities"*
R$_5$: *"The article was very clear and straight to the point"*.

THE:7PS

R$_1$: *"I tried to incorporate all the seven principles of interpretive research"*.

STU:UND

R$_1$: *"To help understand the distinction between the natural reality and social reality"*
R$_2$: *"This assignment helped me understand the topic"*
R$_3$: *"I understood everything that was said, because there are many sources of communication"*
R$_4$: *"I did not understand most if the article, I just tried to base my assignment on the article as best as I can"*
R$_5$: *"The article however helped me understand the process that I will have to follow during my research construction"*
R$_6$: *"I am honestly struggling to understand the whole concept of paradigm research writing"*
R$_7$: *"The use if the infinity symbol and the continuous iteration process that a research is required to go through to gather the data is made easier to understand with the use of this visual representation"*
R$_8$: *"This really served as a guideline in doing my questionnaires, understand better on how to collect data and equally analyse it"*.

STU:SUG

R$_1$: *"I think it would be a great idea to combine all the four categories that we were placed in by selecting each student from the category to join with other students from other categories to form a group of four members so that the grounded theory processes can be fully understood whereby each student can focus on an individual task/duty"*

4.2   Discussion

Apparently the infinity approach paper assisted the students in forming a picture of how to structure their study <ART:STR>. It also provided insight in how students may approach the research problem.

The students experienced that the infinity approach material assisted them with the analysis of interpretive data <STU:EXP>. In addition, it provided an understanding of the interpretive research paradigm. It provided a process for students to follow. An understanding of data collection and how to do data analysis were also gained. Finally, it prompted a student to indicate that an understanding of the difference between natural reality and social reality was gained <STU:UND>.

One student felt that the article was very clear and straight to the point; this is also evident in the assessed assignment where a mark above of 90% was achieved by that particular student <STU:EXP>.

An encouraging result is that one student attempted to apply the seven principles for conducting interpretive research suggested by [11] as discussed in the infinity approach material made available. This student received a mark of 90%. This assignment provided a clear explanation on what was done for the assignment and how an understanding about the selected topic were gained. The student's response in answering the questionnaire supports this <THE:7PS>. The student identified a complicated investigation question (If a bakery wanted to increase sales by having events, what would appeal to their customers?) for the assignment and linked some of the 7 principles to the study. Afterwards the student reminisced that it was not easy to implement them all due to the limited scope of the assignment; this is a valid point.

One student made a suggestion for improvement of the assignment on the infinity approach. This improvement suggests that groups of students should collaboratively work on the assignment to include more than one approach as covered by the assignment <STU:SUG>. This suggestion may be of value, especially when considering it along with the feedback provided in the previous paragraph regarding the limited scope of the assignment. Students working in groups may complete a more complex assignment.

Upon reflection, the assessors concluded that the students over-complicated the assignment unnecessarily. They tried in some instances to formulate a research problem or question rather than gain an understanding about their given topic. Many students could not provide a clear picture of what they have done during the study. They also struggled to express their thoughts. Taking into consideration that honours students are novice researchers, this result may be expected.

5   Conclusion

When conducting research from an interpretivist stance, a researcher attempts to understand a particular phenomenon or situation. To do this a series of repetitive steps, grounded in hermeneutics, need to be conducted. These steps are not always clear and understandable to young researchers, especially those whose undergraduate studies relate to the positivist stance. In an attempt to address this issue, the infinity approach [5] was developed to guide young researchers in conducting interpretive research.

This paper reviews and evaluates the infinity approach through a pilot study where students enrolled for a post-graduate honours course were required to apply interpretivist techniques to an educational case study. Afterwards a questionnaire with open-ended questions was completed by the students. From the feedback discussed in the foregoing paragraphs, our research question—*does the infinity approach assist students in gaining and understanding of interpretive research and the process of doing interpretive research?*—may be considered to be answered positively, since most participants answered in an affirmative tone. In summary, the following conclusions are drawn from our pilot case study:

– The suggested infinity approach helps in structuring an interpretive research study.
– The suggested infinity approach assists students in gaining an understanding about the interpretive research process.

Although this study is (only) a pilot study, the evaluation of the suggested approach provided valuable feedback. It also provided an indication that the infinity approach assist students indeed in understanding interpretive research and the processes involved in conducting such research.

The intervention with the intention to guide young IT researchers already implemented a first phase where a lecture series provides support to young and inexperienced researchers. The second phase is on-going and has as purpose to develop material that may facilitate research methods challenging to the young IT researcher. The infinity approach is one attempt to produce such material. The third phase entails the evaluation and refinement of suggested material. With this pilot evaluation of the infinity approach, the first step of a two-step third phase is completed. The second step entails a study involving an in-depth explanatory case study, (see [24] for comparison).

Improvements regarding the preparation of honours students for research may include lectures offered as full day(s) in a workshop format. This may allow more in-depth scaffolding of concepts. Such an arrangement may allow enough time to facilitate assignment completion to ensure that students do not over-complicated matters.

Improvements regarding future assignments on this topic include grouping students in pairs. This may help the students to obtain a higher level of understanding with regard to interpretivism since a peer will provide a much-needed sounding board. Also, collaboration among students may allow a larger assignment scope, a point highlighted by the student who endeavored to link some of the seven principles to the given problem. It may also be considered to include a debriefing session upon completion of a similar assignment—where groups of students may share their knowledge with one another.

References

1. ATLAS.ti: The Qualitative Data Analysis & Research Software (2014)
2. Boehm, B.: A spiral model of software development and enhancement. ACM SIG-SOFT SEN **11**(4), 14–24 (1986)
3. Bogdan, R., Biklen, S.: Qualitative Research for Education: An Introduction to Theory and Methods. Allyn & Bacon, Boston (1992)
4. Botes, R., Goede, R., Smit, I.: Demonstrating an interpretive data collection and analysis process for a DSR project. Technical report, North-West University (2014)
5. Botes, R., Smit, I.: Taking the mystery out of interpretive research: crossing the infinity approach. Technical report, North-West University (2015)
6. Burrell, G., Morgan, G.: Sociological Paradigms and Organisational Analysis: Elements of the Sociology of Corporate Life. Routledge, Abingdon (2017)
7. Calitz, A.P., Greyling, J., Glaum, A.: CS and IS alumni post-graduate course and supervision perceptions. In: Gruner, S. (ed.) SACLA 2016. CCIS, vol. 642, pp. 115–122. Springer, Cham (2016). doi:10.1007/978-3-319-47680-3_11
8. Droysen, J.G.: Grundriß der Historik. Frommann, Hamburg (1858)
9. Guest, G., Bunce, A., Johnson, L.: How many interviews are enough? An experiment with data saturation and variability. Field Methods **18**(1), 59–82 (2006)
10. Hughes, J.A.: The Philosophy of Social Research. Longman, Harlow (1980)
11. Klein, H.K., Myers, M.D.: A set of principles for conducting and evaluating interpretive field studies in information systems. MIS Q. **23**(1), 67–93 (1999)
12. Myers, M.D.: Qualitative Research in Business and Management. SAGE, Thousand Oaks (2013)
13. Myers, M.D.: Qualitative research in information systems. MIS Q. **21**(2), 241–242 (1997)
14. Myers, M.D., Newman, M.: The qualitative interview in IS research: examining the craft. Inf. Organ. **17**(1), 2–26 (2007)
15. Noor, K.B.M.: Case study: a strategic research methodology. Am. J. Appl. Sci. **5**(11), 1602–1604 (2008)
16. Oates, B.J.: Researching Information Systems and Computing. SAGE, Thousand Oaks (2006)
17. ResearchWare: Technical report
18. Ryan, G.W., Bernard, H.R.: Techniques to identify themes. Field Methods **15**(1), 85–109 (2003)
19. Saunders, M., Lewis, P., Thornhill, A.: Research Methods for Business Students. Prentice Hall, Upper Saddle River (2009)
20. Schnädelbach, H.: Philosophy in Germany 1831–1933. Cambridge University Press, Cambridge (1984)

21. Schwartz-Shea, P., Yanow, D.: Interpretive Research Design: Concepts and Processes. Routledge, Abingdon (2013)
22. Straus, A., Corbin, J.M.: Basics of Qualitative Research: Techniques and Procedures for developing Grounded Theory. SAGE, Thousand Oaks (1998)
23. Takhar-Lail, A.: Market Research Methodologies: Multi-Method and Qualitative Approaches. IGI Global, Hershey (2014)
24. Yin, R.K.: Case Study Research, Design and Methods, 5th edn. SAGE, Thousand Oaks (2014)

Mapping a Design Science Research Cycle to the Postgraduate Research Report

Alta van der Merwe, Aurona Gerber, and Hanlie Smuts[✉]

Department of Informatics, University of Pretoria, Pretoria, South Africa
{alta.vdm,aurona.gerber,hanlie.smuts}@up.ac.za

Abstract. Design science research (DSR) is well-known in different domains, including information systems (IS), for the construction of artefacts. One of the most challenging aspects of IS postgraduate studies (with DSR) is determining the structure of the study and its report, which should reflect all the components necessary to build a convincing argument in support of such a study's claims or assertions. Analysing several postgraduate IS-DSR reports as examples, this paper presents a mapping between recommendable structures for research reports and the DSR process model of Vaishnavi and Kuechler, which several of our current postgraduate students have found helpful.

Keywords: Design science research · Students' report writing · Design science report · Postgraduate education · Research reports

1 Introduction

Design science research (DSR) adopts a pragmatic research paradigm to develop artefacts that are innovative and solve real-world problems [12]. DSR is relevant for information systems (IS) research because it directly addresses two of the discipline's key aspects, namely the central role of the IS artefact in IS research and the perceived lack of professional relevance of IS research [9]. The notion of 'design as research' in the IS domain is relatively new. The adoption of DSR in IS research is mainly due to [10], wherein an IS research framework (ISRF) is provided that emphasises the rigour and relevance of the research. At the same time, [23] introduced a process model for DSR with awareness, suggestion, development, evaluation and conclusion as subsequent phases. This model is discussed in more detail in Sect. 2. In IS, research often includes the construction of some kind of artefact. Exactly what such an artefact entails is often also the topic of rigorous debate [19,22]. Although [7] provides a framework for reporting in a research project, it do not discuss the process model of DSR in the same way as [23]. Independently of the research approach followed for a research project, students find the process of structuring the research report in such a manner that it forms a valid argument to be a challenge. This is often especially true in postgraduate studies that include the construction of an artefact in the research project. In such cases, students need to develop a document structure

© Springer International Publishing AG 2017
J. Liebenberg and S. Gruner (Eds.): SACLA 2017, CCIS 730, pp. 293–308, 2017.
https://doi.org/10.1007/978-3-319-69670-6_21

that supports the research contribution, and which includes an artefact comprising more than one component. The purpose of our paper is to consider the DSR process model as in [23], and propose possible document structures to support the research contribution in the research report. Using an analysis of several postgraduate research reports that successfully adopted DSR, this paper develops a mapping between the proposed structure of a research report and the DSR process model of [23]. The mapping is presented in four scenarios and validated with two examples of completed research reports. The mapping and scenarios were subsequently validated in a workshop with doctoral degree candidates. The feedback results indicated that the mapping was useful for all DSR students who needed to structure their research reports.

2   Background: DSR

DSR is primarily concerned with research on design as science [1,6,7,13–15,23]. The intent of DSR is to create an artefact through a balanced process that combines the highest standards of rigour with a high level of relevance. One of the measures of DSR is whether the research resulted in a relevant artefact, but also whether the process was rigorous [8,24]. Figure 1 depicts the ISRF of [8,10]. Using this framework, DSR is described as research building and evaluating computing artefacts designed to meet identified needs [8]. The goal of the artefact is the fulfilment of a specific need or utility. The description of the needs would provide the requirements for the artefact. In the building of the artefact, knowledge from the applicable knowledge base is used. During evaluation, the artefact is measured against the needs to evaluate its utility [22]. One of the central discussions in DSR is what is recognised as an 'artefact' in the DSR paradigm [19,22]. One of the reasons is that the artefact is not always tangible, but more often intangible (such as a model, software, a framework or architecture). In DSR, it is accepted that the artefact embodies or is part of the design theory [21]. The following three discussions on DSR artefacts appear in [18]:

- Constructs, models and methods are DSR 'artefacts' in [25], whereas constructs, models, methods, instantiation and better theories are DSR 'outputs' in [23].
- DSR 'outputs' were analyzed in [21], identified as constructs, models, methods and instantiations, as well as social innovations or new properties of technical, social or informational resources. Accordingly, an 'artefact' is any designed object with an embedded solution to an understood research problem.
- In [19] we can find an exhaustive list of acceptable DSR artefacts that included software, algorithms, methods, models, frameworks and architecture, grouped into eight types: system design, method, language or notation, algorithm, guideline, requirements, pattern and metric.

For the purpose of this paper, it is accepted that the IS DSR artefact is anything that is delivered by a rigorous research and development process and that can be shown to fulfil an identified need.

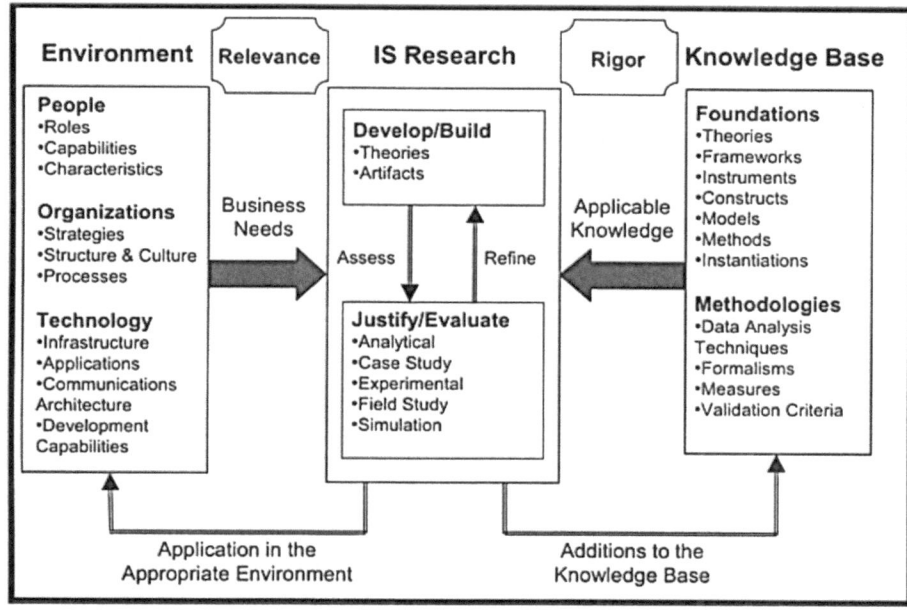

Fig. 1. Information systems research framework according to [10]

2.1   DSR Methodology

One of the most-cited methods accepted by DSResearchers is the design cycle or process model of [23], (see Fig. 2). The method was derived from other DSR advocates such as [8,9,16]. In [23] the following five phases for the execution of a typical DSR project are recommended:

1. **Awareness of the problem:** The awareness could be generated from practical experience or from related disciplines. The output from this phase is a proposal.
2. **Suggestion:** The suggestion is closely related to the awareness of the problem (as indicated by the dotted line). The suggestion is often included as a tentative design in the complete proposal as output. However, an approach to develop a suggestion might be included in the proposal if a possible solution is not immediately evident.
3. **Development:** The tentative design is implemented during this phase and the technique for implementation will differ depending on the artefact.
4. **Evaluation:** When the artefact has been developed, the evaluation of the artefact is mandatory, usually according to requirements and criteria specified during the suggestion phase (as part of the proposal). The result of the evaluation should be carefully noted and explained. This phase may result in the refinement of an awareness, a suggestion or a development, especially if the result of the evaluation is not satisfactory.

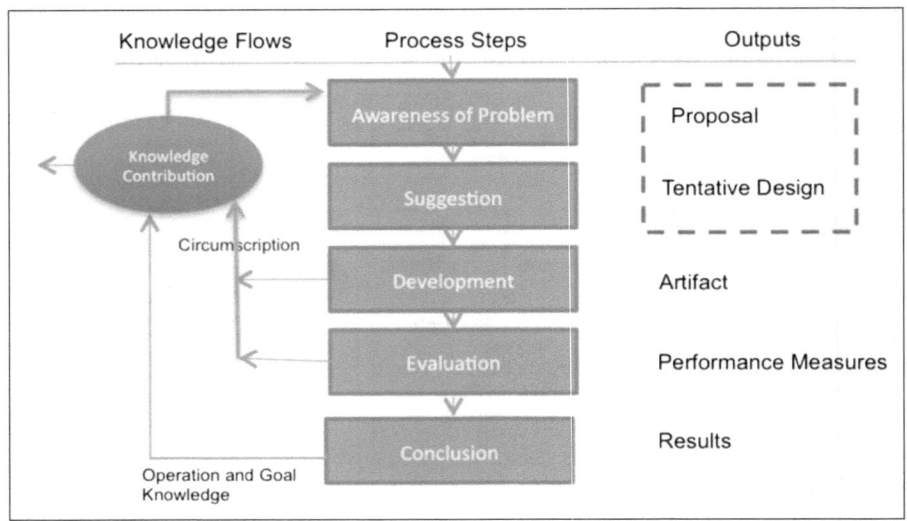

Fig. 2. Design science research process model (DSR cycle) according to [23]

5. **Conclusion:** This is the final phase when the research results and contribution are identified. This not only includes the artefact, but all additional knowledge with regard to the process, construction and evaluation that were acquired. The output of this phase is an acceptable research contribution.

The notion of 'iteration', as indicated by the arrows on the left in Fig. 2, is embedded in the DSR method. It is possible to branch back to awareness during the execution of development, evaluation and conclusion. Several cycles of the abovementioned notions are often executed during the construction of a DSR artefact. Circumscription is due to the discovery of constraint knowledge about the theories gained through the detection and analysis of contradictions, but in practice these cycles also occur because development, evaluation and conclusion in DSR often expose new problems that could be entered into the DSR cycle at the awareness stage [23].

The DSR process model of [23] summarises the phases that are necessary to execute a DSR project. Nowadays, the approach is often adopted by researchers and postgraduate students in IS as an acknowledged, repeatable process for the construction of a useful artefact and research contribution. Given the adoption of this approach, it is possible to argue that the specific process model for DSR plays a valuable role in ensuring that computing research is more rigorous and repeatable, but also relevant and useful. For the remainder of this paper, reference will be made to the DSR process model of [23] (Fig. 2) simply as 'the DSR process model'.

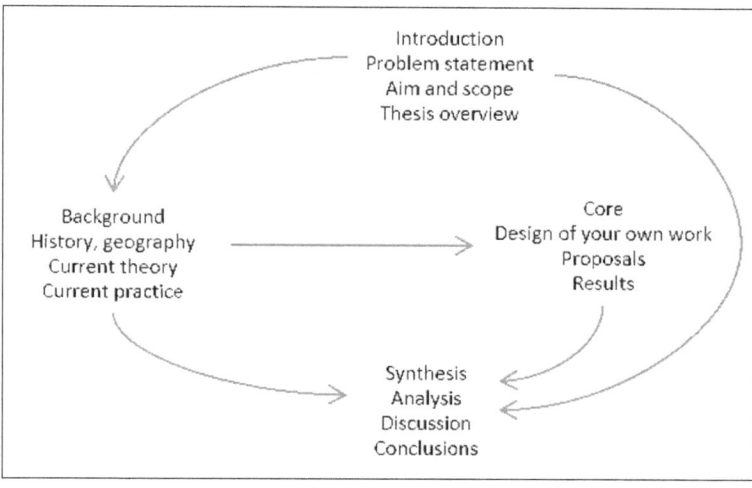

Fig. 3. Structure of a research report in [4]

3   Typical Structure of a Research Report

Multiple considerations, such as subject chosen and research approach, drive the planning around the structure of a postgraduate study or research report [20]. In theoretical and textual studies the introduction is normally followed by chapters that reflect the analysis of secondary literature and the conclusion. In contrast, the structure of an empirical study is largely dictated by the methods utilised.

Generally, for most studies, the student will have to include a chapter that explains the research design, another on the analysis and presentation of findings, as well as a discussion and recommendations [17, 26]. Therefore, the structure of a research report (dissertation or thesis) followed by students in most disciplines usually consists of an introduction, background, body (core) and conclusion [4, 11]. As mentioned in [4] (see Fig. 3), all sections in a research report are connected. The report's conclusion connects to the goal, and the background acts as input to the core or body of the work.

In [11] it was proposed that the text consists of an introduction, literature review, method, body and conclusion, (see Fig. 4). The purpose of the introduction is to give an overview of the report, and includes sections on the research questions, objectives, a problem statement and a brief chapter overview. The second section in this document structure is the literature review, where the theory base for the study is introduced: from a high-level or broad perspective, as well as an in-depth discussion of relevant topics. This section might consist of more than one chapter. The method chapter, which provides an overview of the research design, including the research instruments for data analysis and data collection, follows the literature review. The body section of the research report usually consists of the evidence of how the student either proved his or her hypothesis if working deductively, or how he or she derived at a contribution

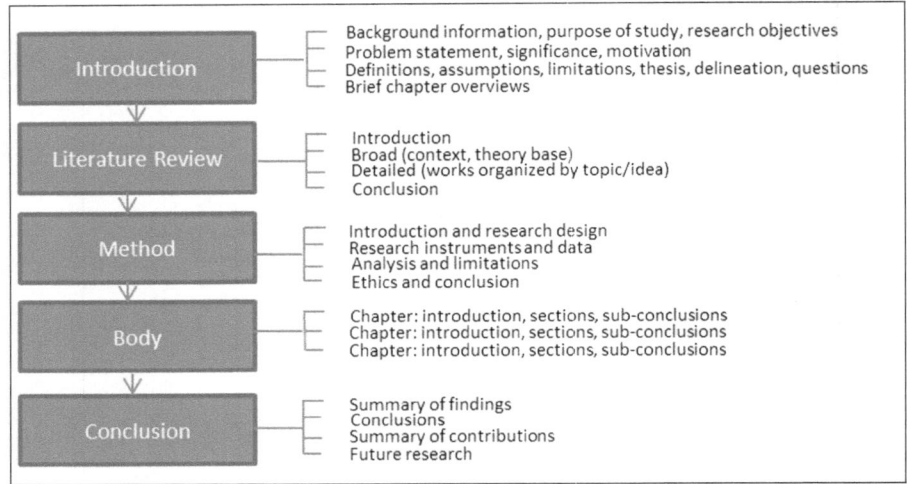

Fig. 4. Structure of a research paper or research report in [11]

if working inductively. The body might be more than one chapter, depending on the format in which the student presents the work. Lastly, the conclusion of the work consists of a summary of the findings and the contribution. The questions that guide this structure, although not explicitly mentioned, are the following: what do I want to know? (Sect. 1); what do I know? (Sect. 2); what will I do? (Sect. 3); what did I find when I executed my plan? (Sect. 4); what is the solution or contribution? (Sect. 5).

Despite the proposed structure of a research paper or research report discussed above, *students often find it difficult to structure a study* that adopts a DSR method. It is, for instance, not clear how to integrate the DSR process model into the proposed document structure. Awareness of a problem could typically fit in Sect. 1 of the document structure, but where would the suggestion then fit in? The literature review should support the problem and therefore the awareness. So does the literature review follow the awareness or does the awareness description follow the literature review? How would the development phase be included given the proposed document structure, especially if more than one development cycle was included in artefact development? Including different theory sections for different development cycles to accommodate the rigour requirement of [10] is even more confusing.

4   Research Report Structures for DSR Process Models

Since the acceptance of DSR as an acceptable research approach for IS research, several discussions about the structure of a DSR research project have commenced. For example, [7] proposed the publication scheme summarised in Table 1 for DSR studies. Due to the wide-spread adoption of the process model of [23],

Table 1. Publication scheme for DSR studies according to [7]

Section	Contents
1. Introduction	The introduction should include the problem definition, significance or motivation, an introduction to key concepts, research questions or objectives, scope of the study, an overview of methods and findings, theoretical and practical significance, as well as the structure of the remainder of the paper.
2. Literature review	The literature review includes prior work that is relevant to the study, including theories, empirical research studies and findings or reports from practice.
3. Method	The method section includes the research approach that was employed.
4. Artefact description	The artefact description should be a concise description of the artefact at the appropriate level of abstraction to make a new contribution to the knowledge base
5. Evaluation	The evaluation is evidence that the artefact is useful.
6. Discussion	The interpretation of the results includes stating what the results mean and how they relate to the objectives stated in the introduction section. The discussion can include a summary of what was learnt, a comparison to prior work, limitations, theoretical significance, practical significance, and areas that require further work.
7. Conclusions	The concluding paragraphs restate the important findings of the work

the purpose of our paper is to extend the publication scheme by integrating the DSR process model and mapping it to the proposed research report structure (as discussed in the previous section). Postgraduate research reports (theses and dissertations) of students that successfully adopted the DSR process model were analysed and four different scenarios were identified. In the first scenario, the student only had one DSR design cycle, and a single artefact was constructed during the design. In the second, third and fourth scenarios, the students developed composite artefacts that consisted of more than one component. The DSR process therefore included more than one cycle.

4.1 Scenario 1: One Cycle of Design

Often in students' curricula, especially at master's degree level, a student might be involved in the design of a simple artefact with a single function, as opposed to a composite artefact that consists of more than one component. Examples of such artefacts include a system with a single defined function, such as a mobile application. It is also possible to have one single cycle when the artefact consists of more than one component, but the components are predeveloped and the designer is only involved in the assembly of existing components in the DSR process. Presenting research of this nature can then be presented using a single mapping from the DSR process model, as illustrated in Fig. 5. In Fig. 5 the mapping between the DSR process model and the research report structure is indicated with numbered arrows (1–8).

Mapping 1: *Introduction and Awareness of the Problem.* In Sect. 1 of a research report (the introduction), the student should already introduce the problem. This this correlates with the awareness phase of the DSR process model.

Mapping 2: *Introduction and Suggestion.* In Sect. 1 of the research report (the introduction), the student already provides an indication of the type of solution (artefact) for the problem (as discussed in Mapping 1).

Mapping 3: *Literature Review and Awareness of the Problem.* In the literature review, the student provides proof of the identified problem by, for instance, discussing a problem that is experienced in practice (relevance) and/or providing proof that the problem has not previously been resolved in literature. Usually, this section should include a comprehensive discussion of the existing related literature and should indicate the lack of a solution in literature.

Mapping 4: *Literature Review and Suggestion.* The student could already suggest an artefact that could provide a solution for the problem at the end of the literature review. This artefact could be a construct, model, method, instantiation or better theory. At this stage, the student might also introduce any theories that will be used if applicable (to address the rigour requirement of [10]). It is also possible to only provide the suggestion in the body of the research report.

Mapping 5: *Method and Development.* In the method section of the research report, the student includes a description of the adopted and adapted DSR process model, which includes the planning of the development of the proposed artefact. Depending on the type of artefact, the student includes how the artefact will be constructed. For an experimental study, the plan may include how the artefact will be built and tested. For a qualitative study, the student may include the questionnaires and analysis methods to be used to collect data to build the artefact, such as in the case of a construct, method or conceptual model. The study may also outline how the artefact will be tested. Depending on the scope of the study, this may include a proof of concept and not a full test.

Mapping 6: *Body and Development.* The student includes all the data relevant to the building of the artefact in the body of the report. For an experimental study, it might include data on the experiment conducted during the development. For the development of a software artefact, the body might include the description of the system itself and the different screens and functionality. For a qualitative study, the results might be the data that was collected, as well as the analysis of the data. The body section includes the results of the study, including the artefact itself. Generally, in a more inductive study, the artefact is presented at the end, while a more deductive study proposes the artefact in the beginning of the body section and then follows a more descriptive process of the development of the artefact.

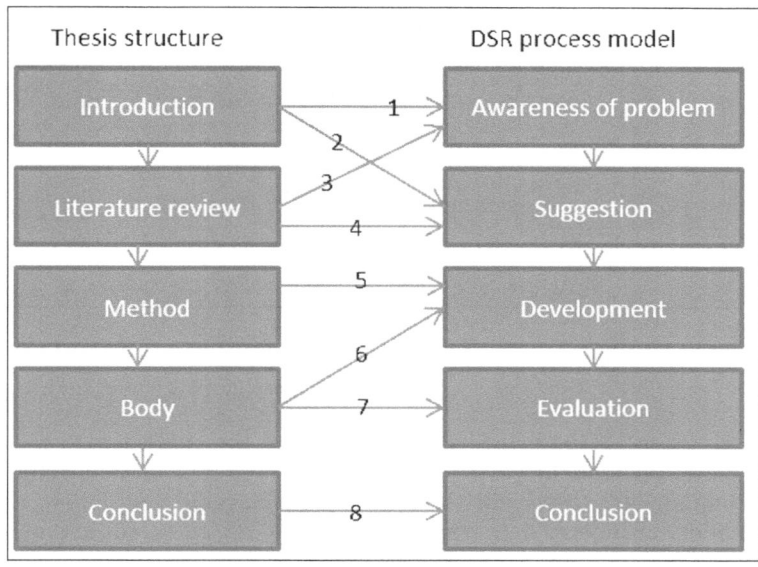

Fig. 5. Mapping for a single design science research process cycle

Mapping 7: *Body and Evaluation.* The body section, usually consisting of several chapters, also includes the results of any evaluation that was done. In a more experimental study, such as the development of an application, this section includes the testing results of the application. For an inductive qualitative study (where the main component of the study consists of constructing the artefact), the evaluation may include a proof of concept or validation using data collected from focus groups or interviews, depending on the scope of the study.

Mapping 8: *Conclusion.* The last section of the research report summarises the study and research contribution, including how the artefact as research contribution has value from a rigour and relevance perspective [10].

4.2   Scenario 2: DSR Process Model with Many Cycles of Design

In doctoral degree studies it is often the case that a composite artefact that consists of more than one component is constructed. In such a scenario, the same general format that was discussed in the previous section could be followed, with extensions in the method (Mapping 5, which influences the development) and the body (Mapping 6, which influences the development): see Fig. 6. As indicated in Fig. 6, the research report will still include a main DSR cycle as the main guiding structure, such as the one discussed in the first scenario. However, the method section might include a description of several subcycles that are then included in the body of the research report. The awareness of the first subcycle will form part

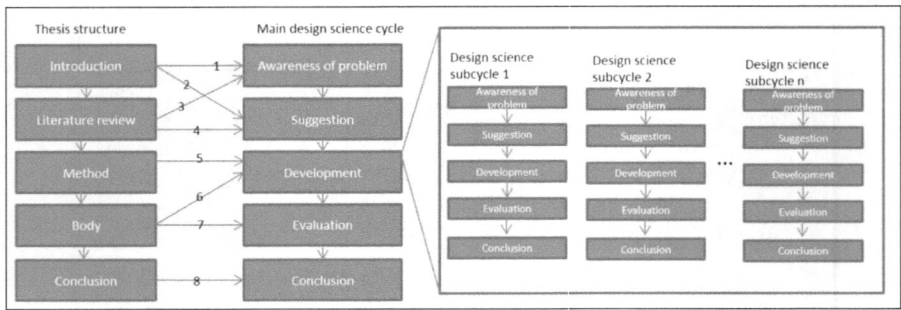

Fig. 6. Design science research process model with many cycles

of the development of the main DSR process when the student realises that the development of the main artefact consists of the development of subcomponents. For the components, there will then be separate cycles that follow the DSR process model, as indicated by the arrows in Fig. 2. These subresearch cycles could also extend into further cycles. Each artefact component may be evaluated separately, or the testing and evaluation could be included in the evaluation of the complete artefact in the main research cycle.

4.3   Scenario 3: Problem Establishment as Part of the Research Process

A variation of Scenario 2 occurs when the student cannot provide sufficient motivation for the research problem from the literature review and needs to provide additional evidence by, for instance, conducting a pre-study as part of his or her research study. In this case, the same structure as in Fig. 5 is proposed, however the awareness and suggestion phases of the DSR process model are included in the body of the research report, after the literature review.

4.4   Scenario 4: Change in the Research Report Structure

As a final scenario for a DSR research report, it is possible to include the method section before the literature review. This structure could be problematic and students are cautioned against its use. This structure might be confusing to the reader (examiner), as he or she will be confronted with the research design in the method section before the awareness of the problem (problem description) and the suggestion are presented.

5   Examples of the Use of the DSR Process Model

As discussed in the previous section, four possible scenarios were identified to map a research report structure to a study that adopted the DSR process model. In this section, some examples are discussed to illustrate the proposed mapping.

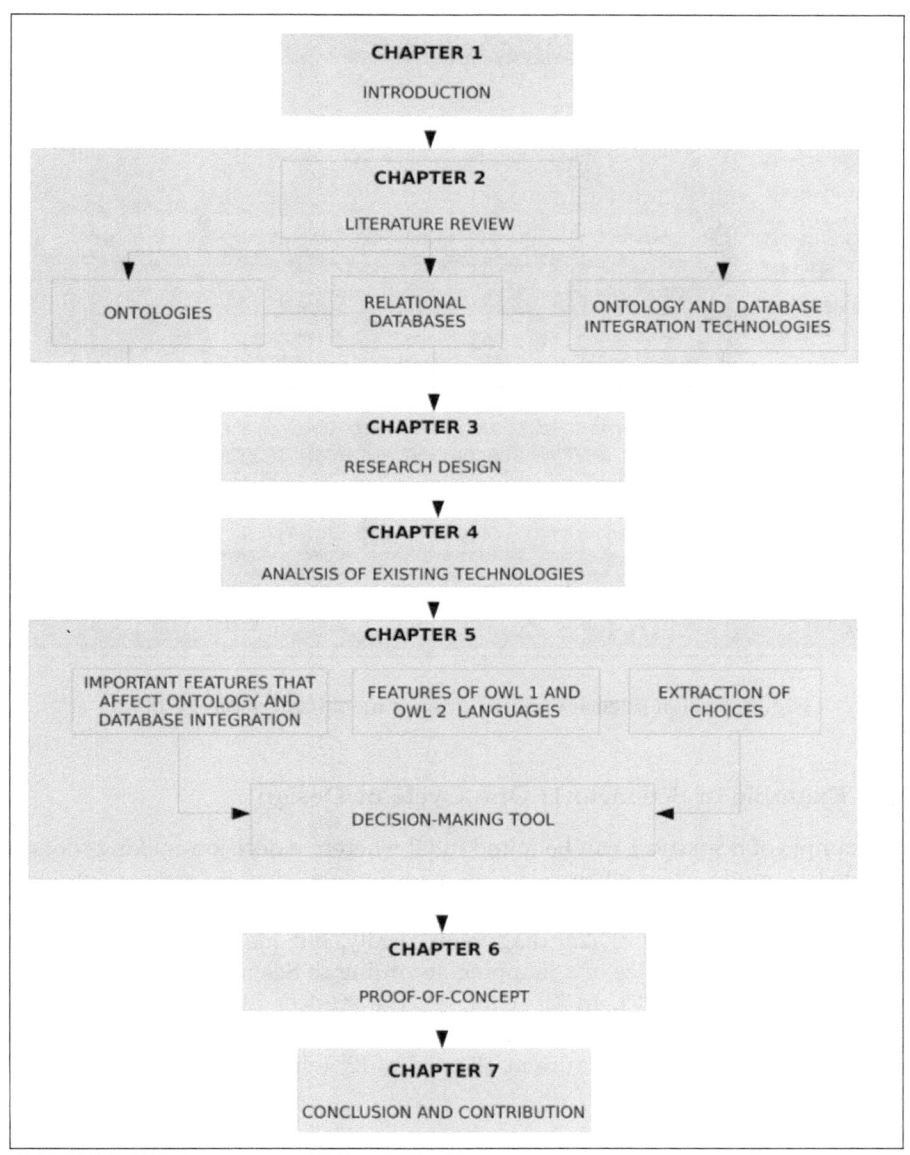

Fig. 7. Design science research process model in [2], with the student's additional comment: *"This research design consists of five steps: awareness of the problem, suggestion, development, evaluation and conclusion. For the awareness of the problem, a literature review was conducted, while theory analysis, theoretical study and artefact building were conducted for the suggestion and development. Verification using a proof of concept was conducted for the evaluation and conclusion"*

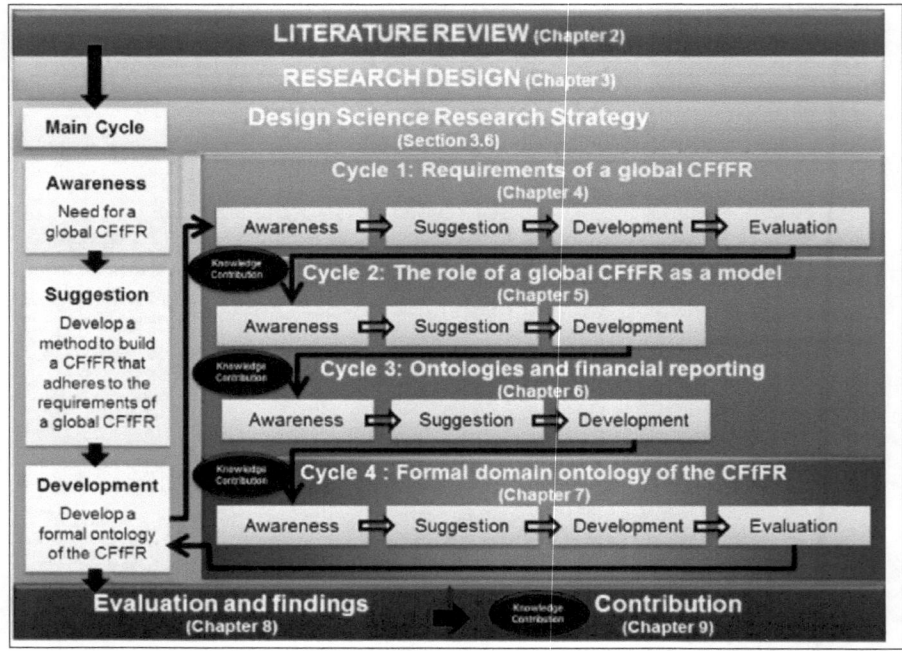

Fig. 8. Design science research process model as a strategy in [5]

5.1   Example of Scenario 1: One Cycle of Design

An example of Scenario 1 can be found in [2] wherein a decision-making tool was develped to guide users when selecting a technology that integrates ontologies with relational databases. In this case the student did not map his chapters to the DSR process model of [23] diagrammatically, but his verbal description in [2] (chpt. 1) gives evidence of a mapping according to Scenario 1: see the caption text (*in italics*) to Fig. 7). In [2] (chpt. 1), the student already introduced the problem (awareness) and proposed a tool (suggestion), but only linked the DSR model to his document structure at the end of [2] (chpt. 1) after discussing the DSR process model.

5.2   Example of Scenario 2: DSR Process Model with Many Cycles of Design

A conceptual framework for financial reporting was developed and presented in [5]. The chapter map of that doctoral thesis is depicted in Fig. 8. The development of the conceptual framework as an artefact consisted of one main research cycle and four subcycles. Figure 8 illustrates that the main cycle is documented in [5] (chpt. 1–3). The development phase branches off into four subcycles, which are reported on individually in the body [5] (chpt. 4–7). Cycles 1 and 4 include

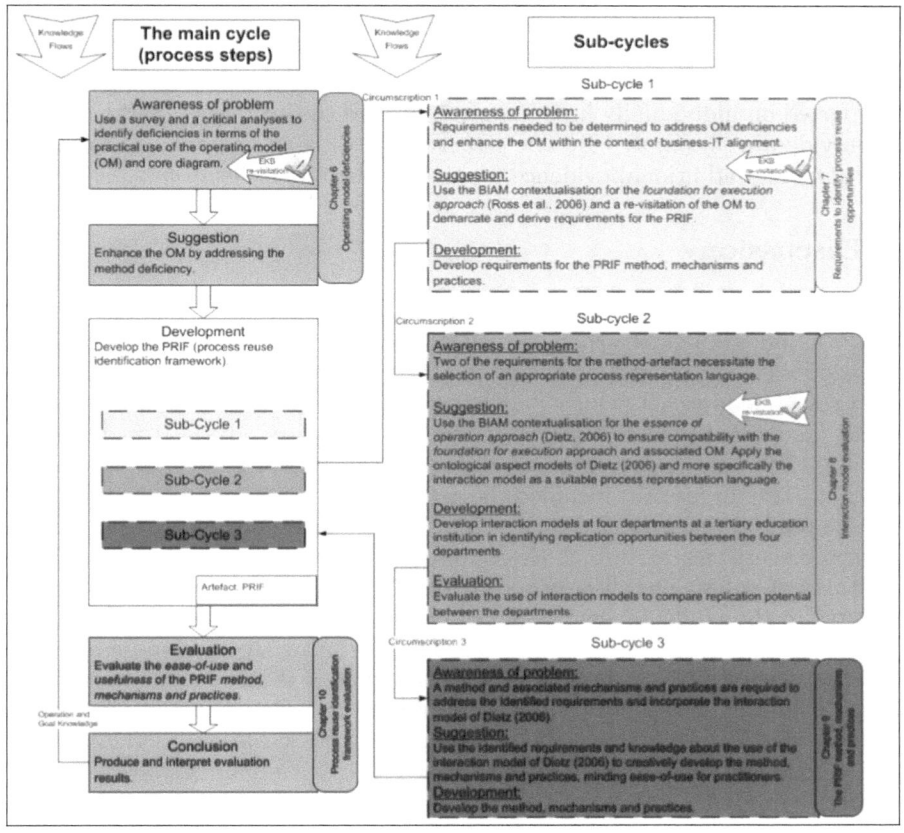

Fig. 9. PRIF developed in [3]

evaluation activities, however cycles 2-3 do not. These components are tested in the main research cycle, which is described in [5] (chpt. 8). The conclusion is in [5] (chpt. 9): it summarises the study's contribution. The method used in [5] aligns with the discussion of Scenario 2 (see above) where a main DSR cycle is proposed with a number of subcycles as part of the development phase.

5.3   Example of Scenario 3: Establishment of the Problem as Part of the Research Process

As mentioned, it is possible to also include the awareness and suggestion in the body of the research report. In the another example [3] we can also find more than one cycle for the development of a process reuse identification framework (PRIF), as illustrated in Fig. 9. In contrast to [5], a survey was conducted in [3] to establish the research problem clearly. This problem is then discussed in the body of [3]. The problem is introduced in the introduction and literature review sections, but since the evidence was not conclusive, the problem was confirmed to align with the relevance requirement of [10] by providing the survey results after the method section in the body of the report [3] (chpt. 5). The subcycles are all reported on individually in [3] (chpt. 7–9), and the evaluation and conclusion are provided in [3] (chpt. 10). The structure in [3] is an example of Scenario 3, which provides additional evidence for the problem identification.

6   Discussion

In the past few years, we encountered several postgraduate students in computing (both under our supervision, as well as research reports examined externally) that adopted DSR, more specifically, the DSR process model of [23] as their preferred research approach. During a PhD workshop conducted with postgraduate second- and third-year DSR PhD candidates who were involved in the structuring of their research reports, the different scenarios presented in Sect. 5 were proposed. The discussions included the presentation of examples from completed research reports that adopted DSR. All postgraduate students ($n = 7$) found the scenario discussions, as well as the examples of other students' work and DSR structures of value. The discussion on the DSR process model of [23] and how it relates to the students' own cycles of design and development of the artefact resulted in enthusiastic interaction. Discussions during the session included the fact that each student's work is unique and that it is necessary for each student to construct his or her own research report workflow. Students specifically emphasized the usefulness of the mapping and the scenarios, as they may be applied to both qualitative and quantitative research, and already accommodate some of the unique nuances of their research. In addition, students emphasised particular advantages of their applications of the mapping and scenarios:

- The first advantage pertained to the approach to their research when executing multiple design cycles. As the challenge for a student often lies in the 'how to', students at the workshop reflected that the scenario of multiple cycles provided clear guidelines for their approach and illustrated how one cycle informed the next in the context of what they know to be (through theory) their required research output.
- The second advantage students identified related to the write-up of their research reports. Students who were in the process of writing up their dissertations or theses, highlighted the fact that a major challenge they experienced was to write up their research in a way that an examiner or external stakeholder may follow the presented argument. In this instance, the mapping and scenarios provided a solid recommendation on structuring a DSR study in order to produce an organised research report that correctly reflected the results and outcomes of their research.
- Lastly, students reflected that research papers, such as journal papers and conference proceedings, in the DSR literature accommodated and referred to multiple stances that were aligned to many research methodologies and approaches where they were applied—an incidence that makes it more difficult

for the students to find their own way. The scenarios and mapping provided a perfect vantage point—like a one-stop shop—from where the students could make sense of the DSR body of knowledge and particular application in their research study before embarking on a study of the broader DSR domain. This pointed to an inside-out consideration, rather than an outside-in approach, which, with such a focused starting point, provided clear direction and saved time when dealing with DSR research reports.

7 Conclusion

We acknowledge that there will be slight variations in the structure, depending on the unique research problems and approaches. However, in general, it is possible for studies that adopt DSR to map the requirements of a research report to the DSR process model. Given the feedback that was discussed above, the scenarios presented in this paper should provide valuable guidance for students who need to structure a DSR research report.

Acknowledgments. Thanks to *Neels van Rooyen* for his help with the type-setting of this paper.

References

1. Baskerville, R., Vaishnavi, V.: Pre-theory design frameworks and design theorizing. In: HICSS Proceedings 49th Annual Hawaii International Conference on System Sciences, pp. 4464–4473 (2016)
2. Chimamiwa, G.: Using ontologies to structure information in a web information portal. Dissertation: University of South Africa (2011)
3. de Vries, M.: A process reuse identification framework using an alignment model. Doctoral Dissertation: University of Pretoria (2012)
4. Evans, D., Gruba, P., Zobel, J.: How to Write a Better Thesis. Springer, Cham (2014). https://doi.org/10.1007/978-3-319-04286-2
5. Gerber, T.: The conceptual framework for financial reporting represented in a formal language. Doctoral Dissertation, University of Pretoria (2015)
6. Gill, T.G., Hevner, A.R.: A fitness-utility model for design science research. In: Jain, H., Sinha, A.P., Vitharana, P. (eds.) DESRIST 2011. LNCS, vol. 6629, pp. 237–252. Springer, Heidelberg (2011). https://doi.org/10.1007/978-3-642-20633-7_17
7. Gregor, S., Hevner, A.: Positioning and presenting design science research for maximum impact. MIS Q. **37**, 337–355 (2013)
8. Hevner, A.: A three-cycle view of design science research. Scand. J. Inf. Syst. **19**(2), 87–92 (2007)
9. Hevner, A., Chatterjee, S.: Design science research in information systems. Design **22**, 209–233 (2010)
10. Hevner, A., March, S.: Design science in information systems research. MIS Q. **28**, 75–105 (2004)
11. Hofstee, E.: Constructing a Good Dissertation. EPE (2006)

12. Hovorka, D.S.: Design science research: a call for a pragmatic perspective. In: Proceedings of SIGPrag Workshop, Sprouts Working Papers on Information Systems (2009)
13. Kuechler, B., Vaishnavi, V.: Extending prior research with design science research: two patterns for DSRIS project generation. In: Jain, H., Sinha, A.P., Vitharana, P. (eds.) DESRIST 2011. LNCS, vol. 6629, pp. 166–175. Springer, Heidelberg (2011). https://doi.org/10.1007/978-3-642-20633-7_12
14. Kuechler, B., Vaishnavi, V.: On theory development in design science research: anatomy of a research project. Eur. J. Inf. Syst. **17**, 489–504 (2008)
15. Kuechler, W., Vaishnavi, V.: The emergence of design research in information systems in North America. J. Des. Res. **7**(1), 1–16 (2008)
16. March, S., Storey, V.: Design and natural science research on information technology. Decis. Support Syst. **15**(4), 251–266 (2008)
17. Mouton, J.: How to Succeed in Your Masters and Doctoral Studies. Van Schaik (2008)
18. Naidoo, R., Gerber, A., van der Merwe, A.: An exploratory survey of design science research amongst South African computing scholars. In: Proceedings South African Institute for Computer Scientists and Information Technologists, SAICSIT 2012, pp. 335–335 (2012)
19. Offermann, P., Blom, S., Schönherr, M., Bub, U.: Artifact types in information systems design science – a literature review. In: Winter, R., Zhao, J.L., Aier, S. (eds.) DESRIST 2010. LNCS, vol. 6105, pp. 77–92. Springer, Heidelberg (2010). https://doi.org/10.1007/978-3-642-13335-0_6
20. Olivier, M.: Information Technology Research: A Practical Guide for Computer Science and Informatics. Van Schaik (2004)
21. Peffers, K., Tuunanen, T., Rothenberger, M.A., Chatterjee, S.: A design science research methodology for information systems research. J. Manag. Inf. Syst. **24**(3), 45–77 (2008)
22. Vahidov, R.: Design researcher's IS artifact: a representational framework. In: Proceedings of 1st International Conference on Design Science Research in Information Systems and Technology, DERIST 2006, pp. 19–33. Claremont (2006)
23. Vaishnavi, V., Kuechler, B.: Design science research in information systems. Association for Information Systems, Technical Report (2004)
24. Venable, J.R.: Design science research post hevner et al.: criteria, standards, guidelines, and expectations. In: Winter, R., Zhao, J.L., Aier, S. (eds.) DESRIST 2010. LNCS, vol. 6105, pp. 109–123. Springer, Heidelberg (2010). https://doi.org/10.1007/978-3-642-13335-0_8
25. Winter, R.: Design science research in Europe. Eur. J. Inf. Syst. **17**(5), 470–475 (2008)
26. Zobel, J.: Writing for Computer Science, 3rd edn. Springer, London (2014). https://doi.org/10.1007/978-1-4471-6639-9

A Topic-Level Comparison of the ACM/IEEE CS Curriculum Volumes

Linda Marshall[✉]

Department of Computer Science, University of Pretoria, Pretoria, South Africa
lmarshall@cs.up.ac.za

Abstract. Curricula are not static, especially in Computer Science. Curricula specifications are updated and it is not always clear what the exact updates are nor the impact these updated may have on a curriculum which has been developed to comply with the specification. In this paper, the results of a comparison on the knowledge unit and topic-level between the ACM/IEEE Computer Science curriculum volumes of 2001, 2008 and 2013 is presented.

Keywords: Computer Science · Curriculum design · Post-secondary education · Modelling · Digraphs · Visualisation

1 Introduction

Given the rate at which technology changes, it not surprising that the ACM/IEEE Computer Science (CS) Curriculum volume needs to change too. Approximately every ten years, an updated version of the volume is released. Along with the new curriculum specification, high level changes between the new volume and the previous volume(s) are provided as part of the curriculum volume report.

With the release of CS2013 in 2013, institutions who have previously modelled their CS curricula on CC2001 may be wondering what the differences between the two curricula volumes are. This paper the differences between the 'core' aspects between the curriculum volumes of 2001, 2008 and 2013. CS2013 outlines differences between CC2001 and CS2013 especially on the *knowledge area* level [2]. However it is harder to determine whether a curriculum that was designed using the *knowledge units* and *topics* of CC2001 compares favourably with CS2013. By means of a framework [8] that is used to facilitate the comparison on a finer level of granularity than 'knowledge areas', this paper presents a comparison of CC2001 and CS2008 with CS2013.

2 Overview of the ACM/IEEE Curriculum Volumes

Prior to the ACM/IEEE Computing Curricula of 1991, the ACM and the IEEE-CS worked independently on computing curricula recommendations. In 1988,

© Springer International Publishing AG 2017
J. Liebenberg and S. Gruner (Eds.): SACLA 2017, CCIS 730, pp. 309–324, 2017.
https://doi.org/10.1007/978-3-319-69670-6_22

the Joint Curriculum Task Force, comprising of the ACM and the IEEE-CS, was formed. The outcome of this partnership was the *Computing Curricula 1991* report [1]. Ten years later, the *Computing Curricula 2001* (CC2001) report was released. This was followed by the 2005 overview report which identified different disciplines within computing [6]. The *Computing Curricula 2001* was included in the overview report and renamed *Computer Science 2001* to better reflect the discipline it represented. In 2008, an update to the 2001 *Computer Science* report was released and is referred to as the *Computer Science 2008 update* (CS2008) report [4]. In late 2013, the latest version of the Computer Science curriculum volume (CS2013) was released.

As from CC2001, the computing curriculum was defined in terms of *Knowledge Areas* (KAs), *Knowledge Units* (KUs) and *Topics*. KAs represented the highest level of the curriculum specification. Each KA has KUs associated with it. These KUs are further described by topics and learning outcomes. Topics could further be broken into subtopics.

In CC2001, KUs were identified as being either *core* or *elective* [3]. Core KUs and their respective topics are regarded as base line requirements in any curriculum that is developed for the discipline. The minimum number of hours that should be spent on each KU are also specified. Elective KUs are optional. CS2013 moved the designation of *core* or *elective* from the KU level to the topic level. A distinction was made between two types of core topics, *core tier 1* (CT1) and *core tier 2* (CT2). A KU may therefore contain a mixture of CT1, CT2 and elective topics. A further requirement in CS2013 is that all topics in CT1 must be included in a curriculum, while at least 90% (with an 80% bare minimum) of CT2 is considered essential for inclusion [2]. In this paper, no distinction is made between CT1 and CT2 topics.

In CC2001, 14 KAs representing the Computer Science Body of Knowledge were identified. These KAs increased to 18 in CS2013 [2,10]. A tabular summary of these changes is given in [8]. The highlights of the changes are: the KAs *AL, AR, DS, HC, IM, IS, OS, PL and SE* in CS2013 remain the same as they are in CC2001 and CS2008; KAs *CN, GV, NC and SP* changed focus, but not enough to drop them entirely; and the *PF* knowledge area was dropped in CS2013 and *IAS, PBD, PD, SDF and SF* were introduced.

As with the KAs, the number of core KUs and topics also increased from 2001 to 2013. Figure 1 shows the number of KAs in which core topics reside. Finally, the information shown for CS2013 reflects KUs that contain one or more CT1 or CT2 topics, as well as the KAs associated with these KUs.

3   Modelling Curriculum Volumes as Digraphs

To be able compare the curriculum volumes with one another, they need to be modelled to reflect the hierarchical nature of their information contents. This hierarchical structure cannot be as represented as a tree structure because one KU, topic or subtopic may belong to one or more KA, KU or topic.

A curriculum volume is modelled in layers like an onion. In the centre is the discipline, in this case Computer Science (CS). The next ring represents the KAs,

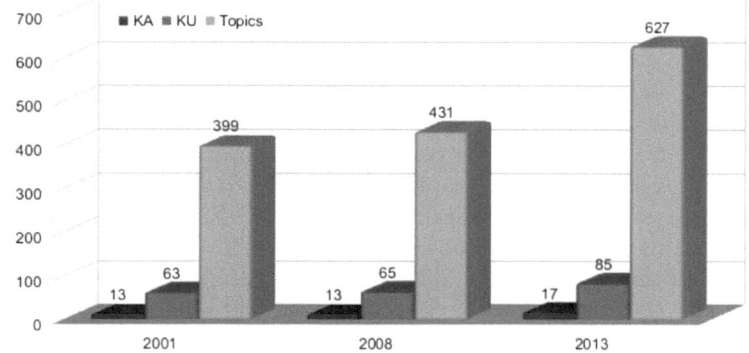

Fig. 1. Core content of ACM/IEEE-CS curricula

followed by the KUs. The outer rings represents topics and subtopics. For simplicity, subtopics and topics are seen as the same. KAs, KUs and topics modelled as vertices and the relationships between them by the edges of a directed graph. Once the digraphs are defined, the structure and the content of the curriculum volumes can be compared and similarity and differences measured.

4   Comparison of CC2001, CS2008 and CS2013

Comparing digraph structures visually, refer to Fig. 2, often reveals differences (and similarities) intuitively. Such inspection can expose interesting areas for further investigation.

The Curriculum Exemplar spreadsheet that accompanies CS2013 makes provision for Curriculum design to the KU level [2] making it possible to use the spreadsheet to compare CC2001, CS2008 and CS2013 to this level. This paper explores the differences to the topic level.

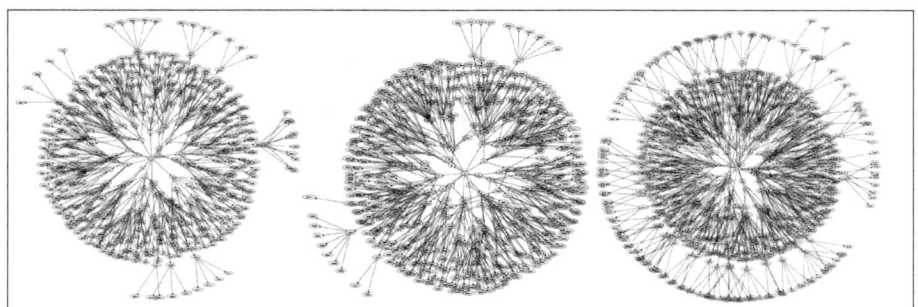

Fig. 2. Visible structural differences between CC2001, CS2008, CS2013

4.1   Algorithmic Difference Comparison

The algorithmic difference comparison uses the *Graph Trans-morphism Algorithm* to build a comparison profile. The algorithm accepts two graphs as input. These two graphs will be referred to as the *ideal* and the *model*. The ideal, I, is the specification to which the compliance of the model, M, is to be measured. The algorithm builds a graph, the complier (C), using the information presented in M and the structure of I. C is therefore a subgraph isomorphism of I that contains as many vertices from M as possible. The differences between the graphs and the ratios of these differences in relation to I provide a means to semi-automate the comparison of relatively large digraphs [8,9].

With this method, the graphs of CS2008 and CS2013 respectively will be compared to CC2001; and CS2013 will be compared to CS2008. The results of the comparisons are given in Table 1. The major columns in the table represent the curricula volumes being compared. Each of the major columns are further divided into two columns. The first column represents the number of vertices, $|V|$, for the quantity comparison in the row. The second column represents the number of edges in the digraph, $|E'|$, for the corresponding quantities. The quantities I, M and C give the cardinality of the ideal, model and complier for the particular combination of volumes. Calculating the sum of the values representing KAs, KUs and topics in Fig. 1 will result in an off-by-one number of vertices for each of the volumes as given in Table 1. This is as a result of the CS vertex (Sect. 3) being added to each of the digraph representations.

The quantities $I\backslash C$, $I\backslash M$, $M\backslash C$ and $C\backslash M$ give the cardinalities of the differences between the digraphs considered in the comparison. The difference between two digraphs A and B, $A\backslash B$, results in a set of vertices and a set of edges which can be found in digraph A and are not in digraph B. From the table the following observations are made:

Table 1. Set cardinalities for the comparison of the curriculum volumes

Quantity	CC2001(I), CS2008(M)		CC2001(I), CS2013(M)		CS2008(I), CS2013(M)	
	$|V|$	$|E'|$	$|V|$	$|E'|$	$|V|$	$|E'|$
I	476	477	476	477	510	519
M	510	519	730	729	730	729
C	356	356	185	184	178	181
$I\backslash C$	120	121	291	293	332	338
$I\backslash M$	138	248	310	356	353	417
$M\backslash C$	172	290	564	611	573	627
$C\backslash M$	18	127	19	66	21	79

Observation 1: *Increase in cardinality (number of vertices or edges) of digraphs I and M.* The cardinality of the digraphs across the curriculum volumes has increased from CC2001 to CS2013. The cardinality of CC2001 in terms of vertices is 476. Cardinality increased to 510 in CS2008 and 730 in CS2013. This indicates that the core aspects required in a curriculum in order to comply with CS2013 has increased from CC2001.

Observation 2: *Cardinality of digraph C.* The cardinalities associated with the respective compliers, C, decrease from left to right in the table. Since C reflects the extent to which M corresponds to I in each column, the table reflects the change that has taken place in the 2013 volume relative to its predecessors.

Observation 1 concurs with the increase in density in the digraphs seen in Fig. 2. *Observation 2* indicates that changes have taken place in CS2013. The comparisons of CS2013 with CC2001 and CS2008 both result in a relatively small cardinality of the complier. One of the changes that has been made has clearly been documented in CS2013 and was reproduced in Table 1. By taking the KA changes into account and modelling equivalences in the digraphs, a better indication of compliance may be achieved.

Equivalences are modelled by including in all digraph representations, that is CC2001, CS2008 and CS2013, an equivalence 'fan'. For example, from Table 1 it is clear that *Net-centric Computing* changed to *Networking and Communication*. Including the digraph given in Fig. 3 in all digraphs representing curricula results in a better complier being built by the algorithm. All edges in the existing digraphs representing the curriculum volumes with a destination of either *Net-centric Computing* or *Networking and Communication* must be altered to have a destination of *Fan_1_in*. Similarly, all edges where the source was either *Net-centric Computing* or *Networking and Communication* must change to have a source of *Fan_1_out*. The addition of equivalences will bloat the cardinalities for the respective digraphs. The equivalences will however not have an influence on the difference cardinalities as the entire "fan" has been included in all digraphs and therefore will be removed by the difference operation. Table 2 presents the cardinalities with equivalent KAs.

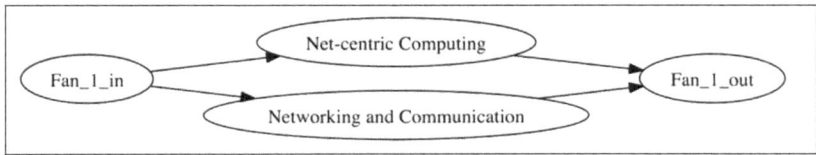

Fig. 3. KA equivalence modelling

A comparison of the cardinalities between Tables 1 and 2 gives an indication of the bloating caused by including the 'fan'. The bloat for I and M across all

Table 2. Set cardinalities with KA equivalences

Quantity	CC2001(I), CS2008(M)		CC2001(I), CS2013(M)		CS2008(I), CS2013(M)	
	$\lvert V \rvert$	$\lvert E' \rvert$	$\lvert V \rvert$	$\lvert E' \rvert$	$\lvert V \rvert$	$\lvert E' \rvert$
I	491	497	491	497	525	539
M	525	539	745	749	745	749
C	371	376	201	205	194	202
$I \backslash C$	120	121	290	292	331	337
$I \backslash M$	138	248	305	351	348	407
$M \backslash C$	172	290	559	603	568	617
$C \backslash M$	18	127	15	59	17	70

comparisons should be exactly the same, namely 15 vertices and 20 edges. This is confirmed by a comparison of the values in the two tables.

The values for C for the comparison of CC2001 with CS2008 results in the same bloat as was seen in I and M. The difference quantities for the comparison of CC2001 with CS2008 yields the same values for the cardinality of vertices and edges.

The comparison of CC2001 and CS2013 results in a C with 16 vertices and 21 edges more for the comparison with equivalent KAs than with the original comparison. C is therefore a better representation of M and a better match to I. Due to the introduction of equivalences, the cardinalities of the difference quantities in Table 2 declined slightly compared to their counterpart entries in Table 1. The comparison of CS2008 with CS2013 follows a similar trend, leading to the following observations.

Observation 3: *CS2013 differs significantly from CC2001 and CS2008.* A noteworthy change has taken place in the specification of CS2013. CC2001 and CS2008 are not that different from each other.

Observation 4: *Modelling KA equivalences improves comparison.* From the difference quantity values in Tables 1 and 2, it is difficult to determine directly whether the KA equivalences result in an improvement with regards to the comparison of the curriculum volumes. The results for each of the comparisons should be normalised so that a comparison can be made. Each ratio is uniquely identified by a ratio descriptor, $R(\mathcal{X}, \mathcal{Y})$. $\mathcal{X}$ represents the quantity for which the ratio is being described and $\mathcal{Y}$ the digraph relative to which the ratio is being calculated. The ratio is calculated by dividing the cardinality of $\mathcal{X}$ with the cardinality of $\mathcal{Y}$ for vertices and edges respectively. This normalisation will take the form of calculating a ratio for each of the cardinalities in the respective tables with respect to the cardinality of I, since I is the representation of the ultimate goal. These ratios can be placed in a table for comparison, but in many instances a general overview of the comparison is lost when considering individual ratios. A method to visualise the ratios is therefore desirable and is discussed next.

4.2   Visualisation of the Ratios for Difference Comparison

The visual representation of the ratios, $R(\mathcal{X}, \mathcal{Y})$, needs to take multivariate data into account. A radar chart (also referred to as a web, spider or star chart) is well suited for this type of data [5]. A radar chart representing a perfect comparison between I and M is given in Fig. 4. Each of the quantities in relation to I is plotted on a 'spoke' of the radar chart. The dotted closed shape in the figure represents the shape of a perfect comparison. Note that the ratios $R(I, I)$, $R(M, I)$ and $R(C, I)$ are all 1 and the difference ratios are all 0.

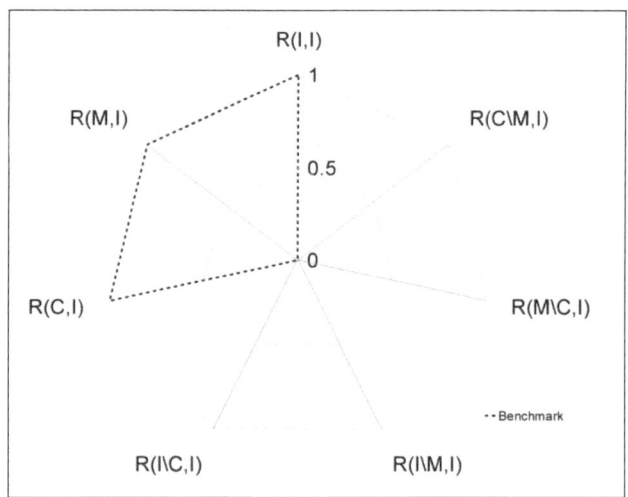

Fig. 4. Radar chart illustrating a perfect comparison of ratios in relation to I

4.3   KU-Level Equivalence Comparison

To determine differences at a finer level of granularity, equivalences in KUs must be added to the digraphs representing the KA equivalences. A manual approach to find KUs that differ can be taken. This approach can be very time consuming and error-prone.

A semi-automated approach that requires that the Graph Trans-morphism Algorithm is applied to the combinations of digraphs and the results of the difference quantity sets for which the cardinalities have been determined are considered. From these sets, all vertices representing KUs that remain in the difference sets need to be considered for equivalences. The difference sets of the difference quantities which will provide the most significant insights into equivalences will be $I \backslash M$ and $M \backslash C$.

For CC2001 and CS2013 with equivalent KAs, the digraphs representing CC2001(I), CS2013(M) and the result of the Graph Trans-morphism algorithm (C), the number of KUs in $I \backslash M$ is 35 and in $M \backslash C$ is 57. Among these KUs, 11 equivalences in KUs can be identified. Further equivalences can be identified by recursively applying the algorithm.

Again, the inclusion of KU equivalences increases the similarities between the digraphs. This can be seen in the ratio values for only KA equivalences and with KA and KU equivalences. *Observation 4* can be generalised to state that the inclusion of equivalences improves comparison. Table 3 gives the ratio values for vertices for the equivalences. In the majority of the cases, the addition of the KU equivalences (second column of each major column) results in better comparison between the digraphs. A radar chart can make this visible. With regards to M, the following observation can be made.

Observation 5: *For all comparisons M is larger then I.* The ratios for quantity M in all cases show that compared chronologically with a previous curriculum volume (I), the digraph representing M is larger than the curriculum volume represented by I.

Table 3. Vertex ratios with (i) KAs equivalent, (ii) KAs and KUs equivalent and (iii) KAs, KUs and topics equivalent

Quantity	CC2001(I), CS2008(M)			CC2001(I), CS2013(M)			CS2008(I), CS2013(M)		
	KAs	KAs and KUs	KAs, KUs and topics	KAs	KAs and KUs	KAs, KUs and topics	KAs	KAs and KUs	KAs, KUs and topics
$R(I, I)$	1.00	1.00	1.00	1.00	1.00	1.00	1.00	1.00	1.00
$R(M, I)$	1.07	1.07	1.07	1.51	1.48	1.29	1.42	1.39	1.21
$R(C, I)$	0.76	0.77	0.94	0.41	0.46	0.81	0.37	0.42	0.75
$R(I \backslash C, I)$	0.24	0.23	0.06	0.59	0.54	0.19	0.63	0.58	0.25
$R(I \backslash M, I)$	0.28	0.25	0.08	0.62	0.56	0.21	0.66	0.61	0.27
$R(M \backslash C, I)$	0.35	0.32	0.15	1.14	1.04	0.50	1.08	0.99	0.48
$R(C \backslash M, I)$	0.04	0.02	0.02	0.03	0.02	0.02	0.03	0.02	0.02

The radar charts in Fig. 5 represent the ratios for vertices and edges of CC2001 and CS2013 with and without consideration for equivalences. The vertex ratio values for the digraphs without equivalences are calculated using the values for the cardinalities of vertices given in the second major column of Table 1 and given in the first column of the second major column of Table 3. The values of the second major column of Table 2 are used to calculate the ratios for vertices given in the second column of the second major column of Table 3.

The values on the radar charts show that by mapping equivalences, the comparison of the digraphs improves, if only marginally. It is also be seen that the digraph representing M is larger than the digraph representing I, as reflected in the ratio $R(M, I)$. This confirms *Observation 5*. For both vertices and edges the ratio $R(C\backslash M)$ tends towards 0. This is an indication that C is a good representation of M which results in *Observation 6*.

Observation 6: *For all comparisons C represents M well.* The closer the ratio $C\backslash M$ is to 0, the better the representation of M in terms of C with respect to I is.

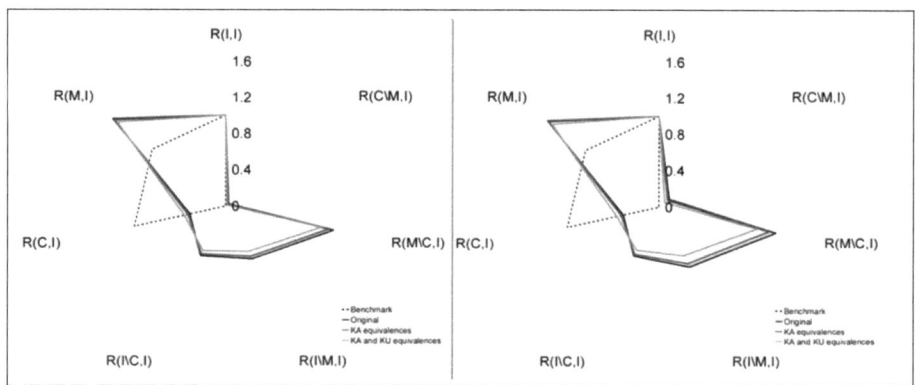

Fig. 5. CC2001 as I compared with CS2013 as M; left-hand-side: vertices, right-hand-side: edges

The comparisons in the second major column of Table 3 compare CS2013 as 'model' with CC2001 as 'ideal'. *Swapping* the 'model' and 'ideal' yields *reciprocal difference* quantities. The vertex set for $I\backslash M$ for CC2001 as I and CS2013 as M will be the same as the vertex set for $M\backslash C$ for M representing CC2001 and I representing CS2013. A further side-effect of the swapping is that the integrity of the digraph representations can be determined. If these reciprocal sets are not the same, then there is an error in one or both of the digraphs representing I and M.

4.4   Topic-Level Comparison Numbers

By including KA and KU equivalences in the digraphs representing CC2001, CS2008 and CS2013, the differences in topics can be determined. Applying the algorithm to the same combinations of digraphs as previously and representing the resultant ratios as radar charts will give an indication to what extent the topics between the digraphs differ.

The topic counts, reflected in Fig. 1, account for the cardinality of the vertices that represent topics in each of the curriculum volumes. From these values it can be seen that the number of topics in the curriculum volumes has steadily increased. This, however, does not indicate how the topics have increased *in relation* to the curriculum volume. Rather than comparing the exact counts representing topic vertices for each of the digraphs, a *ratio* is calculated by dividing the count by the cardinality of the corresponding digraph. This gives an indication of how much of the curriculum volume specification is represented by topics. In all 3 curriculum volumes under consideration the ratio was $\approx 80\%$, indicating that the relative topic count has increased proportionally to the increase in the curriculum volumes.

Table 4. Topic counts for $I \backslash M$ and $M \backslash C$ per comparison

	CC2001(I),CS2008(M)	CC2001(I),CS2013(M)	CS2008(I),CS2013(M)
C	278	126	121
$I \backslash C$	121	273	310
$I \backslash M$	113	270	312
$M \backslash C$	153	503	508
$C \backslash M$	0	2	2

By applying the same technique used to determine KU equivalences, that is, searching for equivalences in the sets representing the difference quantities, to determine topic equivalences, the results in Table 4 appear. From the comparisons, the number of topics in the complier decreases. This indicates differences between the curriculum volumes in terms of topics. It also implies that a good comparison between the volumes is not likely.

The topics in C that are not in M ($C \backslash M$) are insignificant in comparison to the other values. There are no topics in C when CS2008 is compared with CC2001. This means that no topics were inferred by the algorithm. Only 2 topics were inferred in the other comparisons. An investigation into these topics reveals that for the comparison of CC2001 and CS2013 these topics in encoded form are: T00491 and T00722. Similarly for CS2008 and CS2013 the topics are: T00613 and T00722. T00722 is common to both comparisons. On closer inspection, the topic encoded as: T00491 is *Philosophical questions*, T00613 is *Risk analysis*, and T00722 is *Team management*. T00491 falls under the KU *Fundamental issues in intelligent systems* encoded as U0051 and has subtopics *The Turing test, Searle's "Chinese Room" thought experiment*, and *Ethical issues in AI* in CC2001. In CS2008 the topic resides in the same KU, but without the subtopics. The subtopics as listed CC2001 are listed as topics of U0051 in CS2008. In CS2013 the topic T00491 has been replaced with *What is intelligent behaviour?* and *The Turing test* is a subtopic of this topic. T00613 and T00722 are related and fall into the *Software project management* KU. A detailed unravelling of the KU will reveal that the topics have been grouped in terms of *Teams, Effort* and *Risk* in CS2013, while in the previous volumes this grouping was not as explicit.

Considering the topics in each of the difference sets and finding equivalences for them can be seen as a daunting task. Finding the differences between the difference sets representing only topics can be semi-automated by taking the difference between the difference sets of topics. This difference will highlight the topics for which there are no matches in the difference sets. Looking at these topics will provide a starting point for finding equivalences. Once equivalences have been found, they can be included in the digraph representations and the framework can be iteratively applied until no more equivalences can be found. This technique is similar to the one illustrated for KUs and to a lesser extent, the KAs.

From the results, presented in the last column of each major column in Table 3, the addition of topic equivalences improves the ratios for each of the comparisons. In some of the cases, this can be quite significant. As the values for ratio $R(M, I)$ is already above 1, it needs to come closer to 1. This it does for all cases except in the comparison of CC2001 as I and CS2008 as M. The ratios for $R(C, I)$ should also tend towards 1. In all comparisons, the addition of the topic equivalences improves the ratio by from 17 to 36 percentage points. The difference ratios should tend towards 0 to indicate better matching. In most cases, when topic equivalences are added, the ratios approximately halve. The only case were the difference is minimal is for $R(C \backslash M, I)$. This means that C, the representation of M in terms of I, is a good representation of M.

The radar charts in Fig. 6 compare the *KAs and KUs* equivalence ratios with those for *KAs, KUs and topics* for CC2001 and CS2013. This is done so that a comparison can be made between Figs. 5 and 6.

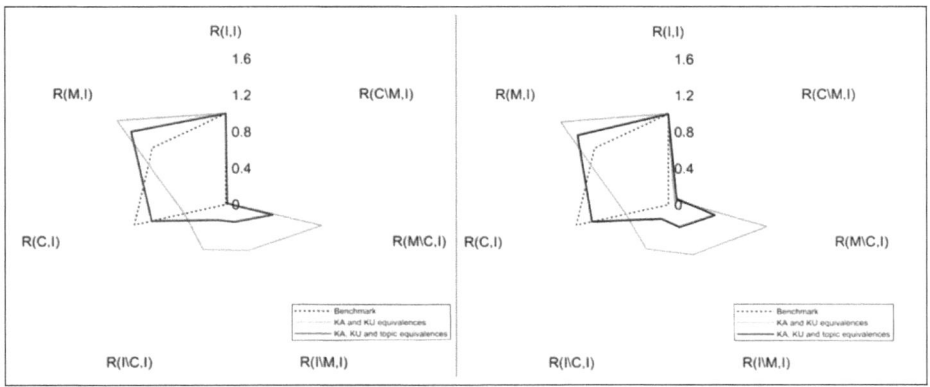

Fig. 6. CC2001 as I compared with CS2013 as M, KA and KU against KA, KU and topics; left-hand-side: vertices, right-hand-side: edges

Figure 6 shows that the inclusion of the topic equivalences results in better matches between the curriculum volumes. This supports *Observation 4*. Similar radar charts can be drawn for the other curricula comparisons using the ratios from Table 3.

4.5   Topic-Level Comparison in More Detail

By considering the topics from the difference set $I \backslash C$, a list of topics in I which are not in C results. The list of topics for the comparisons of the following combinations of I and M: CC2001 with CS2008; CC2001 with CS2013; CS2008 with CC2001; CS2008 with CS2013; CS2013 with CC2001; and CS2013 with CS2008, results in a total of 1333 topics which are in I and not in the respective C. For illustration, consider the topics for each of the comparisons for the *Algorithms and Complexity* KA in Table 5.

The first and second columns in Table 5, specify the curricula selected to represent I and M in the comparison. The KU column, lists the KU in I for which there are topics. This is followed by topics for that KU. There are no topics for the comparison of CC2001 with CS2008 nor for the reverse. Therefore it can be said that CC2001 and CS2008 include exactly the same topics for the *Algorithms and Complexity* KA. The differences are between CC2001 or CS2008 with CS2013. The comparisons with I as CC2001 and M as CS2013 are the same as when I is CS2008. This indicates that these topics are no longer included in CS2013 or specified in the same way. Remember, equivalences have been included in the comparison and therefore these may not have been deemed sufficiently equivalent enough. The reverse, I representing CS2013 and M either CC2001 or CS2008, provides topics which are included or ellaborated on in CS2013 which were not explicitly (nor in terms of equivalences) in either CC2001 or CS2008. Interestingly, for these comparisons with CS2013, the list of topics for both CC2001 and CS2008 are identical. This is not always the case. Due to space constraints, differences for all KAs for each of the comparisons are not provided in this paper[1]. Instead, an overview of the differences is given in Table 6.

The topic difference presented in Table 5 appears in Table 6 for the corresponding KA (*Algorithms and Complexity*). There are zero values for the comparison of CC2001 and CS2008 and *vice versa* indicating no difference. The other differences in Table 6 correspond the detailed differences shown in Table 5. The percentages have been calculated using the total number of topics for the KA in the respective I. This means that even though the total number of topics in comparisons may be the same, the percentages will differ in relation in the topics in the KA for the particular curriculum. The KAs introduced to or removed from CS2013 are therefore visible in the table. Where topics have been newly introduced in CS2013, such as for the KA *Computational Science*, there is no overlap with topics in CC2001 or CS2008. Other KAs may have a lower percentage, such as *Software Development Fundamentals*. In these cases, topics have been moved from KAs which may no longer exist, or due to a better fit in the updated KA.

From Table 6, *trends* in curricula specifications can be identified. In CC2001 and CS2008, *Programming Languages* is more prominent than in CS2013. CS2008 had a strong focus on *Architecture and Organisation* which has subsided in CS2013. The focus on *Information Assurance and Security* issues is

---

[1] A comprehensive list of topics per comparison is available from the author.

Table 5. Topics in $I\backslash C$ per comparison (Algorithms and Complexity KA)

I	M	KU	Topic
CC2001	CS2008	None	None
CC2001	CS2013	Algorithmic strategies	Numerical approximation algorithms
		Basic computability	Implications of uncomputability; Tractable and intractable problems; Uncomputable functions
		Distributed algorithms	Consensus and election; Fault tolerance; Stabilisation; Termination detection
		Fundamental computing algorithms	Topological sort; Transitive closure (Floyd's algorithm)
CS2008	CC2001	None	None
CS2008	CS2013	Algorithmic strategies	Numerical approximation algorithms
		Basic computability	Implications of uncomputability; Tractable and intractable problems; Uncomputable functions
		Distributed algorithms	Consensus and election; Fault tolerance; Stabilisation; Termination detection
		Fundamental algorithms	Topological sort; Transitive closure (Floyd's algorithm)
CS2013	CC2001	Algorithmic strategies	Dynamic programming; Reduction: transform-and-conquer
		Basic analysis	Analysis of iterative and recursive algorithms
		Basic automata Computability and Complexity	Regular expressions; Introduction to P and NP classes and P vs NP problem; Introduction to NP-complete class and exemplary NP-complete problems (e.g. SAT, Knapsack)
		Fundamental data structures and algorithms	Common operations on binary search trees such as select min, max, insert, delete, iterate over tree; Heaps
CS2013	CS2008	Algorithmic strategies	Dynamic programming; Reduction: transform-and-conquer
		Basic analysis	Analysis of iterative and recursive algorithms
		Basic automata Computability and Complexity	Regular expressions; Introduction to P and NP classes and P vs NP problem; Introduction to NP-complete class and exemplary NP-complete problems (e.g. SAT, Knapsack)
		Fundamental data structures and algorithms	Common operations on binary search trees such as select min, max, insert, delete, iterate over tree; Heaps

Table 6. Percentage difference of topics in $I \backslash C$ per comparison

I	CC2001	CC2001	CS2008	CS2008	CS2013	CS2013
M	CS2008	CS2013	CC2001	CS2013	CC2001	CS2008
Knowledge area	Comparison $I \backslash C$ percentages					
Algorithms and Complexity	0	27	0	27	19	19
Architecture and Organisation	48	9	70	70	18	62
Computational Science					100	100
Discrete Structures	4	12	2	14	30	32
Graphics and Visual Computing	0	91	0	91	86	86
Human-Computer Interaction	40	30	69	46	72	61
Information Assurance and Security					91	88
Information Management	6	24	20	30	35	30
Intelligent Systems	0	39	0	50	54	54
Net-centric Computing	36	64	41	74		
Networking and Communication					88	88
Operating Systems	3	15	8	11	19	10
Parallel and Distributed Computing					76	76
Programming Fundamentals	6	34	28	47		
Programming Languages	5	70	3	73	63	63
Social and Professional Issues	0	53	18	62		
Social and Professional Practice					70	67
Software Development Fundamentals					37	37
Software Engineering	6	44	35	60	57	55
Systems Fundamentals					88	88

significant in CS2013, yet some of the topics – more in CC2001 than in CS2008 – are in the previous curriculum volumes. A similar comparison can be made for *Networking and Communication* and *Parallel and Distributed Computing*.

The implications, which these changes have on curriculum development are significant. A curriculum which complies with a previous curriculum volume, will not by default comply with a more recent curriculum volume. Table 6, provides an indication of where changes in a curriculum presented at an institution of higher education ought to be considered. For instance, if a presented curriculum complied 100% in the *Algorithms and Complexity* KA for CC2001 and CS2008, there are another 19% of topics which have not been considered in both instances (as identified in Table 5).

5   Conclusion and Future Work

From the comparisons presented in this paper, two observations can be made. The first is that the ACM/IEEE curriculum volumes have changed rather significantly from CC2001 to CS2013 in content and structure. Considering the vertices and edges of the difference sets gives an indication of where the differences are. This was illustrated in Sect. 4.4 for a brief overview and Sect. 4.5 for an illustration of the differences between the curriculum volumes for the *Algorithms and Complexity* KA.

For the difference quantities, $I \backslash C$ and $I \backslash M$, the differences between CC2001 (I) and CS2013 (M) [7] are smaller than those between CS2008 (I) and CS2013 (M). This indicates that CS2013 is closer in content to CC2001 than it is to CS2008. The difference ratios for $R(I \backslash C, I)$ and $R(I \backslash M, I)$ between CC2001 (as I) and CS2008 and CS2013 (as M) respectively, are similar in proportion. This indicates that C is a good representation for M. The discrepancy between the comparisons for $R(M \backslash C, I)$, in Fig. 7, highlights that CC2001 and CS2008 are more similar than what CS2013 is to either CC2001 or CS2008 (as in Table 6).

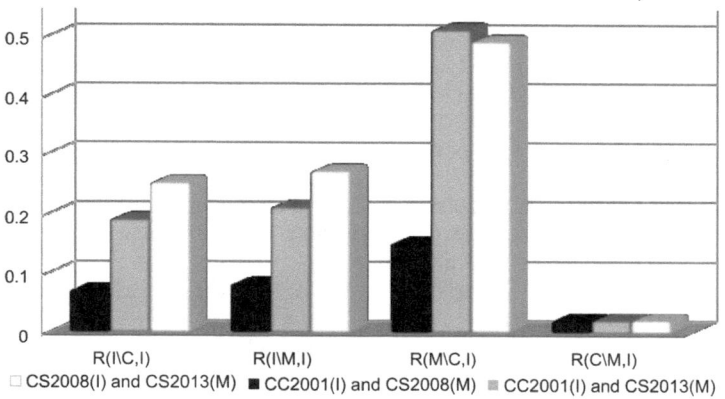

Fig. 7. Comparison of the difference ratios

The second observation relates to the addition of the equivalences. It is clear from *Observation 4* and the subsequent references to the observation in terms of KUs and topics, that the addition of the equivalences does improve the comparison. It is therefore necessary to include the equivalences on the topic level [8, Chapter 11.3].

For *future work* it is envisaged that for all KAs, KUs and topics, equivalences will be modelled taking aspects that are classified as 'core' and 'elective' into account. A comparison which takes core and elective topics into account may result in more accurate comparisons between the curriculum volumes, as some aspects (mostly KUs and topics) may have swapped their status between being 'core' and 'elective' (or *vice versa*). A further aspect that needs to be included in the digraph model of CS2013 is the relationships between KUs as specified in the curriculum volume. It would be interesting to see what the effect these may have on the results presented in this paper would be.

Last but not least, it should be emphasised that the method presented in this paper is not specific to the application on the ACM/IEEE CS curricula alone. It can be applied to any data set that is similarly structured and similarly representable by directed graphs. By modelling a curriculum presented at some higher education institution, such curriculum can be compared with a modelled 'ideal' curriculum specification, and the compliance to the envisaged 'ideal' can be measured. A further application of the technique includes the possibility of comparing curricula presented different institutions of higher education against each other.

References

1. ACM/IEEE-CS Joint Curriculum Task Force: A Summary of the ACM/IEEE-CS Joint Curriculum Task Force Report - Computing Curricula 1991. Commun. ACM **34**(6), 68–84 (1991). http://doi.acm.org/10.1145/103701.103710
2. ACM/IEEE-CS Joint Task Force on Computing Curricula: Computer science curricula 2013. Technical report. ACM Press and IEEE Computer Society Press, December 2013. http://dx.doi.org/10.1145/2534860
3. ACM/IEEE-Curriculum 2001 Task Force: Computing Curricula 2001: Computer Science (2001). http://www.acm.org/education/curricula.html. Accessed 22 Aug 2007
4. ACM/IEEE-Curriculum CS2008 Joint Task Force: Computer Science Curriculum 2008: An Interim Revision of CS 2001 (2008). http://www.acm.org/education/curricula.html. Accessed 22 Aug 2007
5. Heerand, J., Bostock, M., Ogievetsky, V.: A tour through the visualisation zoo. Queue **8**(5), 20–30 (2010)
6. Joint Task Force for Computing Curricula: Computing Curricula 2005: The Overview Report covering undergraduate degree programs in Computer Engineering, Computer Science, Information Systems, Information Technology, Software Engineering (2005). www.acm.org/education/curric_vols/CC2005-March06Final.pdf. Accessed 22 Aug 2007
7. Marshall, L.: A comparison of the core aspects of the ACM/IEEE Computer Science Curriculum 2013 Strawman report with the specified core of CC2001 and CS2008 Review. In: Computer Science Education Research Conference, CSERC 2012, pp. 29–34. ACM (2012)
8. Marshall, L.: A graph-based framework for comparing curricula. Ph.D. thesis, University of Pretoria, South Africa (2014)
9. Marshall, L., Kourie, D.: Deriving a digraph isomorphism for digraph compliance measurement. In: Proceedings of the 2010 Annual Research Conference of the South African Institute of Computer Scientists and Information Technologists, SAICSIT 2010, pp. 160–169. ACM, New York (2010). http://doi.acm.org/10.1145/1899503.1899521
10. Sahami, M., Roach, S., Cuadros-Vargas, E., Reed, D.: Computer science curriculum 2013: reviewing the strawman report from the ACM/IEEE-CS task force. In: Proceedings of the 43rd ACM Technical Symposium on Computer Science Education, SIGCSE 2012, pp. 3–4. ACM, New York (2012). http://doi.acm.org/10.1145/2157136.2157140

Author Index